Rethinking Coal

Rethinking Coal

Chemicals and Carbon-Based Materials in the 21st Century

Harold Schobert

OXFORD
UNIVERSITY PRESS

Oxford University Press is a department of the University of Oxford. It furthers the University's objective of excellence in research, scholarship, and education by publishing worldwide. Oxford is a registered trade mark of Oxford University Press in the UK and certain other countries.

Published in the United States of America by Oxford University Press
198 Madison Avenue, New York, NY 10016, United States of America.

Library of Congress Cataloging-in-Publication Data
Names: Schobert, Harold H., 1943- author.
Title: Rethinking coal : chemicals and carbon-based materials in the 21st century / Harold Schobert.
Description: New York, NY : Oxford University Press, [2022] | Includes bibliographical references and index.
Identifiers: LCCN 2022018864 (print) | LCCN 2022018865 (ebook) | ISBN 9780199767083 (hardback) | ISBN 9780199912377 (epub) | ISBN 9780197646809 (UPDF) | ISBN 9780197646816 (online)
Subjects: LCSH: Coal trade—History—21st century.
Classification: LCC HD9540.5 .S343 2022 (print) | LCC HD9540.5 (ebook) | DDC 338.2/724—dc23/eng/20220708
LC record available at https://lccn.loc.gov/2022018864
LC ebook record available at https://lccn.loc.gov/2022018865

DOI: 10.1093/oso/9780199767083.001.0001

1 3 5 7 9 8 6 4 2

Printed by Sheridan Books, Inc., United States of America

Dedicated to the memory of Dr. Frank R. Eshelman—educator, scientist, inventor, and friend

Contents

Preface

This book has been written four times. To begin at the beginning, about 35 years ago I wrote a book on coal called *Coal: The Energy Source of the Past and Future* (American Chemical Society, 1987). At that time the world was not far removed from the oil embargoes and price shocks of the 1970s and from the nuclear accidents at Chernobyl and Three Mile Island. There seemed good reason to believe that oil and natural gas supplies would become less and less available as we moved into the twenty-first century. Climate change was not much talked about. As for renewable energy sources, well, maybe someday. . . . Until "someday" arrived, we could rely on coal.

In 2010, the previous book was nearing 25 years of age, and it seemed to me that it might be time for an update. Coal was still the major energy source for electricity generation in most parts of the world. There were plans afoot in many places for synthetic fuels plants based on coal. "Plan A" had been to do a thorough rewriting of the 1987 book, incorporating the advances in coal combustion and conversion to gaseous and liquid fuels and the improvements in our understanding of coal as a substance. So, I got started.

Various life events intervened, so a fresh start—"Plan B"—was made in mid-decade. It was already apparent that the ground was starting to shift. The coal industry seemed to be going into decline, synthetic fuel plants were being postponed or canceled outright, and much of the financial support for research or development activities in coal had evaporated. And at about the same time progress on the manuscript had evaporated. Finally, in early 2020, came "Plan C," which was, once and for all, to finish the book at all costs. By that time, it was very clear that, over the course of these past ten years, much had changed for coal.

Now, climate change is a major concern for many people and indeed many governments. The hydraulic fracturing (fracking) technology for recovering gas and oil from shales has radically changed the global situation for those fuels. Some countries, including the United States, suddenly had the unexpected good fortune of newly abundant and relatively inexpensive oil and gas. We weren't running out after all—at least not now. Steady improvements in the technologies of wind and solar generation of electricity have led to dramatically lower costs.

Natural gas and wind and solar energy became less expensive and more competitive with coal. Record heatwaves and an increased incidence of severe weather became more frequent, which is consistent with most predictions of global climate change models. Direct chemical evidence, based on the chemistry and physics of carbon isotopes and explained in Chapter 15, indicates that use of fossil fuels is a major contributor to increased carbon dioxide in the atmosphere. When compared on the basis of equal amounts of energy produced, coal combustion emits more carbon dioxide than use of other fossil fuels. So now we see coal mines and coal-fired power plants planned to be phased out or actually closed in many countries. Some major financial institutions have announced that they will no longer provide funding for coal projects. Perhaps the most stunning development was the announcement in the spring of 2021 by the United Mineworkers Union—the labor organization for mineworkers in the United States—that it will accept a switch to use of renewable energy in exchange for creation of jobs for displaced coal miners. These jobs could well be in the renewable energy industry.

The vision of the future for coal has changed greatly in these past ten years. Some see it as getting darker. There has been much talk, sometimes loudly, of a "war on coal." The whole problem, it seemed, was that the coal industry was hamstrung by governmental policies and regulations that placed coal at a competitive disadvantage. The hypothesis was that everything would be fine if we could reduce the regulatory burden and have "pro-coal" persons in government positions. In science, hypotheses call for experimental tests. In the United States, we conducted that experiment for four years. The hypothesis was shown to be false.

There is indeed a war on coal. The problem is that it is not the war that it's been thought to be. The real war on coal is a pincers movement waged vigorously on two fronts. First, the greatly increased availability of inexpensive natural gas and the increasingly attractive costs of solar and wind energy make a switch away from coal seem increasingly worthwhile. Second, the increasing perception that serious action is needed on reducing carbon dioxide emissions make coal less attractive *as a fuel*, as a result of the inexorable rules of chemistry and arithmetic. Even if these factors were not important—though they certainly are—steady work by coal scientists and engineers has given us continual improvements in power plant and blast furnace technology so that we can produce the same amounts of electricity and steel with less and less coal. I continue to believe strongly that there is a future for coal, but that future does not involve wasting it by burning it. The worst and most wasteful thing we can do with coal is to burn it. The future lies in making chemical products and carbon-based materials. In fact, we have the possibility of building the infrastructure for the renewable energy industry with carbon materials from coal.

These considerations led to the "Plan C" manuscript. As things turned out, when the presumptive masterpiece was finished, it was twice as long and had three times as many illustrations as Oxford University Press was expecting. "Plan D" was radical emergency surgery, which has created this book as it now exists.

This book has four parts. The first discusses what exactly we are talking about when we mention coal—how coals form in nature, what they are composed of, and what properties they have as solids. This sets the foundation for the second part, which discusses the present uses of coals—in other words, what we are doing with coals now. The third part discusses the factors that have caused such a shift in the vision of the future. These two factors are competition from other energy sources and the growing appreciation of the human contribution to climate change. The final section presents some examples of where I think the future for coal lies in chemicals and carbon-based materials. If there is to be a future for the coal industry, that's where it has to be.

Coals are found on every continent, including Antarctica. Coal science and technology are international endeavors. I have tried to reflect those facts by including, where appropriate, information from many countries. Many of the practical examples are from the United States, only because those are the ones I know the best.

This book is not primarily intended to be a textbook, nor is it intended to be a monograph for professionals working in coal science and technology. I have tried to tell the story of coal to the general public to explain how coal is formed, how scientists study it, how it is mined, the ways that we use it, environmental issues surrounding coal utilization, and how coal might be used in the future. Most chapters have numerous citations as a guide for readers who wish to explore further. There is also a Glossary and a Bibliographic Essay for additional guidance. I have assumed that most readers will have had at least an introductory course in chemistry as well as some modest exposure to the terminology and principles of other sciences. Those who have had no science beyond high school can still read this book with profit, and hopefully with some enjoyment. I hope that some of my colleagues will enjoy it as well.

Finally, all of us are well aware of the rapid succession of profound changes occurring currently and in the recent past—pandemic, economic dislocations, supply chain issues, the Russian invasion of Ukraine. Statistics or current event information may have changed since this manuscript was completed in the Spring of 2021, but the composition, properties, and reactions of coals have not.

Acknowledgments

Thomas Mann has said that a writer is someone for whom writing is more difficult than it is for other people. He must be right. This book would not exist without the assistance of my dear wife Nita, an indefatigable searcher of the literature, who also wrestled with much of the hard work of getting my drafts into final shape. It is a special pleasure to acknowledge all she has done to make this book possible. I am also happy to acknowledge Jeremy Lewis of Oxford University Press for his advice and particularly for his extraordinary patience through the long gestation period of this book. It would be wrong to say that he has the proverbial patience of Job—several generations of the Job family would be more like it. I sometimes reflect on the comment by John Kenneth Galbraith: those who yearn for the defeat of their enemy are said to wish that he might write a book.

Throughout my career I have been blessed with a great many former professors, colleagues, students, postdoctoral associates, technicians, consulting clients, lab rats, and friends in all walks of life with whom I have exchanged ideas and from whom I have learned much. It would be impossible to list them all, but I most certainly extend my sincere gratitude to all. Collectively, they represent at least a dozen countries. Nita and I are grateful to those who have helped us with permissions for illustrations or with the actual illustrations themselves. In particular, Lindsay Findley patiently and professionally converted my pencil sketches from old lecture notes into finished drawings. At the eleventh hour, Professor John Bunt of North-West University came through with a figure and the kind permission to use it.

Nita and I are especially grateful to the inimitable Cora Forrester, our granddaughter, for useful advice on handling some of the electronic images and on using drop boxes. Having worked for a half-century since receiving a PhD, it is somewhat humbling to be bailed out by an eleven-year-old fifth-grade student.

This book was written on my own time and, except for the illustrations acknowledged separately, entirely with our own resources. I am pleased to acknowledge and thank the assistance mentioned here, but opinions, mistakes, and omissions are entirely my own responsibility.

1

Introduction

The 2016 Nobel Laureate in Literature[1] has eloquently characterized the coal situation today: "The times they are a-changin." On or about December 2010,[2] the coal industry worldwide seemed to be in great shape. Global demand for coal had not only recovered from the 2008 recession, but was by then expanding at annual rates of about 5%. Coal prices were high and poised to head even higher. And then. . . .

By the middle of the 2010s, world coal consumption was declining, even in countries such as the United States and China, big consumers of coal. By the end of 2015, coal prices had dropped by one-third to a half from their near-record levels in 2010 (Focus Economics 2016). Major coal producers were filing for bankruptcy (Moritz-Rabson 2019). The past few years have since brought a steady stream of bad news. In the United States, natural gas first surpassed coal as the principal fuel for electricity generation in 2015 (Saha 2019). India proposes to build no new coal plants after 2022 (Mathiesen 2016). France aims to phase out coal use by 2023 (England 2016), with the United Kingdom following in 2025 (Evans 2016), and Canada in 2030 (Cotting and Potter 2018). India expects to increase its electricity generation from renewable resources by a factor of five, also by 2030. In 2018 came the last piece of hard coal to be mined in Germany (Dobuch 2018). That same year was the first time that the number of coal-fired power plants decreased worldwide (Evans and Pearce 2020). Major financial sources have announced that they will no longer invest in or make loans to coal projects (Mufson and Grandoni 2020). On June 10, 2020, Britain, whose coal had fueled the Industrial Revolution, had gone for two months without using coal to generate electricity (Rowlatt 2020).

But what is it that we are talking about when we discuss coal? As with many complicated questions, the answer depends on whom you ask. Consider these various and conflicting sentiments: Coal provides a vital energy component

---

[1] Bob Dylan.
[2] This is a paraphrase of Virginia Woolf's phrase "on or about December 1910," from her essay "Mr. Bennett and Mrs. Brown." The complete phrase is "on or about December 1910, human character changed." In December 2010, it was the coal industry that was about to change.

of the economies of many nations. Coal mining and use have caused environmental damage as well as human misery and suffering. Coal could reduce dependence on imported energy supplies for countries lacking domestic petroleum or natural gas resources. Coal's predominance as an energy source passed decades ago, and its importance is rapidly waning. Coal could support a transition from a petroleum-based economy to one based on renewable, nonpolluting energy sources by mid-century. Coal offers a source of clean synthetic fuels and valuable chemicals and materials. Coal represents a fascinating, complex, scientific jigsaw puzzle; the more pieces we put into place, the more we learn about our Earth's history. Coal is an uninteresting black rock; since all we do is dig it up and burn it, why bother about what it's made of or how it came to be that way? Coal is the worst offender in the human contribution to global warming. Coal is just the black stuff shoveled into one end of a plant so that electricity, gas, or steel comes out the other end.

Where is coal headed? It's good to start with the useful reminder that, "It's difficult to make predictions, especially about the future."[3] With that caveat in mind, the coal industry of the future—if there is one—and the ways—if there are any—in which we use coal likely will be much different from the boom times of the past. Despite some people who argue that we must stop using coal, or indeed all fossil fuels, right now, preferably by this time tomorrow morning, coal will continue to be used for decades to come. Almost certainly it will not be used on the scale of years past, and likely will be used in ways different from those used in the past. Despite those who argue that if only we could get rid of noxious government regulations and red tape, King Coal would be back in business like the good old days, the American experiment of 2017–2021 shows that hypothesis to be false.

Every energy source, definitely including coal, has some technical advantages and some disadvantages. Every energy source impacts the environment to some extent. Some might possibly provide environmental benefits. Every energy source has economic advantages and economic penalties. The challenge for all of us—not just for those in the coal industry or for the scientists and engineers who study coal—is, in a way, like trying to steer a course down a stream with rocks on both sides, hoping to arrive at an optimum role for each source of energy, whether coal, natural gas, solar, wind, or something else entirely. Adding to this challenge is that, along this metaphorical river, the positions of the rocks keep changing. In 1977, solar cells cost $76.67 per watt; by 2016, the price had dropped to $0.26 per watt.

---

[3] This comment has been attributed to sources as diverse as Niels Bohr, a Nobel Laureate in physics, and Yogi Berra, a Hall of Fame baseball player.

Figure 1.1 For many decades, smoke billowing from the stacks of a coal-burning plant
was a sign of industrial activity and prosperity.
Courtesy of iStock.com/DWalker44.

From 2010 to 2020, the cost of solar energy dropped by 82%, from $0.38 per
kilowatt-hour to $0.07 (Rollet 2020). Fifty years ago, only a few prescient indi-
viduals cared about carbon dioxide in the atmosphere. Seventy-five years
ago, railroads ran on coal. A century ago, windmills for electricity generation
powered only a single home or farming operation.

Much of the world has come to expect a lifestyle based on abundant electricity, all day, every day; ready transportation, public or private; an array of
consumer products made of metals or plastics; comfortable temperatures
in homes and commercial buildings; more heat for industrial processes;
and plenty of chemical products, such as medicines and fertilizers. All these
needs can be satisfied by coal and actually have been satisfied by coal in years
past. Coal was the energy source that drove the Industrial Revolution. It provided the fuel that produced the steel. It powered transportation networks of
railways and ships. It heated houses, stores, offices, and factories. By-products
from coal were converted into hundreds of useful chemicals and materials.
For a century or more, chimneys and smokestacks served as symbols of a
prosperous economy.[4] By about 1900, coal dominated energy production in
most industrialized nations (Figure 1.1).

---

[4] The plume of smoke from chimneys and smokestacks is, in a way, a curious image. The smoke suggests
that the factory or power plant is busily active, doing tremendous amounts of work. In actuality, billowing

The world still has abundant reserves of coal, possibly nearly a trillion tonnes (IEA 2020), despite our having burned untold billions of tonnes in past centuries. Though many countries are closing coal-burning power plants, in 2018, almost 40% of the world's electricity came from such plants (EIA 2017). At least in the near future, coal will continue to maintain a role in electricity generation. It has to. There is neither enough manufacturing capacity nor financial wherewithal in the world to replace all these plants in short order. Coke production, mainly for smelting iron ore, remains the second-most important use of coal.[5] Liquid fuels produced from coal are equivalent to about one-third of South Africa's petroleum refinery capacity (EIA 2017). Coal combustion still provides heat for industrial and steam for industrial use. In some places, coal continues to be burned in homes. Economics and geopolitics may favor, or indeed require, a country relying on its indigenous energy resources rather than on imports. Coal likely will continue to be an important energy source in those countries where it is cheaper than oil, natural gas, or other sources in terms of cost per unit of energy.

It can't be denied that coal mining and consumption, as it was practiced worldwide until well into twentieth century, brought with it a lot of problems. Coal burning polluted the environment with smoke, cinders, soot, fogs of sulfuric acid, and ash. Coal is recognized as a major factor in the human contribution to climate change. Underground mining was a dangerous occupation, with employment conditions often tantamount to serfdom. Unrestricted surface mining caused extensive environmental disruption. Change came at different times in different countries, though it was the 1970s that saw the rise of environmental consciousness in many countries. Legislation and regulations that focused on control of emissions substantially reduced potential pollution caused by burning coal. Labor laws, along with workplace health and safety regulations, have greatly improved working conditions for coal miners in most industrialized countries. Worldwide, mining is less hazardous than such occupations as farming, deep-sea commercial fishing, or lumbering (Trujillo 2020). Mined-land reclamation laws now require restoration of the landscape to a condition as it was before surface mining began.

---

clouds of smoke are a sign of inefficient burning—coal is being wasted, and the smoke is contributing to pollution. The ideal image would be a perfectly clean stack, but then the viewer would think that nothing was happening.

[5] Coke, more specifically metallurgical coke, is made from certain types of coal, as will be discussed in detail in Chapter 10. The iron and steel industry represents the second-largest market for coal, though small in comparison to the demand for coal in electric power generation. Except for iron ore itself, coke is the most important raw material in the iron and steel industry.

To try to see where coal is going, we need to know something about coal: what it is, how it was used in the past and is used today, and factors likely to impact its future. Aristotle tells us that we can only have proper knowledge of an object or substance—coal, for example—when we understand what Aristotle would call its "cause" (Aristotle 1952). A deeper investigation of cause leads to defining four kinds of causes. The first identifies what processes, or what maker, produced the substance being investigated. The second identifies its shape or composition. The third cause covers physical characteristics and qualities of the substance. The final cause addresses the purpose or function of the substance.

The first chapters of this book address Aristotle's first three causes: What are the processes in nature that led to the formation of coal? What is coal made of? What are its properties? The second group of chapters addresses the current uses of coal. Then, following chapters discuss factors that can impact how coal might (or might not) be used in the future: the extraordinary change, especially in North America, caused by the availability of large quantities of inexpensive petroleum and natural gas by hydraulic fracturing and concerns about climate change and how coal use contributes to carbon dioxide emissions. Finally, the point is made that it is a waste of a valuable resource to burn coal. Rather, it is a hydrocarbon resource that could be used much more profitably and beneficially to make chemical products and carbon materials. That will be the future of coal: chemicals and materials.

References

Aristotle. 1952. "Physics." In: *Aristotle, Volume I. Great Books of the Western World*, edited by Robert Maynard Hutchins, 259. Chicago: University of Chicago Press.

Cotting, Ashleigh, and Ellie Potter. 2018. "Canada Finalizes Plan to End Most Coal-Fired Power Generation by 2030." *S&P Global*, December 13, 2018. https://www.spglobal.com/marketintelligence/en/news-insights/trending/vCIXwAFTZs-04k7Dhy705g2.

Dobuch, Grace. 2018. "Germany Closes Its Last Active Coal Mine, Ending 200-Year-Old Industry." *Fortune*, December 21, 2018. https://fortune.com/2018/12/21/germany-closes-last-coal-mine/.

EIA. 2017. "South Africa." *U.S. Energy Information Agency*. https://www.eia.gov/international/analysis/country/ZAF.

England, Charlotte. 2016. "France to Shut Down All Coal-Fired Power Plants by 2023." *Independent*, November 17, 2016. https://www.independent.co.uk/news/world/europe/france-close-coal-plants-shut-down-2023-global-warming-climate-change-a7422966.html.

Evans, Simon. 2016. "Countdown to 2025: Tracking the UK Coal Phase Out." *Carbon Brief*, February 10, 2016. https://www.carbonbrief.org/countdown-to-2025-tracking-the-uk-coal-phase-out.

Evans, Simon, and Rosamund Pearce. 2020. "Infographics. Mapped: The World's Coal Power Plants." *Carbon Brief*, March 26, 2020. https://www.carbonbrief.org/mapped-worlds-coal-power-plants.

Focus Economics. 2016. "Thermal Coal Price Outlook." https://www.focus-economics.com/commodities/energy/thermal-coal.

IEA. 2020. "Coal: Fuels & Technologies." *International Energy Agency*. https://www.iea.org/fuels-and-technologies/coal.

Mathiesen, Karl. 2016. "India to Halt Building New Coal Plants in 2022." *Climate Home News*, December 16, 2016. https://www.climatechangenews.com/2016/12/16/india-to-halt-building-new-coal-plants-in-2022/.

Mortiz-Rabson, Daniel. 2019. "Eleven Coal Companies Have Filed for Bankruptcy Since Trump Took Office." *Newsweek*, October 30, 2019. https://www.newsweek.com/eight-coal-companies-have-filed-bankruptcy-since-trump-took-office-1468734.

Mufson, S., and D. Grandoni. 2020. "JPMorgan Pulling Back from Coal Companies, Arctic Drillers." *Minneapolis Star Tribune*, February 26, 2020.

Rollet, Catherine. 2020. "Solar Costs Have Fallen 82% Since 2010." *PV Magazine*, June 3, 2020. https://www.pv-magazine.com/2020/06/03/solar-costs-have-fallen-82-since-2010/.

Rowlatt, Justin. 2020. "Britain Goes Coal Free as Renewables Edge Out Fossil Fuels." *BBCNews*, June 10, 2020. https://www.bbc.com/news/science-environment-52973089.

Saha, Devashree. 2019. "Natural Gas Beat Coal in the US: Will Renewables and Storage Soon Beat Natural Gas?" *World Resources Institute*. https://www.wri.org/blog/2019/07/natural-gas-beat-coal-us-will-renewables-and-storage-soon-beat-natural-gas.

Trujillo, Noelia. 2020. "18 of the Most Dangerous Jobs Around the World." *Reader's Digest*, January 6, 2020. https://www.rd.com/list/dangerous-jobs-around-world/.

2
Classification

Coals can be found in many places all over the world, from Svalbard to Antarctica, from China all the way around the globe and back again. The stuff called "coal" has a great deal of variation, from a moist, crumbly brown substance looking like exceptionally old wood, to a hard, lustrous, black material. Coals vary so much that Peter Luckie, at Penn State University, recommended that we should never use the word "coal" in the singular, but rather refer to "coals." Because there is so wide a range of properties and compositions, there has to be a system to describe and classify these various coals, to provide producers, sellers, buyers, and users of coals a common set of terms so that all parties know exactly what they are buying, selling, and using.

Successful commercial trade has to be based on a way for the buyer and seller to agree on exactly the properties and characteristics of the material that is being bought and sold. Efficient use requires that fuel engineers know the characteristics of the specific kind of coal they will be using. Equipment can be designed and built to process almost any kind of coal, but specifications and operation have to be mated to the composition and properties of the specific coal to be used. A system of standards means that a coal of a particular name, such as anthracite or bituminous coal, or lignite, fits within the range of some agreed-upon list of properties. Then the buyer, equipment designer, or user can confidently expect that that coal will conform to that list of properties, within reasonable tolerances.

Coals can be classified in three ways. We will see in the next chapter that coals were formed from plants that lived tens to hundreds of million years ago. The components of those plants experienced, over ages, various biochemical and geochemical changes that converted the original plant substance into a form of coal. These changes will be reflected in the chemical composition and physical properties of the specific coal being studied. Coal *rank* expresses the extent to which a coal changed, or progressed away from, the original plant material. (Throughout the book, words that are italicized and bold-faced when first used will be found in the glossary.)

Plants are composed of a variety of components, such as wood, bark, and pollen, each with its own characteristic composition. When substances of different chemical composition—plant components in this case—are subjected to a common set of conditions (the biochemical and geochemical changes) they will almost certainly be changed in different ways. Components of ancient plants retain a semblance of their identity in a piece of coal and will have characteristic appearances that can be discerned under a microscope. These remnants of original plant matter are called *macerals*. They will be discussed in coming chapters. The relative proportions of different macerals in a specimen of coal determine the coal *type*.

All coals contain some amount of non-combustible inorganic materials. The amount of mineral matter, or the amount of ash produced from it, provides a measure of coal *grade*. In most processes ash is something to be collected, dealt with, and disposed of in an environmentally acceptable manner—in other words, a nuisance. Though ash occasionally has commercial uses or can be used as a raw material to extract valuable elements (Chapter 16), users of coal would generally prefer the lowest possible ash yield.

There is no single, universally adopted system for classifying coals. Classification systems have evolved along two parallel tracks. One track includes systems useful mainly for studies of coal chemistry and geology based on chemical composition—especially carbon content—and optical properties. The other track relates to classification for practical use, mostly how coals behave when heated or burned. In countries in which coal has had significant importance to the national economy, standards-setting organizations have established procedures for classifying coals by rank. In the United States, the American Society for Testing and Materials, commonly known as ASTM, sets the standards (ASTM 1984, 247). The ASTM classification system finds wide acceptance even outside the United States. Other bodies, such as the British Standards Institution (BSI) and the Normenausschuss in Germany (DIN), also have established standards for rank classification of coals. Other countries or organizations have done so as well.

This chapter describes the ASTM system for classifying coals. This system will provide the way of identifying or naming coals for use in discussions throughout the rest of the book. The ASTM classification has existed since 1934 (M. Marcinowski, personal communication, 2021). The tests forming the basis for the system relate to the applications that were most important when the standard was established: burning lumps of coal on a grate and converting coal to metallurgical coke for the iron and steel industry. Today, they represent only a small portion of coal use. Nonetheless, the standards haven't changed. Partly this comes from a deservedly conservative and cautious reluctance to change standards that were agreed upon through extensive, painstaking committee work and laboratory testing and that have been in use for many decades. In addition, over these many decades, coal scientists and engineers have built up an extensive empirical, but very useful, body of practical knowledge relating rank classification to composition, properties, and likely behavior of coals when they are used. Why tinker with what works?

By far the dominant uses of coals involve burning them, either to use the heat of burning coal directly or to use the heat to make steam. The paramount question in using any fuel, including the coals, is how much heat can be produced when a known quantity is burned: the heat of combustion, also called the *calorific value* or the heating value.[1] To make a fair comparison of one fuel with another, the heat released during combustion has to be on the basis of the same quantities of fuel in each case. For solid fuels such as coals, calorific value is expressed as heat released per unit weight. Throughout most of the world, units of calorific value are megajoules per kilogram (MJ/kg) or, sometimes, kilocalories per kilogram (kcal/kg). The United States still uses British thermal units per pound (Btu/lb). Knowing the calorific value makes it possible to calculate how much of a specific coal must be burned to generate a needed quantity of heat. Other factors being equal (which they seldom are in the coal business), the highest available calorific value would be desired, which means that the least amount of coal would have to be purchased to provide the heat required.

Measuring the calorific value involves burning a carefully weighed quantity of coal in pure oxygen inside a reaction chamber, which itself is immersed in a known quantity of water. The experimental apparatus is called a calorimeter. The surrounding water captures the heat released when the sample burns. The amount of heat that will raise the temperature of 1 gram of water by 1 Celsius degree is very accurately known. The mass of water and its temperature increase give us the information needed to determine the amount of heat released from the coal. Knowing the mass of the coal burned provides the information needed to calculate the calorific value.[2] Calorific values of the

---

[1] Strictly speaking, the term "calorific value" applies only to solid or liquid fuels and is questionable for use with highly volatile liquids, according to ASTM definition. We will generally follow this usage; when it's not applicable, we will refer to the "heat of combustion."

[2] Already there is a complication: the reporting of two apparently different calorific values for the same coal. When a coal sample burns, carbon dioxide and water are the dominant products. At the combustion temperature, water will be present in its gaseous form (i.e., steam). But when considering calorific value at normal temperatures, say 25°C, water will be a liquid. Condensation of steam to liquid water releases a large amount of heat, 2,267 kJ/kg. A reported numerical value for the calorific value of a coal in which the combustion products are taken to be carbon dioxide and liquid water will include the heat released when the sample was burned plus the heat of condensation of steam. In comparison, the reported calorific value when the products are taken to be carbon dioxide and steam does not include heat of condensation; it will necessarily be the smaller of the two values. In coal science and technology, the term *higher heating value* (HHV) reports the value when the products are carbon dioxide and liquid water. The *lower heating value* (LHV) applies when the products are taken to be carbon dioxide and steam.

world's coals vary widely, from less than 14 to more than 30 MJ/kg, as received in the analytical laboratory. Coals have lower calorific values than natural gas or petroleum fuels, but wood and many other biomass-derived fuels are lower still.

Proximate analysis gives indications of how a coal sample likely will behave when heated and then burned. It's unfortunate that the word "proximate" sounds much like the more common word "approximate." There is nothing at all approximate about a proximate analysis. Each step follows rigorous standards. The term "proximate" refers to the analytical practice of reporting the amounts of several components as if they were a single generic substance. This practice is not limited to coal analysis.

Proximate analysis of coals directly measures three quantities: moisture, volatile matter, and ash. Their values are used to calculate a fourth: fixed carbon. The test procedures of the proximate analysis are well established and carefully followed, but it is likely nowadays that no process or piece of equipment using coal anywhere in the world operates at the exact conditions of the proximate analysis. Why not replace it with something more modern? The same argument can be made as applied to the rank classification itself. More than a century of scientific and engineering practice has established a great many useful relationships that tie the expected composition and behavior of a particular coal to its rank classification or to one or more of the parameters of the proximate analysis. Some of these relationships provide quantitative predictions, others serve as qualitative indicators; all can be useful at times.

All coals have some amount of water associated with them. Some geologically young coals may have more than 50% water associated with the coal substance itself. They are more water than coal. Heating any sample of coal gently at 105–110°C causes a loss of weight. Since 100°C is the normal boiling point of water, material removed at the test temperature is presumed to have been water, in which case the weight lost represents the moisture content of the sample. In keeping with the term "proximate," the test does not call for condensing and collecting the material driven off to see if it contains anything other than water or even to verify that it is water. By definition, the measured weight loss at these conditions is said to be due to *moisture*.

Heating coal to 950°C in an inert atmosphere for a standardized length of time causes another weight loss. The inert atmosphere assures that the coal sample will not catch fire during the test. Some of the material that escapes in this experiment is a mixture of gases. The rest escapes as vapors that condense to an oily or tarry material—a mélange of organic compounds. Here again, the test does not require collecting the gases, oils, and tars, nor does it call for further analysis to determine specific compounds that volatized or to measure their amounts. Rather, all of the material that escapes at the conditions of this experiment are lumped together and called *volatile matter.* Strictly speaking, the term "volatile matter" applies only to material driven off at the specific conditions of the volatile matter test. In the ASTM test for volatile matter (ASTM 1984, 410), the sample is heated at 950°C for 7 minutes. Materials driven off at other conditions of temperature and time are called *volatiles.*

All coals, as mined, contain a variety of inorganic substances. We will discuss them in Chapter 5. When any coal burns, it leaves behind a residue of an inorganic, non-combustible solid ash. The amount of ash collected from a carefully weighed sample of a coal, burned under standardized conditions (ASTM 1984, 406), represents the ash value, or ash yield, of the sample. Ash is a complex mixture, primarily of compounds of silicon and aluminum, but also with compounds of such elements as iron, magnesium, calcium, sodium, and potassium. All are lumped together as *ash.* The proximate analysis does not require determining the chemical composition of the ash. Finding the actual composition of the ash can be done as a separate analytical procedure, and often is, because it provides important indicators of how the ash will behave when coal is burned.

Ash produced from a coal sample results from several chemical reactions and phase changes experienced by the minerals originally present in the sample. Some of those reactions will cause a gain in weight, such as the oxidation of iron disulfide (pyrite) to iron sulfate. Others result in a weight loss, as in the decomposition of calcium carbonate (calcite) to calcium oxide. The ash that forms is not truly a component of the original coal sample. There is no ash in coal. Ash is a product of several simultaneous chemical reactions experienced by the minerals that really were in the coal.

The combustible, carbonaceous residue that did not vaporize as volatile matter is called *fixed carbon.* "Fixed" is an ancient term in chemical nomenclature, applying to a substance or material that is resistant to change. Fixed carbon is fixed because it doesn't volatilize. Years ago, oxygen and nitrogen were often referred to as "fixed gases" because they could not be made to condense to liquids at temperatures normally attainable in a laboratory. The proximate analysis does not measure fixed carbon directly. Rather, it is found by difference, by subtracting the percentages of moisture, volatile matter, and ash from 100.

Coals as they come from the ground contain not only the combustible material that we really want—the "coaly" part of the coal—but also some amount of moisture and some quantity of non-combustible minerals that produce ash. The analytical laboratory can certainly report its results on the sample exactly as it was when it was handed in, called the "as-received basis." Analysis data reported on an as-received basis include the moisture and the ash yield.

The moisture content of any sample of coal can vary somewhat depending on how it was handled and stored before the laboratory got hold of it. To avoid uncertainty or confusion that could result from changes in the moisture content due to how the sample had been handled and stored, it can be more informative to remove—mathematically—the effect of moisture on the analytical results and express the data on a moisture-free basis. The Appendix shows how this, and other corrections discussed here, is done. Because data on coal composition can be reported in a variety of ways, analytical reports need to be treated with caution unless the basis is mentioned on which the data are being reported.

Since neither water nor minerals burn, analysis results can be corrected to a moisture *and* mineral-matter-free (mmmf) basis. Doing so gives information only on the combustible portion of the coal, which is usually the portion we wanted in the first place. While the arithmetic is straightforward, figuring out the amount of mineral matter is not. As a coal sample burns, the minerals transform to ash. Collecting and weighing the ash is not a problem. Knowing the ash yield makes it possible to express coal composition on a moisture-and-ash-free (maf) basis. Neither the weight nor the chemical composition of the ash represents the composition and weight of minerals originally in the coal sample. Consequently ash, though easy to make, collect, and weigh, does not directly measure the mineral matter originally present in the sample.

Ideally, to characterize a sample of coal, we should have an accurate measure of the weight of minerals originally present in the unburned sample. Until the advent of modern, computer-interfaced analytical instruments, isolating and identifying minerals in a sample of coal represented a time-consuming task not suited to routine analytical work. Instead, coal scientists developed formulas to estimate the amount of mineral matter. Such approaches used the weight of ash and other commonly measured properties, such as the sulfur content of the sample, to do this. The Parr formula (found in the Appendix), commonly used in US practice, provides a fairly simple approach for doing this.

Only in recent decades has it been relatively easy to identify and count the minerals originally present in a coal sample. Various techniques, many involving electron microscopy, have been developed to observe, count, and analyze directly the mineral grains in a sample of coal (Creelman and Ward 1996). Equipment for doing this can be very expensive and may require a dedicated operator. Small laboratories may not be able to afford to set up to do such work. Thus, methods for calculating mineral matter, such as the Parr formula, remain useful today. Having a value for the amount of mineral matter, whether measured directly or calculated, allows converting the analytical data to the moisture and mmmf basis.

It's also possible to remove, mathematically, only the contribution of the mineral matter to the analysis results but not remove the moisture content. As shown below, this is actually done in the ASTM system. Doing so provides results on a moist, mineral-matter-free (m,mmf) basis, where the comma should be read as if it meant "but."

Looking into the sample crucible at the end of the volatile matter test would show, for most—but not all—coals, a powdery, black residue of fixed carbon and ash. In some cases, the residue is not a powdery material but instead is a single piece that appears to have formed as if the particles of coal somehow fused or agglomerated together. These coals are called *agglomerating* or *caking coals*. The only coals that exhibit this behavior are in the bituminous rank range (Figure 2.1). Some agglomerating coals not only fuse but also swell to fill much or all of the volume of the crucible. Such behavior is of immense practical importance in using coals in metallurgy, especially in smelting iron ores (Chapter 10).

The calorific value and volatile matter (or fixed carbon), converted to an appropriate basis, provide the information needed for rank classification based on the ASTM system. ASTM rank classification uses volatile matter or fixed carbon reported on an mmmf basis and on the heating value reported on an m,mmf basis. Figure 2.1 displays the ASTM classification system (ASTM

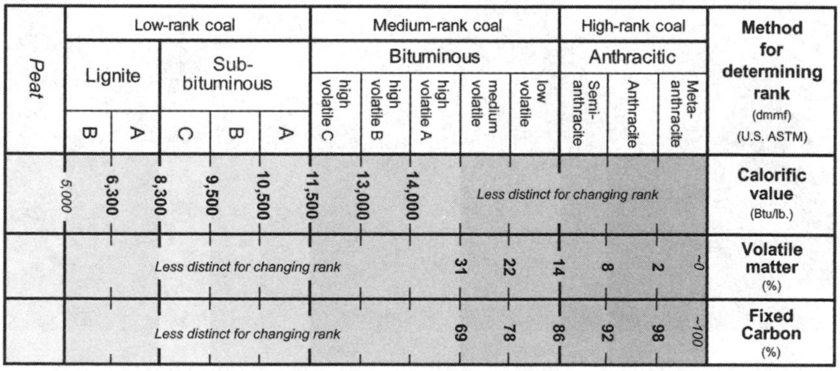

Figure 2.1 The table for the classification of coals by rank.
Artwork by Stephen Greb, Kentucky Geological Survey.

1984, 247). Informally, lignites and subbituminous coals can be lumped together as low-rank coals, and bituminous coals and anthracites as high-rank coals. Figure 2.2 shows a specimen of lignite, at one end of the rank range; at the other end, a specimen of anthracite is shown in Figure 2.3.

The proximate analysis confirms the great variability among the world's coals. Moisture contents range from nearly zero to almost 70%. On a moisture-free basis, ash values run from less than 1% to about 30%, volatile matter from 5% to almost 60%, and fixed carbon from 20% to about 85%. Proximate analysis data also provide useful indications of the behavior and results that can be expected when coals are used.

Moisture has no role in producing useful energy. It dilutes the combustible portion of coal. In fact, moisture is worse than useless because it takes a great deal of heat to raise water to its boiling point and then evaporate it. Burning a high-moisture coal wastes valuable heat just to evaporate water. Either we must accept this inefficiency, or we must find ways to dry the coal before it is used. The quantity of moisture indicates how much drying will be needed in grinding, pulverizing, and handling coal to be put into a boiler in

Figure 2.2 A very young (Pliocene) lignite from Serbia. Woody structures are still evident in these pieces. This is the low end of the coal ranks.
Photograph courtesy of the US Geological Survey. Public domain.

Figure 2.3 The high end of the coal ranks: Carboniferous anthracite from Pennsylvania. It is hard, black, and glossy.
Courtesy of Donna Pizzarelli, US Geological Survey, Public domain.

an electricity-generating station. And, if coal is to be transported any distance, why pay to ship water?

Ash brings its own set of problems. Minerals also dilute the combustible portion of the coal. They don't contribute to releasing useful energy. Ash has to be collected, handled, stored, and disposed of in environmentally acceptable ways. The higher the ash yield, the more collecting, handling, storing, and disposing has to be done. Converting the minerals into ash can release potentially harmful substances, such as mercury or selenium. These, too, have to be trapped and treated in an environmentally acceptable way.

Volatile matter and fixed carbon values help predict how coal will burn. The ratio of fixed carbon to volatile matter represents the *fuel ratio* of the coal. Fuel ratio indicates how difficult a coal will be to ignite, how rapidly it will burn completely, and the quality of the flame (i.e., long and smoky or short and clean). Volatiles vaporize, ignite, and burn as a gas. Fixed carbon burns as a solid. Coals of high volatile matter content can usually be ignited easily but burn quickly. Burned on a grate, such coals have a large, often smoky flame. A high fixed carbon content indicates a coal that may be difficult to ignite but will tend to burn more slowly. A burning bed of such coal likely will have a short, hot, relatively smokeless flame. Anthracites, the premium coals for domestic applications, provide a practical example. On an as-received basis, anthracites typically have about 5% volatile matter and 80% fixed carbon, along with high calorific values, around 30 MJ/kg. A bed of anthracite provides considerable heat and burns for a long time, with virtually no smoke. A problem arises if the fire goes out completely because the low volatile matter content makes it challenging to rekindle an anthracite fire.

The relative amounts of volatile matter and fixed carbon also the affect the rate of heat liberation, sometimes called *calorific intensity*. While a high proportion of volatile matter helps a coal ignite easily and burn with a large flame, the volatile matter does not contribute a great deal to the heat output. On the other hand, coals high in fixed carbon and therefore low in volatile matter burn relatively slowly but yield a large amount of heat, anthracites being examples.

The ASTM system does not require information about the chemical composition of the coal substance. However, knowing the major elements that comprise coals and their relative proportions represents information so useful that such an analysis is commonly performed in conjunction with the proximate and calorific value determinations. In laboratory work on coals, the carbon content, on an maf or mmmf basis, is often used as indication of rank.

The elemental composition of coal is determined by procedures collectively known as the *ultimate analysis*. In this context, ultimate is yet another term borrowed from the early days of chemistry. Here, "ultimate" means that the composition is determined without concern for the exact molecular structure in which the elements occur. In coal chemistry, "ultimate" does not have its more common meaning, which would imply that the sample cannot be analyzed any further or that we have accounted for every last possible element that might be present. Ultimate analysis directly measures percentages of carbon, hydrogen, sulfur, and nitrogen in a sample. The percentage of a fifth element, oxygen, commonly is calculated by summing the other four and subtracting the sum from 100%.

Years ago, an ultimate analysis of coal was a very laborious process, a good bit more complicated than the proximate analysis. This might be another explanation of why classification schemes have tended to rely on proximate analysis results. Nonetheless, skilled chemists could produce highly accurate and precise results from the ultimate analysis. Now, various kinds of automated, computer-interfaced instruments can perform elemental analyses much faster and with much less human effort.

Determination of carbon, hydrogen, and sulfur rely on burning a known amount of coal sample, usually in purified oxygen. The carbon dioxide (CO_2), water, and sulfur oxides produced from burning the coal are absorbed in various reagents. Weighing the amount of absorbing reagent before and after the analysis gives the amount of CO_2 (or water or sulfur oxides) produced from that coal sample. This information, along with the weight of the original coal sample, allows the chemist to work backward to calculate the percentages of the elements in the sample. More sophisticated instruments determine the combustion products directly in the gas stream, usually by spectroscopic methods.

In years past, determining nitrogen required digesting the coal sample in hot concentrated sulfuric acid, producing ammonium sulfate. The ammonium sulfate was subsequently decomposed to ammonia, which would be trapped in an absorbing solution and finally measured in another step. This procedure is known as the *Kjeldahl method*, in honor Danish chemist who developed it. Fortunately for coal analysts, automated Kjeldahl analyzers have been developed.

Simple analytical methods to determine oxygen in coals are not available. The amount of oxygen is commonly determined by adding the percentages of the other four elements and subtracting the sum from 100%. The disadvantage is that all of the experimental uncertainties in the carbon, hydrogen, nitrogen, and sulfur measurements are propagated into the calculated value for oxygen. Calculating, rather than measuring, the oxygen content is known as *oxygen by difference*. Good analytical laboratories that use this method will say so, for example, reporting %O (diff.).

Ultimate analysis measures the total sulfur content of the coal, a useful piece of information in itself. But sulfur occurs in coal in several ways. Some is incorporated in the carbonaceous molecular framework of the coal. Sulfide minerals, most notably pyrite, also occur in coal, along with smaller amounts of sulfate minerals. Total sulfur varies widely, from less than 1% to, in rare cases, more than 10%. Most of the variability in total sulfur content comes from differences in the amount of **pyritic sulfur**. The portion of sulfur chemically bonded to the carbonaceous structure, called **organic sulfur**, is usually less than about 2%. In most coals the **sulfatic sulfur** is quite low. Taken together, the determinations of the pyritic, organic, and sulfatic sulfur constitute a *forms-of-sulfur* analysis (ASTM 1984, 357).

The heat liberated when a coal is burned relates to its elemental composition, particularly to the amounts of carbon, hydrogen, and oxygen. Most of the heat comes from the converting carbon to CO_2 and hydrogen to water vapor. The presence of oxygen atoms in a coal reduces its calorific value. This is because oxygen atoms bonded to carbon atoms have, in effect, already partially oxidized some of the carbon atoms. Consequently, they are less able to generate heat than are carbon atoms bonded only to hydrogen or to other carbon atoms. The relative amounts of hydrogen and carbon affect the total amount of heat generated. Hydrogen burning to water vapor produces 3.7 times as much energy as the same weight of carbon burning to CO_2. The calorific value of a good-quality low-volatile bituminous coal containing 4.6% hydrogen is about 36 megajoules per kilogram, on an maf basis.[3] This represents about the best calorific value one can expect from coals, though no one actually burns a moisture- and ash-free coal. More realistically, the calorific value is about 34 MJ/kg on a moisture-free basis and 33 MJ/kg as-received. In contrast, gasoline, with about 16% hydrogen, has a heat of combustion of 43 MJ/kg and natural gas, 54 MJ/kg. Methane, the principal component of natural gas, contains 25% hydrogen.

Carbon contents of coals run from 70% to 95% (maf basis). Oxygen contents decrease as carbon contents increase, ranging from more than 25% to almost zero. Hydrogen and nitrogen contents show less variability; the hydrogen content being around 5.5% in most coals having less than about 88% carbon, and decreases to about 2% at higher contents of carbon. Nitrogen typically occurs at about 1–2% in most coals.

Though ultimate analysis data are not used in some rank classification systems, they can be used as indicators of rank. Particularly for laboratory work on coals, the carbon content on an mmmf basis is often taken as an indicator of rank: the higher the carbon content, the higher the rank. The hydrogen-to-carbon atomic ratio, calculated as shown in the Appendix, also serves to indicate rank. Coal is formed from plants. The final maturation product of coal formation is pure carbon—graphite. Rank also indicates its extent of geological maturity by assigning the position of that coal in the progression from the original plant material to graphite. Because the carbon content of coal increases as it matures, rank conversely gives a qualitative indication of the amount of carbon in the coal.

References

ASTM. 1984. "Standard Classification of Coals by Rank. D 388-84." *Annual Book of ASTM Standards. Volume 05.05. Gaseous Fuels; Coal and Coke*. Philadelphia: American Society for Testing and Materials: Philadelphia.

ASTM. 1984. "Test Method for Forms of Sulfur in Coal. D 2492-84." *Annual Book of ASTM Standards. Volume 05.05. Gaseous Fuels; Coal and Coke*. Philadelphia: American Society for Testing and Materials.

---

[3] Specifically, these are data for the Pocahontas No. 3 coal from Wyoming County, West Virginia (Newman et al. 1967, 7–4).

ASTM. 1984. "Standard Test Method for Ash in the Analysis Sample of Coal and Coke from Coal. D 3174-82." *Annual Book of ASTM Standards. Volume 05.05. Gaseous Fuels; Coal and Coke.* Philadelphia: American Society for Testing and Materials.

ASTM. 1984. "Standard Test Method for Volatile Matter in the Analysis Sample of Coal and Coke. D 3175-82." *Annual Book of ASTM Standards. Volume 05.05. Gaseous Fuels; Coal and Coke.* Philadelphia: American Society for Testing and Materials.

Creelman, Robert A., and Colin R. Ward. 1996. "A Scanning Electron Microscope Method for Automated, Quantitative Analysis of Mineral Matter in Coal." *International Journal of Coal Geology*, 30: 249–269.

Newman, L. L, W. A. Leech, M. H. Mawhinney, C. R. Velzy, C. O. Velzy, A. J. Tigges, H. Karlsson, and W. E. Lewis. 1967. "Fuels and Furnaces." In: *Standard Handbook for Mechanical Engineers*, edited by Theodore Baumeister and Lionel S. Marks, 7.2–7.16, New York: McGraw-Hill.

3
Coalification

Names that recognize distinctions among the varieties of coals and tell us something about their properties bring up new questions: How did they get to be that way? Is the way in which coals form responsible for their compositions and properties? Basically, where do coals come from?

Coals formed from plants. Evidence can be obtained with the unaided eye. Plant remains that have been converted into coal—that is, that have been *coalified*—can be found in coals of all ranks. Coalified plant parts range from sticks or small branches to parts of tree trunks or stumps. Coalified remains of leaves and spores can also be found.

Coals examined under a microscope reveal the remains of a cellular structure very similar in appearance to plant tissues or wood. Coal microscopy reveals such plant-derived structures as spores and pollen grains. Some grains can be used to identify the kind of plant from which they came. The use of pollen grains or spores to identify the plants from which they came is the basis of the science of palynology. Observations from palynology are very helpful in providing understanding of the kinds of plant communities that gave rise to a particular coal deposit.

Solvents commonly used in organic chemistry will extract various constituents that include ones having molecular structures identical to or clearly derived from ones that can be extracted from modern plants. These compounds, which have survived coalification virtually intact, are sometimes called biological markers, or often *biomarkers*, because they help to identify, or mark, the biological origin of the samples being studied. For example, many plants secrete resins. Resins contain a family of compounds called terpenes, as well as their relatives the diterpenes and triterpenes. Extraction of lignites and subbituminous coals yields di- and triterpenes that likely originated in conifers (Goodwin and Mercer 1983, 460–464).

The ultimate analysis tells us that coals consist of carbon, oxygen, and hydrogen as the major components, along with smaller amounts of nitrogen and sulfur. Carbon predominates in every coal. Coals range from about 65% to 95% carbon moisture-and-ash free (maf). Plant compounds also consist predominantly of carbon, oxygen, and hydrogen along with smaller amounts of nitrogen and sulfur. The similarity in composition suggests that coals derived from plant materials. A trillion tonnes of coal still remain (Rock 2020). The only likely natural source of vast amounts of material rich in carbon and hydrogen and containing smaller amounts of oxygen, nitrogen, and sulfur would be living, or once-living, matter.

Plant components and coals are similar in composition, but not identical. Wood contains a much greater proportion of oxygen relative to lignite, about 51% versus 24% respectively, and somewhat more hydrogen as well, 7% versus 5%.[1] As plant material coalifies, the processes that occur will be ones that remove oxygen atoms and some hydrogen atoms. Because the total percentages of all the elements must always sum to 100, reductions in hydrogen and oxygen contents are reflected in a greatly increased percentage of carbon.

Some special coals clearly reveal their plant origin. *Paper coals* look like coalified paper; layers of the coal appear somewhat like a handful of sheets of paper. The *bark coals* of southern China contain high proportions of a component, provisionally called *barkinite*, which has clearly derived from tree bark (Wang et al. 2014). *Coal balls* are aggregations of plant material that have been preserved by impregnation with minerals such as pyrite or calcium carbonate; they have been petrified rather than coalified.

To understand how coals were formed, it helps to know something of the kinds of plants and environments that contributed the material that eventually became coal. The eras of Earth's history are shown in the geologic time scale (Figure 3.1).

The first plants to grow on land appeared toward the end of the Silurian. They were the earliest examples of *vascular plants*, characterized by tubes or channels that transport nutrients and water throughout the plant structure. Though they did not appear until the end of the Silurian, their spread was so rapid that extensive forests already existed by the middle of the Devonian.

The oldest coal comes from the close of the Devonian, about 360 million years ago. Two groups of plants became increasingly important: one whose descendants are the club mosses; the other, a group including modern-day scouring rushes and horsetails. Many of these early plants inhabited subtropical forests and preferred moist places, suggesting that warm and wet environments favor prolific production of the plants that eventually became coal.

The Carboniferous derives its name from the enormous deposits of coal formed at that time. In coal geology, the Carboniferous has likely been studied more extensively than all of the other periods collectively. Earth's climate

---

[1] These data are on a moisture-and-ash-free basis (Newman et al. 1967).

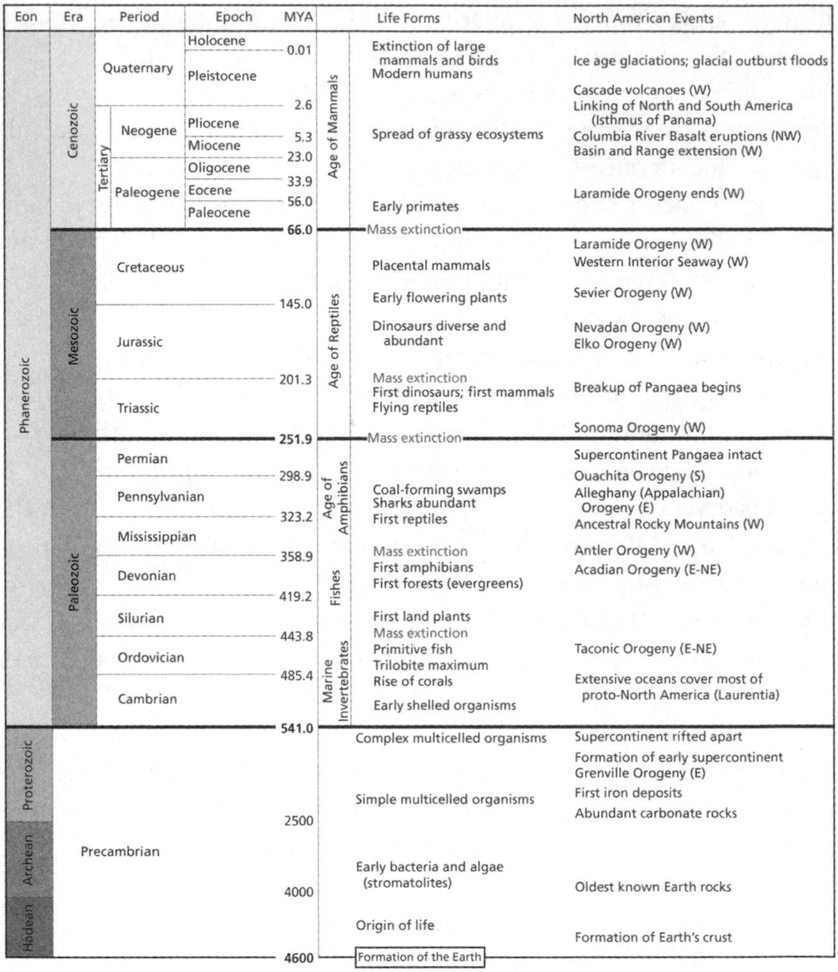

Eon	Era	Period	Epoch	MYA	Life Forms	North American Events
Phanerozoic	Cenozoic (Tertiary)	Quaternary	Holocene	0.01	Extinction of large mammals and birds; Modern humans *(Age of Mammals)*	Ice age glaciations; glacial outburst floods
			Pleistocene	2.6		Cascade volcanoes (W); Linking of North and South America (Isthmus of Panama); Columbia River Basalt eruptions (NW); Basin and Range extension (W)
		Neogene	Pliocene	5.3	Spread of grassy ecosystems	
			Miocene	23.0		
		Paleogene	Oligocene	33.9		Laramide Orogeny ends (W)
			Eocene	56.0	Early primates	
			Paleocene	66.0	Mass extinction	
	Mesozoic	Cretaceous		145.0	Placental mammals; Early flowering plants *(Age of Reptiles)*	Laramide Orogeny (W); Western Interior Seaway (W); Sevier Orogeny (W)
		Jurassic		201.3	Dinosaurs diverse and abundant	Nevadan Orogeny (W); Elko Orogeny (W)
		Triassic		251.9	Mass extinction; First dinosaurs; first mammals; Flying reptiles; Mass extinction	Breakup of Pangaea begins; Sonoma Orogeny (W)
	Paleozoic	Permian		298.9	*(Age of Amphibians)*	Supercontinent Pangaea intact; Ouachita Orogeny (S)
		Pennsylvanian		323.2	Coal-forming swamps; Sharks abundant; First reptiles	Alleghany (Appalachian) Orogeny (E); Ancestral Rocky Mountains (W)
		Mississippian		358.9		
		Devonian		419.2	Mass extinction; First amphibians; First forests (evergreens) *(Fishes)*	Antler Orogeny (W); Acadian Orogeny (E-NE)
		Silurian		443.8	First land plants; Mass extinction	
		Ordovician		485.4	Primitive fish; Trilobite maximum; Rise of corals *(Marine Invertebrates)*	Taconic Orogeny (E-NE)
		Cambrian		541.0	Early shelled organisms	Extensive oceans cover most of proto-North America (Laurentia)
Proterozoic		Precambrian		2500	Complex multicelled organisms; Simple multicelled organisms	Supercontinent rifted apart; Formation of early supercontinent; Grenville Orogeny (E); First iron deposits; Abundant carbonate rocks
Archean				4000	Early bacteria and algae (stromatolites)	Oldest known Earth rocks
Hadean				4600	Origin of life; Formation of the Earth	Formation of Earth's crust

Figure 3.1 The geological time scale.
Courtesy of Rebecca Port, National Park Service.

was warm and humid, similar to tropical or subtropical regions today. These conditions supported extensive proliferation of plant life. Plants grew to enormous proportions compared with their modern descendants. Carboniferous club mosses were about the size of modern trees, while today's club mosses are small plants. Ancient ferns grew some 10 to 30 meters, with stems more than a meter in diameter. The ancestors of ground pines and club mosses, the *lepidodendrons* or scale trees, grew prolifically and often dominated Carboniferous forests. The largest may have reached 40 meters. The name derives from the scars left on the stem where leaves were attached. Stumps of these trees are sometimes found as fossils in the rock beneath a coal seam,

Figure 3.2 A petrified tree stump found at the base of a coal seam.
Photograph courtesy of National Geographic. Public domain.

as shown in Figure 3.2. Modern rushes may grow to about 30 centimeters in height, but their Carboniferous ancestors sometimes attained 30 centimeters in diameter, with proportionately greater heights. These plants all reproduced by spores. The first seed plants of the class of *gymnosperms*, which includes conifers, appeared during the Carboniferous. By the late Carboniferous conifers contributed extensively to the plant life of the period. Expansion of the conifers and other seed plants occurred at the expense of the spore-bearers.

Carboniferous forests were immense. As an example, the Pittsburgh seam of bituminous coal, a single coal deposit dating from the Carboniferous, lies beneath an area of 50,000–65,000 square kilometers. Swamps developed on flat, low-lying ground along the shores of shallow seas or near the mouths of rivers. They occasionally experienced extensive flooding or complete drying. The swampy environment was important for the preservation of plant material, and the flooding or drying had a role in determining the extent of coal deposition.

In the Southern Hemisphere, important deposits of bituminous coal in South Africa and Australia accumulated in the Permian. Plant life in the Permian was much the same as in the Carboniferous. Preserved tissues of plants from the Permian show evidence of growth rings. Plants accumulate growth rings when their biological activity is not continuous but occurs only in favorable periods, with a definite change of seasons. This suggests that the perpetual summer of the Carboniferous had ended and that the cooler climate of the Permian experienced seasonal changes.

Cooler conditions in the Permian favored different plant communities (Falcon and Ham 1988) more suited to cool-to-warm climates rather than the steamy swamps of the Carboniferous. Permian plants necessarily will contain tissues and organs in different relative proportions and of different chemical compositions compared to those of the Carboniferous. When those different kinds of plants undergo coalification, we should expect some differences in the nature of the coals that formed. This is exactly what we see. Permian coals, many of which are found in the southern hemisphere, show characteristics different from the northern-hemisphere Carboniferous coals (Falcon and Ham 1988). Permian-era plant communities contributed different proportions of such components as pollen, waxes, and leaves. The plant components could have contained molecules of different composition and molecular structures relative to those of the Carboniferous plants.

Major episodes of coal formation resumed during the Cretaceous period. Cretaceous coals occur in many parts of the world, from Spitzbergen to Antarctica. Coal occurring in such currently inhospitable places tells us that the climate in those regions was once much milder than it is today.[2] Enormous amounts of low-rank coals of Cretaceous Age occur in the western part of North America, as well as in Australia, New Zealand, Germany, and Siberia.

The Cretaceous period experienced a significant shift in vegetation. During the early Cretaceous, a new kind of plant began to develop, one that enclosed its seeds in a protective covering. Plants with enclosed seeds are called *angiosperms*. As plant evolution continued through the Cretaceous, angiosperms became increasingly important, to the point that they represent the dominant plant form today. Deposition of coal continued into the Tertiary period. Angiosperms were the major contributors to the coal deposits of this period. Some very young coals, such as the lignites in Bulgaria, are of Miocene age.

The trillion tonnes of coal still remaining (Rock 2020) contain about 650 billion tonnes of carbon, based on an assumption that the average carbon content of all the coals in the world is 65%, on an as-received basis. This doesn't take into account the enormous tonnages that have already been extracted

---

[2] We should also take into consideration the possibility that, thanks to continental drift over the eons of geological time, such places were much closer to the Equator during the Cretaceous than they are today.

and burned in previous centuries. Where does this prodigious amount of carbon come from? It accumulated and built up one atom at a time, starting with a very simple molecule—carbon dioxide (CO_2). Because carbon is the primary element in all coals, and because coals occur in nature, a further step in answering the question of where coals come from is to consider the major natural processes that involve carbon.

It must be that some plants passed through their life cycles and accumulated in the environment as organic matter—dead plants—without decaying completely. Formation of coals results from the fact that decay, the process that returns carbon back to the atmosphere, is not perfectly complete. About 98–99% of the organic material does indeed decay. The 1% or 2% of accumulated organic matter that does not decay undergoes a complex series of chemical and geological transformations that eventually lead to the substances that we recognize as fossil fuels—coals, petroleum, and natural gas.

Carbon incorporated in plants ultimately derives from CO_2 in the atmosphere. Plants capture CO_2 during the process of photosynthesis. During photosynthesis, plants convert CO_2 into sugars. Six molecules of CO_2 and six molecules of water are used to produce a molecule of glucose, with oxygen as a second product. Glucose is the most important molecule produced by plants (Schobert 2013a, 19–34). Plants use glucose both for short-term energy storage and as a raw material for the biosynthesis of many other molecules used in their life cycles.

CO_2 is exceptionally stable. Converting it to glucose requires a considerable input of chemical energy to overcome the inherent stability of the CO_2 molecule. The prefix "photo-" indicates that the needed chemical energy comes from light, specifically from the energy in sunlight. For this reason, we can think of the energy that we derive from coals as being a form of stored solar energy.

Sugars formed in photosynthesis are used by plants as nutrients and to synthesize the many other biochemical components needed in their life processes (Schobert 2013a, 19–34). When plants consume their nutrients, some of the carbon in such molecules returns to the atmosphere as CO_2. When the plant eventually dies, its tissues decompose, releasing the remaining carbon, again as CO_2. If the plant is eaten by an animal, CO_2 is returned to the atmosphere by the respiration of the animals while they are alive and by decay of their bodies after their deaths.

Instead of decaying, 1% or 2% of accumulated organic matter undergoes a different process that leads to its conversion to coals (or other fossil fuels). We use these fuels by extracting them from the Earth and burning them. Burning converts the carbon in the molecular constituents of the fuel back to CO_2, which returns to the atmosphere. This process closes the cycle from CO_2 to living organisms to accumulated organic matter and back to CO_2. Possibly *the* major problem confronting humanity today is that some parts of the cycle are running faster than others, leading to increasing concentrations of CO_2 in the atmosphere, which in turn impacts Earth's climate. We return to this in Chapter 15.

Living organisms contain thousands of chemical compounds. Rather than trying to trace the reactions of each one, we can consider decay as a reaction that is the reverse of photosynthesis (i.e., the reaction of glucose and oxygen to produce water and CO_2). In this case a molecule of glucose reacts with six molecules of oxygen, forming six molecules of CO_2 and six molecules of water. Air is the source of oxygen. It is an *aerobic* process. Aerobic decay of glucose is the exact reverse of photosynthesis. It releases carbon from once-living matter, returning it to the atmosphere as CO_2.

In a relatively dry environment, decay starts immediately. Aerobic bacteria and fungi attack the plant tissues, converting their components to CO_2 and water. With enough time, no organic matter will be left to form fossil fuels. Because aerobic bacteria need atmospheric oxygen to live, the principal chemical agent in aerobic decay is oxygen. To shut down the decay process requires finding ways to exclude or eliminate oxygen.

Covering the organic matter with stagnant water, mud, or silt provides effective ways of doing this. This can happen in swamps, marshes, lakes, river deltas, and lagoons. Tropical or subtropical regions favor abundant plant growth, which leads to production of large quantities of organic matter. The more organic matter produced, the better the chances that some of it will be preserved. Organic matter must be buried deeply enough—a meter or more—to eliminate access of oxygen. The very first part of coal formation is a race between the rate of aerobic decay and the rate of burial. Slow burial likely allows aerobic decay to triumph, and carbon goes back to CO_2. Quick burial lets some carbon survive in coals. Modern environments that provide for the preservation of organic matter include, as examples, swamps in Indonesia and the Brahmaputra and Ganges Deltas in India.

Below 1 meter of depth, with oxygen depleted, new chemical processes affect the organic matter. They are also facilitated by bacteria, but ones not needing oxygen for their life cycles: *anaerobic* bacteria. Bacterial attack on organic matter represents the first major phase in the formation of coal, a process sometimes referred to as the **biochemical phase** of coalification. It is also called **diagenesis**, which means the transformation of materials through dissolution and recombination of their components.

Anaerobic bacteria decompose plant material through chemical pathways much different from those used by aerobic bacteria. Anaerobic decay of one molecule of glucose produces three molecules of CO_2 and three molecules of methane. In the comparison of wood and lignite, we saw that coalification will require processes to reduce the proportions of oxygen and of hydrogen relative to carbon. Though carbon is lost in both anaerobic products, much more hydrogen and oxygen are lost than carbon. An atom of carbon lost in CO_2 takes away two atoms of oxygen. An atom of carbon in a methane molecule carries off four atoms of hydrogen. All of the many other chemical components of plants are also potential reactants for the anaerobic processes.

As anaerobic decomposition proceeds, products of the breakdown of plant components combine to produce brown or black solids of high molecular weight and of ill-defined molecular structures, called *humic acids*. Humic acids dissolve in alkaline solutions. If the resulting solution is reacidified, they reappear as a precipitate. This description does not say what humic acids *are*; it simply indicates how we can recognize them by their behavior under certain conditions.

By the time the decomposing organic matter is buried to a depth of about 10 meters, anaerobic bacteria have consumed most of what they are capable of metabolizing, and their action ceases. Also, some of the decomposition reactions produce phenol or its derivatives,[3] which are good bactericides (Schobert 2013b, 103–131). The different components of plants all have their own characteristic molecular compositions and structures. When substances of different compositions and structures are subjected to the same chemical conditions, as in diagenesis, it commonly happens that they react to different extents and possibly in different ways (Schobert 1990a, 27–37). Carbohydrates and proteins react very readily. Fats, oils, waxes, and resins change very little. Trees and shrubs contain a special component, lignin, which forms in cell walls and serves to add mechanical strength and rigidity to trunks and branches. Lignin also undergoes little change in diagenesis.

With about 10 meters of burial, diagenesis has produced a mixture of humic acids with unreacted or partially reacted fats, oils, waxes, resins, and lignin. These substances combine in poorly understood ways to form *kerogens*. Kerogens are brownish-black, high molecular weight solids that do not dissolve in alkaline solutions nor in most acids, nor in common laboratory solvents. Kerogen formation ends the biochemical phase of fossil fuel

[3] Phenol, known by its trivial name carbolic acid, was introduced in surgery as a germicide and disinfectant by the British physician Joseph Lister. A derivative of phenol, commonly called creosol, is an important component of creosote, a widely used disinfectant and antiseptic.

formation. Three types are recognized, which differ depending on the dominant source of the original organic matter. Type I, algal kerogen, comes primarily from algae. Type II, liptinitic kerogen, derives mostly from plankton, small organisms that float and drift freely in water. Type III, humic kerogen, originated mainly from higher plants. Most of the world's commercially important coal deposits derive from humic kerogen. Algal and liptinitic kerogens mainly produce petroleum and natural gas. However, a few coals derive from these kinds of kerogen—boghead coal from type I and cannel coal from type II kerogen.

Diagenesis is not always uneventful. Floods can wash away accumulated organic matter, removing material needed for coal formation. Drying of the swamp can expose organic matter to air, leading to its decomposition by aerobic decay. If flooding or drying persists for long times, coalification comes to a halt. Some other types of sediment—such as sand, clays, or silt—might be deposited on top of the newly forming coal. Possibly, the swamp might be reestablished at a later geological time, leading to formation of a new seam of coal. A repetitive process of swamp formation and destruction eventually results in a series of coal seams being formed, separated by other sediments.

Saline or brackish water in the local environment likely contains sulfate ions. These waters support the growth of bacteria that convert sulfate to hydrogen sulfide or to other sulfur-containing species. Then, sulfur can be incorporated in the plant residues as organic sulfur or as pyrite.

Continued accumulation and compression of type III kerogen forms *peat*. Peat consists of the compressed tissue of newly accumulated plants along with residual, partly decomposed organic matter, complete with sulfur and inorganic materials that either were in the plants to begin with or were washed into the swamp. The weight of upper layers of accumulating organic matter builds up, compacting the lower layers and consolidating them more densely. These effects increase the carbon content from less than 50% (maf basis), roughly typical of wood, to about 60% in peat. It takes about a hundred years to form a layer of peat 5–8 centimeters thick. Some modern swamps have peat layers in excess of 10 meters, suggesting that growth, death, and decay have been proceeding relatively undisturbed for more than 15,000 years. Peat is not usually regarded as a form of coal. It has sometimes been considered as a renewable resource, somewhat like biomass, but it accumulates so slowly that renewing a peat bed would take centuries.

Formation of peat marks the end of the biochemical phase of coalification. The plants died, bacteria attacked, and a layer of peat formed. This may have taken a few thousand years. Now biology stops and geology takes over. The *geochemical phase* of coal formation begins. No distinct demarcation occurs between the biochemical phase and the subsequent events. Peat represents the transition between the two phases. Millions of years will pass as the processes of the Earth slowly but inexorably transform consolidated peat into coal.

Formation of coal from peat results in the peat being compressed tenfold; that is, a 1-meter-thick coal seam came from about 10 meters of peat. Some places in the world have coal seams more than 30 meters thick. If these came from 300 meters of peat, the accumulation of peat must have been going on for well over a million years. Some of the seams of subbituminous coal in Wyoming are close to 80 meters (Schwamle 2019). They must have taken a long time indeed.

Peat, buried increasingly deeply inside Earth's crust, starts a slow transformation to coal. As layers of new sediments begin to build on top of the peat, their weight exerts pressure that slowly compacts the peat and buries it more deeply. Burial exposes the peat to elevated temperatures. The effect of temperature, possibly helped a bit by pressure, transforms peat to brown coals and lignites. The transformation is driven primarily by natural forces inside Earth's crust: temperature, from the natural thermal gradient in the crust, and pressure, from weight of overlying sediments. This is why the term "geochemical phase" is used. It is also called *catagenesis*. Millions of years will pass as geochemical processes slowly but inexorably transform the consolidated peat into coal.

Peats have compositions not much different from those of the major components of higher plants. Conversion of peats to brown coals and lignites decreases their moisture and oxygen contents and increases the carbon contents (maf basis).[4] The apparent reduction in oxygen content accompanying the coalification from peat to lignite results largely from a loss of CO_2 (Antisell 1865, 18). Brown coals and lignites contain recognizable plant fragments.

Subbituminous coals and then bituminous coals subsequently form as temperature and time continue to exert their effects. Compared with lignites, bituminous coals have significantly more carbon, about 80–90% (maf), and much lower oxygen contents, generally in the range of 3–9%. Bituminous coals have the greatest commercial importance of the world's coals.

A carbon content of about 91% marks the beginning of the anthracites. Most traces of the original plant life are gone. Formation of anthracites likely

---

[4] This apparent increase in carbon content is not a result of somehow adding carbon to the system. Because the results of the ultimate analysis are required to sum to 100%, a reduction in the percentage of any one component—in this case, oxygen—necessarily requires that the percentages of the other components increase to keep their sum at 100.

involved geological pressure as well as temperature. Many of the world's anthracites lie in mountainous regions. Mountain building exerts enormous pressures on coal seams and their surrounding rock layers. Heat and pressure generated by the folding and compression of rocks completed the transition to anthracites. At the upper end of the rank range, anthracites contain about 95% carbon and retain only about 2% oxygen (maf basis). The ultimate product is pure carbon—graphite.

Two centuries ago, Humphry Davy, an outstanding chemist, inventor, and poet of his era, examined a sandstone quarry outside Glasgow. There, plant matter that had accumulated close to the surface had scarcely decomposed. About 3 meters down, "it has the appearance of wood slightly charred" (Davy 1980, 95–96). Down where the sandstone lay, the plant matter appeared to have been converted to bituminous coal.

About 60 years after Davy, the German geologist Carl Hilt studied the coals of western Germany, southern Wales, and the Pas de Calais region of France. He found a relationship between volatile matter and the depth to which a sample had been buried, now known as *Hilt's rule*. Originally stated, Hilt's rule said that, "In a given shaft or a given borehole, the volatile matter content decreases with increasing depth" (Hilt quoted in van Krevelen 1993, 92). A decrease in volatile matter is consistent with an increase in rank. More generally, the deeper a coal has been buried, the higher the rank it has attained (Schobert 1990b, 55). It is sometimes extended to say that the deeper coal is of higher rank and geologically older.

More and more sediments building up on top of coal require longer and longer times to accumulate. For a series of coal seams lying over one another, it's reasonable to expect that the one on the bottom is geologically the oldest. The longer and deeper a coal has been buried—the older it is—the more it has been exposed to catagenic processes and the more extensively it has been coalified. These considerations make it reasonable that there should be a relationship among age, depth, and rank.

Hilt's rule works pretty well—most of the time. It is not invariably true. Hilt's rule assumes "normal" geology: that sediments accumulate on top of each other millennium after millennium, and that the Earth's heat never wavers. Unfortunately, there is not an absolute, unvarying three-way relationship among rank, depth, and age. Brown coals found near Moscow formed during the Carboniferous. Never buried deeply, they never experienced the natural temperatures and pressures that could have transformed them to bituminous coals. On the other hand, temperature extremes, as might be caused by the intrusion of magma near a coal deposit, can transform a relatively young coal to high rank. Anthracites in the Canadian Rocky Mountains and in the Peruvian Andes are geologically young—Tertiary or Cretaceous—but have experienced very high temperatures. The temperatures of the magma intrusion caused this coal to be "up-ranked" in a comparatively short geological time.

Charles Lyell, likely the preeminent geologist of the nineteenth century, made an extensive visit to the United States in the early 1840s (Lyell 1845). Part of the trip was devoted to examining the anthracites and bituminous coals in Pennsylvania. Although the rocks across Pennsylvania are of the same geological age, the eastern portion of the state contains anthracites, while from the middle of the state and extending westward into Ohio are great quantities of bituminous coals.

Lyell concluded that the processes of mountain building were responsible for the transformation of bituminous coals to anthracites. The forces responsible for the mountain building were most intense in eastern Pennsylvania, where the rocks were the most deformed by faults and folds.[5] In an anthracite mine in Pottsville, Lyell saw that the strata, including the anthracite vein, had been uplifted to the extent that they were nearly vertical (Lucier 2008, 91–94). One wall of the mine, originally the clay bed on which the organic matter had been deposited, contained fossilized underground roots of the Carboniferous plants that had contributed to coal formation. The opposite wall had fossilized impressions of stems and leaves. This evidence indicated that the coalified Carboniferous environment had been shoved so hard as the mountains were raised that it had been turned on end—pushed up through 90 degrees.

Lyell's contemporary, the American geologist Henry Darwin Rogers, also recognized that the bituminous coals and anthracites of Pennsylvania were once part of a single gigantic coal deposit, the eastern portion of which had been most affected by mountain building. Lyell and Rogers recognized that anthracite was bituminous coal that had lost its "bitumen." *Bitumen* is a generic term that applies to mixtures sometimes described as a "black, sticky mixture of hydrocarbons occurring naturally or pyrolytically" (Morris 1992, 268). Essentially, anthracite was de-bitumenized bituminous coal. Heating bituminous coals in an oxygen-free atmosphere—so that they don't catch fire—can yield significant amounts of tar and oily hydrocarbons, while this does not occur with anthracites.

---

[5] In structural geology, a *fold* is a curve or bend in a structure, such as a layer of rock, which normally would be planar. It takes a lot of force to bend rocks. A *fault* is a fracture, or a whole zone of fractures, in which one side has been displaced relative to the other side.

Lyell showed that coals were "most bituminous" near the western end of the Appalachian coalfield, where the strata are level and apparently unaffected by geological forces, and that they become "progressively debituminized" in the direction of rocks that are distorted (Lucier 2008, 91). Rogers pointed out that the Pennsylvania anthracite region had no evidence for the existence of true igneous rocks, so the conversion of bituminous coals to anthracites in this region could not have been caused by intrusions of magma.

Lyell visited what were, at the time, the northern-most coal mines in the United States, in Blossburg, Pennsylvania. Lyell referred to the region as "one of the extreme outliers of the great Appalachian coal field" (Lyell 1845, 87–89). Here there were nine seams of bituminous coal, some of which were 2 meters thick. These seams were horizontal, contrasting with the steeply tilted beds in the anthracite region, only about 100 kilometers away. This was further indication that the conversion of bituminous coal to anthracite was related to the extent of geological disturbance. The horizontal seams of coal at Blossburg had the same *Stigmaria* fossils in the clay underneath and the same impressions of leaves in overlying sandstone as in the near-vertical seams of anthracite in Pottsville.

Exploration for oil in the Mahakam delta region of Indonesia led to the discovery of a sequence of coal seams overlain by peat and going down some 4 kilometers (Given 1984). At the top, peat is still accumulating. The deepest seam of coal contains about 85% carbon (maf basis); this carbon content and its optical properties classify the coal as high-volatile A bituminous. The coals in between show all the gradations of rank. All of the coals are of the same geological age, mid-Tertiary. The principal contribution to the organic matter of all of the sediments was the same, coming from mangrove palms.[6] There is no evidence of magmatic intrusion, and none of mountain building.

The coals of the Mahakam Delta provide a remarkable example of how depth of burial can be the responsible agent for coalification from peat at least through the end of the high-volatile bituminous coals. The Russian geologist Aleksi Petrovich Keppen observed that the deepest coals in the Donets Basin were anthracites; those in the middle were "caking and coking" coals (i.e., bituminous coals); and at the top were "gas coals" and cannel coals (Keppen 1893, 71). Though Keppen of course lacked the analytical equipment that was available almost a century later, his work is in good general agreement with

---

[6] *Nypa fruticans*, also known as the nipa palm.

the study of the Mahakam Delta coals. Keppen's work also ties in with that of Lyell and Rogers on the effects of mountain building. On the eastern side of the Urals, he found anthracites mixed with graphite and, eventually, beds containing only graphite (Keppen 1893, 21).

Charles Lyell had examined a bed of "plumbaginous anthracite"[7] near Worcester, Massachusetts (Lyell 1845, 247), an anthracite that had not fully converted to graphite. Although this material had 3% volatile matter, verging on being a meta-anthracite, it could produce black markings on paper and was used for making pencils. Lyell noted the veins of quartz in this bed, suggestive of some effect of igneous activity on the anthracite.

We have seen that carbon content of coals increases steadily as coalification proceeds: from about 50% in wood, to 60% in peat, 70% in lignites, 80–90% in bituminous coals, up to about 95% in anthracites. No carbon is added during the coalification process; other elements, mainly oxygen and hydrogen, are lost. Time, temperature, and pressure applied progressively, over geological time, eliminate hydrogen and oxygen mainly by forming small gaseous molecules such as water, methane, and CO_2. These transformations cause the oxygen content to drop sharply during the transition from lignites to bituminous coals. A significant reduction in hydrogen begins near the end of the bituminous rank range and continues with the formation of anthracites. These changes in composition imply that a specific coal deposit should experience a regular progression of rank changes over the course of time. As coalification proceeds, increasingly carbon-rich products will inevitably dominate.

Temperature derives from the natural geothermal gradients in the Earth. With typical geothermal gradients, coals will not likely experience extreme temperatures. How can the extensive changes in composition and molecular structures take place at the comparatively low temperatures that accompany coalification? Time is on our side.

Once peat has been covered with inorganic sediments, the very long time periods typical of geological processes can operate. Reactions that would be immeasurably slow on a human scale can occur over the tens of millions of

---

[7] Graphite was once commonly called "plumbago." In the sixteenth century, a deposit of graphite was discovered in England. Presumably, the graphite was taken to be a form of lead, for which the Latin name is *plumbum*. Because of its grayish-black metallic appearance, the graphite was given the name "plumbago," variously translated as "lead ore" or "black lead." The difference should be immediately obvious to anyone who has held a piece of graphite in one hand and an equal-sized piece of lead in the other, since lead is five times denser than graphite. If this weren't enough, the leadworts, a family of decorative plants, are also known as plumbagos.

years that might be needed for catagenesis. Furthermore, time and temperature are somewhat interchangeable. Many cultures have sayings urging haste in business matters, to the effect that "time is money." In coal geology, time is temperature. A long time at a high temperature or a short time at high temperature could have similar effects—thus the occasional ancient brown coals and young anthracites.

Early stages of catagenesis break down some of the components that had survived diagenesis mostly unaltered. Fats, oils, waxes, and lignin react with the removal of oxygen as CO_2 and water. These reactions cause a steady decrease in the amount of oxygen and a small decrease in hydrogen relative to the amount of carbon. Consequently, two families of products form during catagenesis: those with a higher percentage of hydrogen than the parent kerogen and a second set having a higher percentage of carbon. No hydrogen or carbon are added to the system; these changes arise from chemically redistributing or reshuffling hydrogen that was in the kerogen at the start of catagenesis (Schobert 2013b, 103–131).

The most hydrogen-rich, stable carbon compound is methane, CH_4. The ultimate carbon-rich material is carbon itself. The form of carbon stable under most natural conditions is graphite. Catagenesis normally proceeds only part way to methane and graphite. Just as in any other chemical reaction process, how far catagenesis proceeds depends on the reaction conditions, primarily temperature and time.

Since no hydrogen from an external source is added to natural systems, the hydrogen-rich products can experience increasing percentages of hydrogen only by using hydrogen atoms taken from the family of carbon-rich products. Redistribution of hydrogen *already in the system* accounts both for the increasing hydrogen content of the hydrogen-rich products and for the increasing carbon content (or, from the other perspective, the decreasing hydrogen content) of the carbon-rich products. These changes in composition represent a geochemical manifestation of the old saying that "the rich get richer and the poor get poorer."

Internal redistribution of hydrogen is limited by how much hydrogen was available in the kerogen in the first place. If the kerogen was relatively rich in hydrogen, the hydrogen-rich products will dominate. Alternatively, a kerogen with relatively little hydrogen will lead primarily to carbon-rich products. For every hundred carbon atoms, algal kerogen has about 170 hydrogen atoms, and liptinitic kerogen about 140. In comparison, humic kerogen has about 85 hydrogen atoms per hundred carbon atoms. Normally, algal and liptinitic kerogens produce petroleum and natural gas; humic kerogen produces the coals.

We have already seen that some uncommon coals derive from algal or liptinitic kerogen. These sapropelic[8] coals have higher hydrogen contents than the much more common coals formed from humic kerogen. As a rule, humic coals have less than 6% (maf) hydrogen. Boghead coals form mainly from algal kerogen. They typically contain about 11–12% hydrogen. Plant debris such as spores and pollen grains form liptinitic kerogen, which transforms to cannel coals, of about 8% hydrogen. The comparatively high hydrogen contents of sapropelic coals makes them easier to ignite than the humic coals. The term "cannel" derives from the Scots word for "candle," possibly bestowed on these coals because they supposedly can be ignited with an ordinary match, burning with a luminous, smoky flame.

The eventual product on the carbon-rich side of the hydrogen redistribution diagram will be graphite. As coalification proceeds, the percentage of carbon in the coal steadily increases, linking with Hilt's rule and Keppen's observations. Then it should be possible to use carbon content as a way of ordering, or ranking, the world's coals in a logical progression from the least to the most coalified. As we have seen, in many, but not all, cases, this also ranks coals in order of their geological age. Carbon content does make a useful way of ranking coals, especially for laboratory studies.

Coals vary from one rank to another, vary from one mine to another even for coals of the same rank, and vary from place to place within the same seam. Several factors contribute to this variability. The original plants consisted of botanically and chemically distinct parts, such as wood, pollen, bark, seed coatings, resins, and waxes. The proportions of different plant components vary from one plant to another. *Lignins*, major structural components of wood, vary from species to species, vary in composition in the individual parts of a single plant, and in individual cells that are producing lignin. Different quantities of these plant components may have accumulated in different locations in the coal-forming environment. Plants evolved over the course of geological time. The chemical composition of various plant parts likely changed as well. A resin preserved in a Carboniferous bituminous coal may not be chemically the same as one in a Tertiary lignite. Under the same time and temperature conditions of coalification, we cannot expect materials that are chemically different to produce identical products.

Coalification resembles cooking, though on an immense scale for unimaginable times. The flavor, texture, aroma, and appearance of a dish made in

---

[8] The term "sapropelic" refers to the accumulation of organic matter in water environments such as shallow seas, lakes, or lagoons. In such environments we would expect plankton or algae to be the primary contributors to the organic matter.

the kitchen depend on its specific ingredients, the relative amounts of those ingredients, how long we have cooked it, and at how high a temperature. On occasion, we may even resort to a pressure cooker. We can, within limits, arrive at the same culinary result by using a higher temperature for a shorter time. All of these points have analogies in coalification—different proportions of ingredients, different times, and different temperatures.

References

Antisell, Thomas. 1865. *The Manufacture of Photogenic or Hydrocarbon Oils, From Coal and Other Bituminous Substances, Capable of Supplying Burning Fluids*. New York: D. Appleton and Company.

Davy, Humphry. 1980. *On Geology: The 1805 Lectures for the General Audience*. Madison: University of Wisconsin Press.

Falcon, Rosemary, and A. J. Ham. 1988. "The Characteristics of South African Coals." *Journal of the South African Institute of Mining and Metallurgy* 88: 145–161.

Given, Peter H. 1984. "An Essay on the Organic Geochemistry of Coal." In: *Coal Science. Volume 3*, edited by Martin L. Gorbaty, John W. Larsen, and Irving Wender, 63–252. Academic Press: Orlando.

Goodwin, Trevor Walworth, and Eric Ian Mercer, 1983. "Terpenes and Terpenoids." In *Introduction to Plant Biochemistry*, 460–464. Oxford: Pergamon Press.

Keppen, Aleksiei Petrovich. 1893. *Mining and Metallurgy: With a Set of Mining Maps*. St. Petersburg: Mining Department, Ministry of Crown Domains.

Lucier, P. *Scientists & Swindlers: Consulting on Coal and Oil in America, 1820–1890*. 2008. Baltimore: Johns Hopkins University Press.

Lyell, Charles. 1845. *Travels in North America*. London: John Murray.

Morris, Christopher W. 1992. *Academic Press Dictionary of Science and Technology*, 268. San Diego: Academic Press.

Newman, L. L., W. A. Leech, M. H. Mawhinney, C. R. Velzy, C. O. Velzy, A. J. Tigges, H. Karlsson, and W. E. Lewis. 1967. "Fuels and Furnaces." In: *Standard Handbook for Mechanical Engineers*, edited by Theodore Baumeister and Lionel S. Marks, 7-2 to 7-19, New York: McGraw-Hill.

Rock, F. 2020. "What Is Coal and Where Is It Found?" World Coal Association. https://www.worldcoal.org/coal-facts/what-is-coal-where-is-it-found/.

Schobert, Harold H. 1990a. "Diagenesis." In *The Chemistry of Hydrocarbon Fuels*, 27–37. London: Butterworths: 1990.

Schobert, Harold H. 1990b. "Catagenesis." In *The Chemistry of Hydrocarbon Fuels*, 38–74. London: Butterworths: 1990.

Schobert, Harold H. 2013a. "Photosynthesis and the Formation of Polysaccharides." In *Chemistry of Fossil Fuels and Biofuels*, 19–34. Cambridge: Cambridge University Press.

Schobert, Harold H. 2013b. "Formation of Fossil Fuels." In *Chemistry of Fossil Fuels and Biofuels*, 103–131. Cambridge: Cambridge University Press.

Schwamle, W. 2019. Wyoming Is Home to Some of the Largest Coal Seams in the World. https://mycountry955.com/wyoming-is-home-to-some-of-the-largest-coal-seams-in-the-world/.

van Krevelen, D. W. 1993. *Coal: Typology – Physics – Chemistry – Constitution*. Amsterdam: Elsevier.

Wang, Shaoqing, Yuegang Tang, Harold H. Schobert, Di Jiang, Xin Guo, Fang Huang, Yanan Guo, and Yufei Su. 2014. "Chemical Compositional and Structural Characterization of Late Permian Bark Coals from Southern China." *Fuel* 126: 116–121.

4

Structure

The meaning of "structure" depends on the scale over which we want to know structure—macroscopic, microscopic, or molecular. It also depends on the perspective from which we are answering the question. Geological structure includes the arrangement of the coal seams and their surrounding rocks, including faults, folds, and igneous intrusions. Paleobotanical structure seeks to describe the coalified plant remains. Chemical structure describes bonding and spatial arrangement of the atoms. Material or physical structure includes properties such as texture and hardness, as well as the presence and size of pores, cracks, and fissures.

Seams of humic coals may have distinct bands or layers, which can possess different physical and chemical properties. The relative abundances of the bands and how they might occur together—in individual, thick bands or in smaller, thinner bands—has a great impact on the overall characteristics of the mined coal. These bands, called *lithotypes*, can be distinguished on the basis of appearance and how they break. Banding is particularly noticeable in bituminous coals.

Lithotypes formed from different kinds of plant material, possibly under different conditions that existed during the long period of deposition of the organic matter that became coal. For example, fusain looks like charcoal because, in a sense, it once was charcoal. It may have formed as a result of fires sweeping through the swamp in which the coal was deposited. Trees were partially carbonized by the fire; the charcoal-like structure persisted into the coal. Distinct bands can often be recognized over a fairly small vertical distance in seams. There is little, if any, change in rank over seam thicknesses. The lithotypes, with such distinct properties as color and fracture, represent the effects of changing coal type.

Lithotypes provide information on the macroscopic structure of seams of humic coals. Going smaller in scale takes us to the microscopic domain. Examining rocks shows grains of individual minerals. By definition, rocks are aggregates of one or more mineral grains (Bates and Jackson 1987, 573). Microscopic examination of coals shows that they, too, contain smaller components. Coals, too, are rocks—organic rocks. The individual components of organic rocks (i.e., coals) are "organic minerals." They are the macerals, introduced briefly in Chapter 2.

Description and classification of macerals on the basis of their optical appearance and evident relationship to botanical structures forms the basis of the field of coal petrography. The foundations of this discipline were established in the 1920s by Marie Stopes, whose formal training was in paleobotany (Boulter 2017). Her studies demonstrated that ancient plants must have had to adapt to environmental changes. By extension, this meant that ecology has a major role in the evolution of organisms.

Macerals arise from coalification of different kinds of plant materials. Various plant components each have their own characteristic molecular structures. Substances having different molecular structures will participate in a certain kind of chemical reaction in different ways. The reactions responsible for the origin of macerals are those occurring during diagenesis and catagenesis. Trees, for example, contain many chemically distinct components. Each of these kinds of components consists of distinct molecular species: cellulose and lignin provide structure, carbohydrates and fats serve as foods, and waxes and resins do special jobs. Since macerals ultimately come from the molecular species of the plant components, we can expect that their chemical composition is related to those specific components. The maceral groups make up the lithotypes, and the lithotypes form the coal. This is shown conceptually in Figure 4.1.

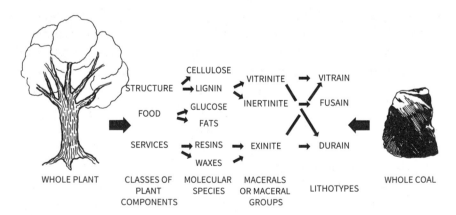

Figure 4.1 The various functions of a plant serve different purposes, so have different kinds of molecular components. During coalification those components eventually become macerals. Maceral groups form the lithotype components of the coal.
Artwork by Lindsay Findley, from the author's sketch.

We expect that the chemical structures in different macerals will be different. Wood forms shiny black macerals called *vitrinites*. Wax contributes to the formation of *cutinites*. Because they formed from plant components of different molecular structures, we can expect differences in chemical structure between vitrinite and cutinite.[1] Given two samples of nominally the same coal, one of which has a high proportion of vitrinite and the other a high proportion of cutinite, we might reasonably expect that they will respond differently when subjected to the same reaction.

Identification of macerals and determination of their relative proportions in a coal sample makes it possible to predict many aspects of the behavior and reactivity of that sample. The identification and determination are done by optical microscopy, a procedure known as **petrographic analysis**. The Appendix provides a table of some of the principal macerals with their origins and properties.

Investigations of the composition, properties, and reactivities of coals largely originated in Europe and Britain. The coals were most often vitrinite-rich bituminous coals of Carboniferous age. This also applied as coal research developed in the United States. Because vitrinites predominate in most northern hemisphere coals and have good chemical reactivities, fundamental studies often focused on vitrinites. Many useful relationships and empirical rules of the behavior and reactivities of coals were generated, but they might be on shaky ground or fail altogether when applied to coals of different ranks, different types, and different geological ages.

There can be significant differences in the composition and properties of coals taken from the same coal field, taken from different seams within that particular field, and even taken from the very same seam (Gessner 1865, 45). It would be hopeless to try to collect and analyze coal samples on a scale of, say, every meter. Given this tremendous variability in coals, how can we make progress in developing a true science of coal?

First, good studies on coals pay great attention to collecting samples that should be representative of the whole. Standards-setting bodies, such as the American Society for Testing and Materials (ASTM), establish methods for collecting samples (ASTM 1984, 336–352). ASTM goes so far as to state that, "Coal is one of the most difficult of materials to sample" (ASTM 1984, 336–352).

---

[1] These names arise from the fact that waxes are in the cuticles of leaves—hence cutinite—and vitrinites often have a vitreous (glassy) appearance.

Francis Bacon, one of the founders of modern science, advocated building up extensive collections, whether of actual specimens, observations of natural phenomena, or experimental results (Principe 2011, 120). After a sufficiently large compilation had been made, it should be possible for scientists to use inductive reasoning to formulate unifying principles. Perhaps several laboratories study the reactivity of one particular coal for which composition and properties are well known. Then a second coal, equally well characterized, is studied in the same kinds of reactions. The differences in chemical behavior between these two coals might appear to relate to some aspect of their compositions. Now data are obtained on a third coal, and a fourth, and so on. One particular characteristic relates to how readily the coals decompose when heated. This provisional explanation will lead to more experiments, on more ranks and types of coals, until an ever-expanding collection of observations and data leads to rules and principles on which we can place increasingly greater reliance.

Isaac Newton, one of the greatest scientists of all time, provided "rules of reasoning in philosophy" (Newton 1992, 398). His Rule III says that if a property can be demonstrated to belong to all bodies on which experiments can be made, it can be assumed to belong to all bodies in the universe (Bauer 2015, 100).[2] If we have samples taken at locations many kilometers apart, and we find that they are all of subbituminous B rank with narrow ranges of carbon, hydrogen, and sulfur contents, we can expect that other samples of this same seam are going to be of the same rank, with similar results from the ultimate analysis.

An excellent comment was once made to me by Frank Karner, at the University of North Dakota: even though the system might be very complex, that doesn't mean we throw away the laws of chemistry and physics. Coals are complex, but, ultimately, at some level we have to be able to understand and explain their behavior in terms of principles of chemistry and physics.

As subjects of laboratory investigation, coals are stubborn. They are heterogeneous. They do not dissolve completely in any solvent, nor do they have a true melting point. Coals cannot be distilled the way that petroleum can. Heating a coal sample produces gases and vapors that can be condensed to liquid products, as in the volatile matter test. The gases and vapors originate from the breaking of chemical bonds in the sample. If an experimenter

---

[2] The original, in Cajori's translation (Newton 1992, 398), is "The qualities of bodies, which admit neither intensification nor remission of degrees, and which are found to belong to all bodies within the reach of our experiments, are to be esteemed the universal qualities of all bodies whatsoever."

captured all of the gases and all of the vapors, saved all of the solid residue, and then mixed them back together, it would not be possible to reconstitute the original sample. Coals do not have a regular atomic arrangement, the way that minerals do, for example, so they can be studied only to a limited extent by x-rays.

Over the past century, chemists have evolved detailed systems of nomenclature for both inorganic and organic compounds. Using the formal name of a compound allows a chemist immediately to write its exact molecular structure. From the structure it is possible to calculate the chemical composition and make good estimates of the expected properties. Going back to the turn of the twentieth century, things weren't quite so straightforward. Then, compounds had names such as "yellow prussiate of soda," "carbureted hydrogen," and "Reinecke's green salt."[3] Even so, it was possible to understand what the names meant, to know the formulas and, often, structures. We don't have that ability in coal science. We work with substances called "Upper Freeport," "Middleburg," and "Lower Kittanning"; "Big Dirty" and the "Better Bed"; "Victoria," "Sophia," and the "Lancashire Ladies"; "Hongai No. 5" and "Illinois No. 6." What are we to make of those names? Not much.

Someone who has experience with coals might remember that Illinois No. 6 coal is usually a high-volatile C bituminous coal. From there, knowledge of the variation of coals with rank could reasonably predict that the moisture- and ash-free carbon content is about 77–79%, hydrogen perhaps 5.5%, oxygen close to 10%. He or she might know that coals of the Illinois Basin tend to be high-sulfur coals. But none of this information is implicit in the name "Illinois No. 6," nor is there any information as to how the atoms are connected into a structure on a molecular level.

Coal scientists concur that coals do not have a structure in the sense either of a polymer or a large biological molecule. Polymers and biological macromolecules have known, identifiable molecular structures. A polymer consists of large molecules, often having molecular weights in the tens or hundreds of thousands of daltons. Polymers have a simple repetitive structure. If we know even a small portion of the structure, we can be confident that a second portion several hundred carbon atoms away from the first would be similar. Biological macromolecules such as DNA can have very complex structures, but ones that can be determined in detail. Once we know the structure, we can be confident that a second sample from the same source is identical—the basis, for example, of DNA analyses in forensics. With coals,

---

[3] Today formally known as tetrasodium hexacyanoferrate (II) or sodium ferrocyanide, methane, and ammonium tetrathiocyanatodiamminechromate (III), respectively.

if the exact molecular structure of a particular sample could be determined, possibly another piece of coal with the same structure might never be found, even a sample taken a few hundred meters away in the same seam, let alone from coals of the same rank. This brings up a crucial question about modeling and proposing molecular structures for coals: Why bother?

> Nor is it necessary that these hypotheses should be true, nor indeed even probable, but it is sufficient if they merely produce calculations which agree with the observations. (Osiander 1993, 22)

Andreas Osiander's statement captures the essential notion of modeling coal structures, though offered in a much different context.[4] The "structure" of coal or an image to represent that structure is a statistical or average model that happens to agree with the results of various analyses. Reaction chemistry and instrumental probes provide information that can be assembled to give a composite picture of the molecular structural relationships in coal. The British statistician George Box reminds us that "all models are wrong, but some models are useful" (Box, Hunter, and Hunter 2005, 440). Although a structural model does not perfectly represent the actual structure of that sample of coal, such models illustrate what might happen when a coal reacts. The best models even allow estimating some of the physical properties of that coal. These models make it possible for chemists to consider likely sites in the molecular structure where reactions might occur and to anticipate the chemical nature of the products.

A good model can also point out what we don't know. A model structure sums up what is known but at the same time provides a starting point for working forward, designing new studies to fill in what isn't yet known. Any model can only be based on the information that was in hand at the time the model was being developed. As new experimental results become available, we can use the model to help us understand them, or we can use the results to help us improve the model.

Coals have often been subjected to reactions that are well understood for simpler organic compounds. Analyzing the products of such reactions applied to coals provides data that allow chemists to infer how these products *might* have been linked together in the coal. At one time or another, coal chemists have probably tried every common reaction of organic chemistry on coals.

---

[4] Osiander's comment comes from the unsigned preface that he added to Nicholas Copernicus's *On the Revolutions of the Heavenly Spheres*. Copernicus himself did not hold this view.

Simple organic compounds having known chemical structures similar to structural features thought to occur in coals, and that can be used to test the expected outcome of various reactions, are called *model compounds*. Many coal researchers study the chemistry of model compounds just as intently as the chemistry of coals themselves.

Oxidation reactions with model compounds are well known and easily studied in the laboratory. For example, ethylbenzene can be oxidized to produce benzoic acid. Other model compounds, such as toluene, also form benzoic acid. This means the molecular structures oxidized in the coal sample could not be determined unequivocally by this experiment. Ethylbenzene contains eight carbon atoms, but its oxidation product, benzoic acid, contains only seven. The "missing" carbon atom escaped as carbon dioxide (CO_2). Analysis of the products at the end of an experiment with a coal may not account for all of the carbon in the original sample. Oxidation of coals does not yield a single product, as ethylbenzene oxidation does. Oxidation products from coals are a complex mixture of many organic acids. Neglecting the "missing carbon" problem, identifying all of the oxidation products would provide the coal chemist with a set of information that could be reassembled in many different ways to obtain many possible structural representations of that coal, all of which would be consistent with the experimental results. This would be like being confronted with a 1,000-piece jigsaw puzzle with no illustration to indicate what the final picture should look like and with pieces that could be assembled in different ways.

These concerns do not imply that chemical reactions of coals and model compounds are useless for providing information about composition and molecular structure. Until instrumental methods were developed for obtaining chemical information from solid, unreacted coal samples, chemical reactions were one of the few tools that chemists had for studying coals. Though the results gave only a partial picture and could produce more than one possible interpretation, they provided useful ideas of what some of the molecular structural features of coals must be like and how they must be arranged.

At some time, in some place, every likely solvent has also been tested on coals. Coals are not truly soluble in any solvent. With a complex, heterogeneous material a solvent might dissolve, or extract, a portion of the material, the process of *solvent extraction*. Alternatively, some solvents can react with material being extracted, the process of *reactive dissolution*.

Solvent extraction provides a solution of coal-like material that can be studied further by various techniques of analytical organic chemistry. Careful selection of solvents can extract specific classes of compounds from coals.

For example, extraction of low-rank coals with benzene yields waxes.[5] The yields of specific classes of compounds, such as waxes, are often small. The components of such extracts represent only a small portion of the original sample.

Heating coals in an inert atmosphere will break chemical bonds, eventually liberating molecules small enough to escape as gases or vapors. The breaking of chemical bonds by the application of heat is the process of *pyrolysis*. The products of coal pyrolysis are very complex mixtures. In some pyrolysis experiments, we cannot be sure that the components first liberated from the coal sample by pyrolysis did not themselves undergo more pyrolysis inside the apparatus, making the interpretation of how the observed products relate to the original sample many times more complicated.

After World War II, chemists developed and improved a variety of instruments to assist in the analysis of complex mixtures or determination of molecular structures. Many do not require the analytical sample to be reacted, degraded, or extracted before it can be studied. They eliminate much of the ambiguity caused by uncertain effects of solvents or chemical reactions. Coal scientists adapted many of the instrumental methods developed by their colleagues working in other fields. These have yielded substantial insights into the nature of coals.

Three instrumental methods have contributed greatly to our understanding of coals at the molecular level. Nuclear magnetic resonance spectroscopy (NMR) applies to elements having an atomic nucleus with an unpaired proton or neutron. Fortunately, this includes isotopes of the two most important elements in coals, hydrogen (^1H) and carbon (^{13}C). Infrared (IR) spectroscopy involves scanning the spectrum of IR wavelengths and recording the wavelengths at which the substance being tested absorbs the radiation. The vibration, stretching, or bending of chemical bonds can be stimulated by IR radiation. Absorption of IR radiation at a particular wavelength shows the characteristic energy required for a particular motion of the specific atoms involved. X-ray diffraction (XRD) is based on the interaction of x-rays with structures in which the molecules or atoms are extremely close together. We will examine some of the key findings of these techniques without delving into the underlying details of how and why they work.

---

[5] Extraction of waxes from lignites has been done on a commercial scale. The products, called *montan waxes*, can be used as components of shoe polish, furniture polish, and waterproof paints. They will be mentioned again in Chapter 16.

Carbon atoms are incorporated into coals in two ways.[6] *Aliphatic* carbon structures, whose name derives from the Greek word for "fat" or "fatty," can occur either as chains or rings of carbon atoms but always tend to be hydrogen-rich. *Aromatic* carbon structures occur in rings, almost always hexagonal, and tend to be hydrogen-poor. Their name comes from the fact that many of the aromatic compounds first studied have characteristic aromas. For example, naphthalene has an odor characteristic of mothballs. Coalification proceeds on a trajectory that leads to increasingly carbon-rich materials. Substances that are increasingly carbon-rich are, necessarily, increasingly hydrogen-poor. We can reasonably conclude that aromatic carbon is likely to be the principal carbon form in coals and that coals should become increasingly aromatic as rank increases.

^{13}C NMR has become the dominant instrumental technique for studying coals. *Aromaticity* of a coal is the fraction of the carbon atoms in aromatic structures relative to the total number of carbon atoms. An aromaticity of 0.6, for example, means that six of every ten carbon atoms are present in aromatic structures. ^{13}C NMR of solid coals shows that aromaticity increases with rank, from about 0.6 in lignites to 0.9 in low-volatile bituminous coals, and just about 1 in anthracites. In most reactions, the chemistry of aliphatic carbon differs from that of aromatic carbon. Knowing the fraction of carbon atoms present in each form helps us understand coal chemistry and helps to predict the behavior of a specific coal in particular reaction conditions. Aromaticity can also be used along with carbon content as an indicator of rank.

A second parameter, ring condensation, also characterizes the aromatic structures. It represents the average number of aromatic rings fused together in the structural units. Benzene, with a single ring of six carbon atoms, has a ring condensation of 1. Naphthalene, with two rings fused together, has a ring condensation of 2. Both aromaticity and ring condensation increase with increasing rank. Ring condensation increases slowly from lignites, where the value would be about 2, through the bituminous coals, with values up to about 6. Above about 90%, ring condensation increases rapidly, rising to some 30–100 in anthracites. Ring condensation is infinite (on an atomic scale) in a perfect crystal of graphite.

As we proceed into the high ranks, particularly into the anthracites, there still remains a need to move hydrogen—chemically—from the carbon-rich

---

[6] This very brief discussion—organic chemistry reduced to one paragraph—represents the shallowest of scratches into the surface of the subject. For readers with little chemistry background, this introduction should suffice. For those who want to probe more deeply, good introductory books are mentioned in the Bibliographic Essay.

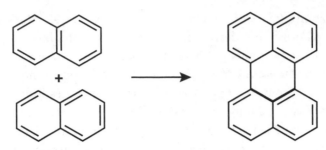

Figure 4.2 The reactions of dehydrogenative polymerization illustrated by the formation of perylene from two molecules of naphthalene. Subsequent reactions of perylene-like structures start the formation of large aromatic "sheets."

material to the hydrogen-rich side on the route to graphite. At values of aromaticity approaching 1, there is not much easy-to-move hydrogen left. We have already picked all of the low-hanging (hydrogen) fruit. But hydrogen can be liberated by the coalescence of aromatic ring structures. This process, *dehydrogenative polymerization*, is illustrated in Figure 4.2 by the reaction of two naphthalene structures to form a molecule of perylene. When this happens, four atoms of hydrogen are "lost," in the sense of their becoming part of the hydrogen-rich products. Nothing prohibits structures such as perylene from undergoing the same process, becoming an even larger aromatic system and even more carbon-rich. Dehydrogenative polymerization is consistent with the rapid, sharp increase in the size of aromatic units in coals with greater than about 90% carbon content.

X-ray studies have shown that, at the upper end of the bituminous rank range, small clusters of rings begin to align in a miniature version of the graphite structure. XRD shows that coals tend to become more like graphite as carbon content increases, especially above about 90% carbon on a moisture-and-ash-free (maf) basis. Virtually all the carbon in anthracites is contained in ring systems; the structures are much larger than in bituminous coals and may consist of large sheets of tens or hundreds of benzene rings fused together. The stacking of these sheets of rings becomes even more ordered than in bituminous coals and is on the way to graphite. The important caveat lies in the words "on the way to." No coal actually is graphite.

A simple polymer, such as polystyrene, has a macromolecular structure consisting of long chains of hundreds or thousands of carbon atoms made by chemically linking together the molecules of its monomer, styrene. In a solid piece of polystyrene, the polymer chains might be entangled with each other on a molecular scale and might experience some weak physical forces of attraction among one another. There are no strong chemical bonds between the chains, only within the chains. Polystyrene prepared in this way readily dissolves in a variety of common organic solvents.

A polymer dissolves in a solvent in a two-step process. Solvent molecules diffuse into the polymer to produce a swollen gel.[7] then the gel dissolves to produce a true solution. We can imagine another polymer that has chemical bonds between the chains in addition to those that normally exist within the chains. Such bonds are called *crosslinks*. A crosslinked polymer exposed to solvents cannot dissolve because solvents cannot break the chemical bonds in the crosslinks. If solvents did break crosslinks, the necessary bond breaking would not be true, reversible dissolution but rather would be a chemical reaction. At most, the interaction of solvent molecules with segments of the polymer chain will cause it to swell to a gel state. This phenomenon is called *solvent swelling*. Crosslinked polymers only swell when interacting with solvents. The amount of swelling that a crosslinked polymer experiences depends on how many crosslinks occur relative to the number of monomer units in the chain, a property called the *crosslink density*.

Coals are not true polymers in that they do not consist of a simple repeating unit that is employed over and over again to form macromolecules. Coals are relatively hard, do not reversibly melt, and do not dissolve completely. These characteristics suggest that the three-dimensional arrangement of aromatic ring systems and the linkages between them could be like a cross-linked polymer. Coals swell when exposed to organic solvents, suggesting that, even though they are not true polymers, we can gain some insights into their structural arrangements by comparing their behavior to that of crosslinked polymers.

Highly crosslinked structures with low values of ring condensation and abundant oxygen exist in resins produced from polymerization of phenol or related compounds with formaldehyde. These phenol-formaldehyde resins have many commercial applications, such as the "board" in circuit boards. They begin to undergo pyrolysis at 300–350°C and decompose without softening. When such resins form with trapped water (the other product of the polymerization process), the water results in pores or "holes" in the structure of the resin. Lignites are porous solids with low values of ring condensation and high amounts of oxygen; they start to experience pyrolysis around

---

[7] A gel remains a coherent mass, being a liquid that is a type of colloidal system, in which very large molecules, such as proteins or polymers, or extremely fine particles, as of clays, are dispersed through the liquid. Gels can experience degrees of swelling ranging from slight to very large upon being contacted with a solvent. Depending on the specific natures of the gel and the solvent, the swelling can be extremely large and eventually transform the swollen gel into a colloidal solution.

300–350°C and do not soften (i.e., are not caking coals). They have high moisture contents as-mined. These points of similarity suggest, but not prove, that the macromolecular structures of lignites likely have some similarity to those of phenol-formaldehyde resins.

Graphite is 100% carbon, has an aromaticity of 1.00, and has essentially an infinite ring condensation on a molecular scale. Anthracites have carbon contents of greater than 91%, aromaticities close to 1, and values of ring condensation likely in the tens or hundreds. This certainly does not say that anthracite is graphite, but it suggests that we should be able to learn more about anthracites by understanding the structure and properties of graphites. XRD shows us this.

Very simple models of coal structures and their variation with rank were developed in the early 1950s by Peter Bernard Hirsch (Hirsch 1954), an eminent materials scientist who spent much of his career at Oxford University. Hirsch's models are not only simple, but often are also accused of being simplistic. They are shown in Figure 4.3, with the open structure of low-rank coals on the left, the liquid structure of bituminous coals in the center, and the anthracitic structure on the right. The straight lines represent aromatic ring systems viewed edge-on. The "squiggles" represent aliphatic, disordered, or amorphous carbon that could be crosslinks between ring systems. Yet looking from the open to liquid to anthracitic structures—as rank increases—the open space (porosity) diminishes, the amount of aliphatic carbon decreases, the aromatic systems get larger and begin to show signs of three-dimensional ordering. Greater porosity likely results in greater moisture-holding capacity. Fewer crosslinks result in greater solvent swelling (neglecting the anthracites, a breed to themselves in this regard). They also result in less aliphatic material to be broken off during pyrolysis and hence less volatile matter. The anthracitic

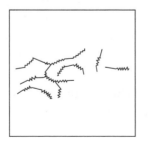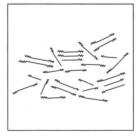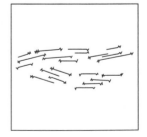

Figure 4.3 The simple Hirsch structural models for coals. The open structure of low-rank coals is shown on the left, the liquid structure of caking and coking bituminous coals in the center, and the anthracitic structure on the right.
Artwork by Lindsay Findley, from the author's sketch.

structure shows hints of graphite. So, simple models are consistent with many experimental observations on coals and their variations rank.

One of my PhD students, Elena Korobetskaya, once proposed a remarkably insightful idea. I never had the resources to follow it up, but hopefully someone will. Virtually all approaches to understanding coalification and coal structure begin with lignin, its conversion to lignites, and on into higher ranks. Lignin itself has a complex structure that varies from one species of plant to another and even from one location to another within the same plant. On the other hand, the structure of graphite is known with exceptional accuracy. Why not start with graphite? Via computational modeling or actual experiments, we could see what happens as atoms of oxygen and hydrogen are added to the structure, eventually incorporating a few atoms each of nitrogen and sulfur. In other words, we can explore how coal structure devolves from graphite, rather than trying to work out how it evolves from lignin.

Increasing the number and size of aromatic ring systems in a material generally results in its becoming stronger and denser, having a higher melting point, and being less soluble in common solvents. The ultimate aromatic compound, the carbon-rich end point of catagenesis, is graphite. Graphite is inert toward virtually all substances except under highly extreme reaction conditions. It is not soluble in any known solvent. It scarcely oxidizes, even at white heat. Its melting point is about 3,600°C. It is denser than any coal.

Because macerals are coalified products of different parts of plants, there will be chemical and molecular structural differences among macerals. Because these differences exist, petrographic analysis of a coal sample provides insights into what we might expect when this particular coal is reacted. Vitrinites dominate in humic coals of the northern hemisphere and display the aspects of composition, structure, and their variation with rank that have been discussed in the previous sections.

Exinites and inertinites are different (Stach et al. 1982, 100–140). Exinites yield more gas and substantially more tar upon carbonization than does the "whole coal" sample that is a mixture of macerals likely dominated by vitrinite. These macerals have higher hydrogen contents and are more aliphatic than vitrinites. On the other hand, *fusinites* show high carbon contents and relatively low hydrogen. Fusinites do not soften to form a fluid phase during production of metallurgical coke. The amounts of different macerals revealed by petrographic analysis can be very helpful in coal selection. In Chapter 13, we will discuss a procedure for making liquid fuels from coals, developed for Utah coals, many of which have good concentrations of *resinites*. With high hydrogen contents and excellent tar yields, resinite-rich coals would be well suited for this process.

Aromatic carbon atoms form only three bonds to neighboring atoms. The fourth valence electron of these carbon atoms is not confined to one specific bond. It is *delocalized*. It participates in bonding systems that essentially spread over the whole aromatic structure. The extent to which light is reflected from a material is dependent on optical properties[8] that contribute to the *reflectance* of that material (Taylor et al. 1998, 371). These properties increase with the number and extent of delocalized electrons. We can measure the reflectance of vitrinites. The greater the percentage of aromatic carbon atoms and the larger the number of condensed rings in vitrinites, the greater will be its reflectance. We have seen that, as rank increases, aromaticity and ring condensation increase, based on measurements independent of optical properties. Therefore, vitrinite reflectance, measured by optical microscopy, is an excellent indicator of coal rank. It might be the best indicator because it depends only on a measurement made on one maceral.

An additional aspect of structure to be considered is the nature of coals as solid materials. Studies of many of the fundamental physical properties of coals have received far less attention than investigations of their chemistry. Coals present two problems deriving from their natural heterogeneity. First, two samples of what is nominally the same coal might contain different proportions of macerals. It seems reasonable to expect that the properties of the macerals taken individually will be different. Second, coals are not just the carbonaceous or "coaly" substance. Coals contain, in addition, embedded particles of minerals, water trapped in tiny pores, and possibly internal cracks that are not apparent in handling the test specimen. Minerals, water, and whatever gas is in an internal crack all will have their own characteristic properties. The measurement made on a test specimen will be influenced by its petrographic composition and all of these inclusions. Coals might also contain so-called *blind pores*. These do not have an opening on the surface of the specimen so could not be penetrated by the gases or liquids used to probe the physical properties.

There are approaches to mitigate these issues. One would be performing the same experimental measurement on a large number of samples of what should be the same coal. To obtain a large number of samples, we can consider the approach of Ivor Evans and C. D. Pomeroy of Britain's National Coal Board: "In some cases, random smashing of larger pieces with a hammer was resorted to in order to obtain greater numbers of specimens" (Evans and

---

[8] These properties are the refractive index and the absorption index. The refractive index is the ratio of the velocity of light in a vacuum to its velocity in the material being studied. The absorption coefficient depends on how much of the light is absorbed in passing through a unit length of the material.

Pomeroy 1966, 55). Measurements can be examined statistically to discard anomalous results. With the remaining data within a statistically acceptable band, the mean value could be taken as a measure of the property being studied. A second approach is to use techniques for testing ever-smaller specimens or areas within a specimen. Likely, the smaller the specimen, the more homogenous it might be and the less probable it might be to have a hidden defect or an embedded mineral particle. A third approach could use noninvasive imaging, such as computed tomography (CT)-scanning,[9] to check for such defects or impurities. The problem is that these approaches are time-consuming, expensive, or both.

Porosity is the percentage of the volume of the substance occupied by void space or pores. To determine porosity, the apparent volume of a sample, pores and all, and the volume of only the truly solid portion itself must be known. Determination of porosity of a coal sample involves measuring its density in mercury, which does not penetrate the pores at ordinary pressures, and measuring the density again in helium, which is assumed to penetrate the pores completely. The volume calculated from the helium density is the volume of the coal substance itself, while the volume calculated from the mercury density is the volume of the coal plus its pores.

Most processes that make use of coals involve a reaction with a liquid or gas and necessarily occur at the surface of coal particles. The extent of reaction is governed by the amount of surface. The total surface is not only the surface of the exterior of the particle, but also includes the surfaces of the pores. The so-called internal surface will be far larger than the external surface. The size and extent of pores affect the ability of reactants to access the internal surface and also affect the ability of reaction products to escape, exposing fresh surface for further reaction. Porosity governs loss of moisture during coal drying and re-absorption of moisture by the dried coal. Porosity can also relate to strength if the pores represent sites of weakness where breakage can occur.

In many applications it is more useful to know the surface area rather than total porosity. *Surface area* measures the space available for reactions to occur. Pore size limits the size of molecules that can penetrate a coal particle and participate in reactions. A high-porosity coal having extremely small pores through which reactants can penetrate or products escape only with difficulty might not participate in reactions as easily as a different coal having lower

---

[9] Computed tomography is a technique that combines a series of x-ray images taken from different angles to create cross-sectional images of the internal structures of the specimen. This technique is enormously helpful in medicine, as many of us have learned.

porosity but larger diameter pores. By knowing surface area and pore volume and making an assumption about the shapes of the pores, the average pore radius can be calculated. This gives an idea of the sizes of molecules that can penetrate or leave the pore system and the relative ease with which molecules can enter or leave.

Measuring the distribution of pore sizes involves forcing mercury into the pores at high pressure. Mercury does not penetrate the pores of coal at atmospheric pressure but does so at elevated pressures. The higher the pressure, the smaller the pores that mercury will penetrate. Measuring mercury penetration as a function of pressure makes it possible to calculate the amount of porosity attributable to the various pore sizes. For many coals, most of the porosity is contained in pores of a width smaller than 2 microns (2000 nanometers),[10] including some of less than 50 nanometers. To put these figures in perspective, the nearly invisible tungsten-wire filament in a 60-watt incandescent light bulb is about 40,000 nanometers in diameter.

The common approach to measuring surface area relies on the adsorption of gases on the coal surface. The measured amount of gas adsorbed at various pressures and the known area of one gas molecule can be used to calculate the coal surface area. The gases generally used for this are nitrogen and CO_2. However, tests of the same sample using the two gases show drastically different results. This means that surface area data reported without any mention of the gas and how they were calculated are useless because we don't know what is actually being measured and reported. The apparent surface areas of coals measured by CO_2 adsorption generally decrease from lignites into the bituminous ranks and then increase again at least to low-volatile bituminous coals (Gan, Nandi, and Walker 1972).

In lignites, internal surfaces contribute about 90% of the total surface area. Lignites have apparent surface areas of $\approx$200 square meters per gram measured by adsorption of CO_2. If all this surface area could somehow be unfolded and spread out, a 40-gram sample of a lignite could cover a soccer pitch. Because pores can fill with water, the moisture content of lignites determined by proximate analysis often ranges from 35% to 45% (as-received). Pore volume could be estimated from moisture determination if it were certain that the pores were full and that no moisture was present on the surface. Doing this in the laboratory involves saturating the sample with water and storing it at 100% relative humidity until it comes to a constant weight.

---

[10] Pore sizes are also given in micrometers (often called microns). A 50-nanometer pore would be 0.05 microns.

Then, the moisture in that sample is considered to be held entirely in the pore system. The moisture content measured in this way is sometimes referred to as the *equilibrium moisture*.[11] Lignites have the highest equilibrium moisture values of any rank, in the range 20–40%. The value drops as rank increases, reaching a minimum at about 88% carbon (maf) and then slowly rising again among the anthracites.

Density-measurement methods depend on why we need the data and what we intend to do with it. *Bulk density* reports the weight of lump coal that can be held in a given volume. This helps us calculate, for example, the size of a storage bin needed to contain a given weight of a coal or the weight of a coal that could be held in a truck, railway car, or reaction vessel of certain size. Bulk density depends on the size of particles in the sample and how efficiently those particles can pack together in a container. Many small particles of coal can be packed into a volume that would hold only a few large, irregularly shaped lumps. Therefore, bulk density conveys no fundamental information about the structure or properties of coals. A rough, rule-of-thumb estimate is that the bulk density of coal is about 800 kilograms per cubic meter. Packed in bulk, a given volume of coal weighs approximately 80% of what the same volume of water would weigh.

The measurement most often used to determine coal density is the *apparent density*. Measurement is done by liquid displacement: the volume of liquid displaced by a known weight of coal must be equal to the volume of the coal. Apparent density depends on the liquid used for the measurements because different liquids penetrate the pores of the coal to different extents. Apparent densities of coals are roughly 1.5 times greater than water. Regardless of the liquid used, a minimum occurs in the range 85–90% carbon. Most kinds of wood have densities less than 1 gram per cubic centimeter, while that of graphite is 2.25 gram per cubic centimeter. The facts that coals have densities intermediate between these extremes and that these densities rise rapidly as rank increases above 90% carbon are consistent with the notions that coals have derived from woody plant material and could conceivably transform all the way to graphite.

True density is measured in helium. No one knows whether the values really are "true." Presumably helium, having the smallest possible molecular size, should penetrate every pore, down to the very tiniest. In that case,

---

[11] This property is also referred to as the *capacity moisture* content because it represents all the water a particular coal can hold when it is fully saturated at 100% relative humidity. It is presumed that the coal within an undisturbed seam is likely at its capacity or equilibrium moisture content. Therefore, this same measurement has a third name, the *bed moisture content*.

density measured in helium really reflects only the coal substance itself and does not inadvertently include pores that could not be accessed by the material used for the density measurement. Values of helium densities of coals are in the same range as apparent densities and follow the same pattern of variation with rank. Helium densities decrease slowly as rank increases, reaching a minimum at about 88% carbon, after which they increase rapidly among the anthracites (Berkowitz 1979, 80).

We need to be able to anticipate the effects of coal hardness in the practical use of coals. Mining operations produce large pieces of coal, too big to be used in modern coal combustion or processing equipment. Coals need to be crushed, ground, and pulverized. Insight into the ease or difficulty of reducing the size of coal pieces is essential for designing and specifying the appropriate equipment to do these things. Tests for quantitative measurement of hardness rely on indentation or penetration of a sample of the material. They involve a probe having a tip of a known shape, such as a sphere or cone. The probe, pressed against the surface with a known force, leaves an indentation. The size of the indentation provides an indication of the hardness of the material (Callister 1994, 130–132). From lignites, microhardness appears to increase with rank, passing through an apparent maximum at about 80% carbon (maf), slowly declining, and finally increasing very rapidly in the anthracite rank range (Berkowitz 1979, 90).

The challenge that still awaits us is for someone—or some research group— to make the "grand synthesis" of reaching the point at which the material properties of coals as solids can be explained in terms of rank, type, and molecular frameworks.

References

ASTM. 1984. "Standard Methods for Collection of a Gross Sample of Coal. D 2234-82." *Annual Book of ASTM Standards. Volume 05.05. Gaseous Fuels; Coal and Coke.* Philadelphia: American Society for Testing and Materials.

Bates, Robert L., and Julia A. Jackson. 1987. *Glossary of Geology.* Alexandria, VA: American Geological Institute.

Bauer, Susan Wise. 2015. *The Story of Western Science.* New York: W. W. Norton.

Berkowitz, Norbert. 1979. *An Introduction to Coal Technology.* New York: Academic Press.

Boulter, Michael. 2017. *Bloomsbury Scientists.* London: UCL Press.

Box, George E. P., Stuart Hunter, and William G. Hunter. 2005. *Statistics for Experimenters.* New York: John Wiley & Sons.

Callister, William D. 1994. *Materials Science and Engineering.* New York: John Wiley & Sons.

Evans, Ivor, and Charles Duncan Pomeroy. 1966. *The Strength, Fracture and Workability of Coal.* Oxford: Pergamon Press.

Gan, H., S. P. Nandi, and Philip L. Walker. 1972. "Nature of the Porosity in American Coals." *Fuel*, 51: 272–277.

Gesner, Abraham. 1865. *A Practical Treatise on Coal, Petroleum, and Other Distilled Oils.* New York: Bailliere Brothers.

Hirsch, Peter Bernhard. 1954. "X-ray Scattering from Coals." *Proceedings of the Royal Society of London. Series A. Mathematical and Physical Sciences* 226, 1165: 143–169.

Newton, Isaac. 1992. *Principia Mathematica*. Translated by Florian Cajoari. Norwalk, CT: Easton Press.

Osiander, Andreas. 1993. "To the Reader on the Hypotheses in this Work." Preface to Copernicus, N. *De Revolutionibus Orbium Coelestium.* Translated by A. M. Duncan. Norwalk, CT: Easton Press.

Principe, Lawrence M. 2011. *The Scientific Revolution: A Very Short Introduction.* Oxford: Oxford University Press.

Stach, E., M-Th. Mackowsky, M. Teichmüller, G. H. Taylor, D. Chandra, and R. Teichmüller. 1982. *Coal Petrology*. Berlin: Gebrüder Borntraeger.

Taylor, G. H., M. Teichmüller, A. Davis, C. F. K. Diessel, R. Littke, and P. Robert. 1998. *Organic Petrology*. Berlin: Gebrüder Borntraeger.

5
Minerals

Specks of minerals can be seen in coals down to the smallest sizes that we can observe with the most powerful microscope. *Minerals* are discrete pieces of substances that have a characteristic composition, crystal structure, and physical properties (Bates and Jackson 1987, 424). Low-rank coals also contain inorganic ions associated with carboxylic acid[1] sites in the coal structure. Ions of other elements can be associated as coordination compounds[2] with nitrogen, sulfur, or oxygen atoms in the coal structure. The term "inorganic constituents" is inclusive of minerals, ion-exchangeable elements, and coordination compounds.

Ash is formed by reactions of the *inorganic constituents* in the sample. Ash was not a constituent of that coal. It is a product of high-temperature reactions that occur as the coal is consumed. The inorganic components and the ash that they produce are usually a small portion of coals, but formation and behavior of ash are important in the use of coals. Sometimes the behavior of ash can have a greater effect on utilization than does the carbonaceous part of the coal. The inorganic constituents establish the composition and mineralogy of ash that forms during utilization. This is also important in issues relating to disposal of the ash in environmentally acceptable ways or in finding uses for the ash.

Inorganic constituents became incorporated in several ways. To all of the sources of complexity of the carbonaceous part of coals we now add the variability in inorganic constituents, deriving from differences in regional geology, water flow, and ion-exchange capacity of the organic matter.

Some of the plants that contributed to coal formation accumulated inorganic matter from the soil in which they grew. Organic matter is preserved by

---

[1] Carboxylic acids are organic oxygen-containing group commonly shown as R–COOH, in which R represents an organic structure, and which are capable of behaving as weak acids. The most familiar carboxylic acid is acetic acid, CH_3COOH, the acidic component of vinegar.

[2] Coordination compounds are based on coordinate covalent bonds, in which the nitrogen, sulfur, or oxygen atom contributes two electrons to form a bond with another atom. The common covalent bonds involve each atom in the bond contributing one electron. Usually, but not exclusively, the "other atom" in the coordinate covalent bond is an atom of a transition metal, such as iron, nickel, or copper.

being covered by the water of the swamp in which the plants were growing. In addition to protecting organic matter from decay, water serves as an agent that transports dissolved inorganic ions and bits and pieces of mineral grains through and into the accumulating organic matter. Water can carry small mineral particles which settle and mix with the coalifying plant material. Sedimentation of these particles will be affected by the velocity and turbulence the moving water. Eventually, they become incorporated into the coal as discrete grains of minerals, called *detrital minerals*. Clays and quartz are examples. Wind also transports detrital minerals. Eventually, the suspended particles fall out of the air, accumulating on land or water surfaces. Ash from volcanic eruptions can travel thousands of kilometers from the volcano. It accumulates as thin partings, on a scale of millimeters to centimeters, between seams of coals.

Some anaerobic bacteria metabolize sulfate ions in water, extracting their oxygen atoms and leaving behind various reduced sulfur species, including sulfide or disulfide (S_2^{-2}) ions. If ferrous (Fe^{+2}) ions happen to be in the water as well, pyrite will precipitate.

Brown coals and lignites can capture ions from the water by *ion exchange*.[3] The amount of exchangeable ions in these coals likely varies over geological time as the composition and flow patterns of water moving through the seam change. Consequently, such coals can exhibit great variability in their inorganic compositions, both vertically and laterally through the seam.

After coal has formed, water can still percolate through vertical cracks— called *cleats*—or horizontal joints in the seam. *Epigenetic* minerals precipitate from the water and deposit in the cleats. Calcite is an example. In some cases, mineral precipitates fill the cavities of coalified remnants of plant tissues.

Characteristics of the minerals provide information useful in anticipating aspects of the behavior of coals when they are used. The apparent hardness or abrasiveness of a coal sample could be affected by its minerals. Quartz is one of the hardest and most abrasive of the minerals occurring in coals. A coal with abundant quartz might cause problems when it is being crushed or ground. Sulfur-containing minerals, such as pyrite, produce sulfur oxides when a coal is burned. Great effort is devoted to reducing the amount of mineral matter in coals before they are used (Chapter 7). To accomplish this, the mineral particles must be liberated from the carbonaceous matrix. How easily

---

[3] The process of water softening is a good practical example of ion exchange. "Hard" water contains ions of such elements as calcium, which impedes the action of soap and leaves deposits in cookware and water heaters. Passing hard water through a bed of resin that has been impregnated with sodium ions "softens" the water as the resin accumulates calcium ions and releases sodium ions. The resin must be recharged by periodic flushing with a highly concentrated solution of sodium ions, removing the accumulated calcium ions.

this can be done is determined by the sizes and shapes of the mineral particles, as well by how extensively they might be intergrown into the carbonaceous matrix or with each other.

At least 60 minerals occur in coals (Gluskoter et al. 1981). The list of the common ones is much shorter, as shown in the Appendix. Most are in a small number of mineral families: oxides, aluminosilicates, sulfides, and carbonates. The minerals are also grouped according to the ways in which they became incorporated in a coal deposit. Detrital minerals originated elsewhere, such as by the weathering of existing rocks, and were then transported by water or wind into the coal-forming environment. *Syngenetic minerals* formed at the same time as the coal itself. Those minerals that came into a coal seam after coalification are said to be epigenetic.

Quartz occurs in virtually all coals (Berkowitz 1979, 46). Quartz may be detrital or epigenetic (Schobert 1987, 41). It is by far the most important of the oxide minerals that occur in coals. Quartz has a high melting point and is comparatively unreactive in ash-forming processes, so may survive largely intact from the original coal sample into the ash. Quartz is harder than most other minerals in coals and can contribute to hardness and abrasiveness in coals and their ashes.

Only three clay minerals are usually important in coals: kaolinite, illite, and montmorillonite. These minerals have structures based on aluminum, silicon, and oxygen atoms, so are called *aluminosilicates*. Kaolinite consists entirely of these three elements. Illite also contains small amounts of potassium, iron, and magnesium. Sodium, potassium, calcium, and magnesium can be incorporated in montmorillonite. Clay minerals are often detrital, formed by weathering of rocks and transported into the coal deposit from outside the immediate area. Clays can also be formed from decomposition of volcanic ash. Clays undergo structural changes during ash formation, contributing to ash-related problems, such as forming tenacious deposits of ash inside combustors.

Sulfur occurs in coals in several ways: incorporated in the molecular frameworks of coals, by carbon-to-sulfur bonds, and also incorporated in sulfide and sulfate families of minerals. Each form of sulfur behaves differently in processes intended to reduce the amount of sulfur. It is often helpful to have knowledge of the amounts of each of the forms of sulfur, the determination of which has its own standard procedures (ASTM 1984a, 357–361). The most important sulfide mineral in coals is pyrite, an example of which is shown in Figure 5.1. In many coals, sulfide minerals provide the largest share of the total sulfur. Some pyrite can form from sulfur produced by bacterial action

Figure 5.1 Pyrite in bituminous coal from West Yorkshire. The pyrite is evident as the light-colored areas, particularly the large piece on the left side.
Courtesy of iStock.com/Joe Peacock.

on sulfur compounds. Additional pyrite can be produced after the coal has formed, when waters containing sulfide or sulfate percolate through the seam. Pyrite formed in this way deposits in cleats or other cracks in the coal. Iron sulfate minerals, such as jarosite, likely form as a result of oxidation of coal. High concentrations of iron sulfates signal that the particular coal being analyzed may be extensively oxidized.

The principal carbonate minerals include calcium and iron carbonates—calcite and siderite. Calcite likely formed by precipitation from waters percolating through a coal seam (Mraw et al. 1983). It can also be liberated by adjacent rocks or recrystallize in the coal (Schobert 1987, 41). At high temperatures, carbonates decompose, forming their related oxides. These oxides then react with clay minerals, forming new, relatively low-melting phases that can cause problems with ash handling.

Coals that formed in marine or brackish environments can contain chlorides, such as sodium chloride. Generally, most coals in Europe, the United States, and South Africa do not have significant amounts of chlorides. However, there are some coals in Britain and Germany that merit the name *salt coals* (Mackowsky 1982) because of high concentrations of sodium chloride. When they do occur, chlorides are undesirable because they can contribute to corrosion of metal or refractory brick surfaces in combustion equipment, and they can contribute to formation of ash deposits.

A useful approach to isolating and identifying minerals in coal is the seemingly oxymoronic *low-temperature ashing* (LTA). The carbonaceous coal matrix can be burned away to liberate the minerals, but at so low a temperature that the minerals would not be altered in the process. The LTA apparatus heats a pulverized sample to about 100–150°C—far below normal ashing temperatures—in an atmosphere of oxygen at reduced pressure. A radio-frequency discharge generates a plasma[4] of oxygen atoms, which react with the coal matrix even at these low temperatures. LTA proceeds without significant change to most minerals. Once the low-temperature ash has been obtained, the minerals in it can be identified by various methods, such as x-ray diffraction. At nominally 150°C, LTA can be excruciatingly slow, especially with relatively unreactive coals, such as anthracites. LTA is a useful research tool but not likely effective in routine analyses or quality-control testing.

The most useful approach to identifying and quantifying the minerals in coals involves scanning electron microscopy interfaced with and controlled by a computer. As the beam moves across the surface of the sample, each mineral particle is analyzed and its size is measured. The composition of that particle is compared with a database of compositions stored in the computer. A mineral particle is taken to be identified if its measured composition agrees, within some reasonable tolerance, with a known composition of a mineral in the database. The report includes the list of minerals identified and percentages of each mineral found in various size fractions. These results provide insights for coal cleaning operations and for studying processes of ash formation when the coal is used. However, the results depend on the database in the computer. A mineral having a composition that doesn't match any minerals in the database is classified as unidentified.

Especially in low-rank coals, some elements can be incorporated in more than one form, not always in minerals. As an example, a portion of the total amount of calcium could be present as exchangeable Ca^{+2} ions in the coal matrix and another portion as the mineral calcite. Because those two portions of calcium are in chemically different environments, they could react in different ways when the coal is used. Our understanding of the inorganic constituents can

---

[4] In this context, the term "plasma" refers to a special state of gaseous matter consisting of ions and electrons, the behavior of which is largely determined by electromagnetic interactions between the various charged particles.

be improved by finding how an element of interest is distributed among the possible ways in which it might occur in the sample.

Sorting out these modes of occurrence can be done with simple chemical treatments, in the process of *chemical fractionation*. Different laboratories have their own procedures for chemical fractionation, but all are of the same essence. First, treating the sample with a solution of a very mild ionic compound—ammonium acetate, for example—removes the exchangeable ions. Second, reacting the residue from the first step with an acid, such as hydrochloric acid, removes acid-soluble materials, such as carbonate minerals. Then, treatment with stronger acids dissolves the silicate and aluminosilicate minerals.

Analyzing the solutions produced in each step and determining the total amount of each element in the sample allows a calculation of the percentage in each of its different modes of occurrence. As a hypothetical example, it might be possible to say that 60% of the calcium is on ion-exchange sites, 30% is acid-soluble (most likely calcite), and 10% is in a silicate mineral, which might be wollastonite ($CaSiO_3$). Chemical fractionation cannot, by itself, identify specific minerals, but it provides information that supports reasonable guesses. Continuing this example, carbonate minerals dissolve in hydrochloric acid: calcite is the dominant calcium carbonate mineral in coals, so the calcium removed from coal by hydrochloric acid is probably calcite—but not positively, without corroborating evidence from other analyses. The solutions from the various steps in the fractionation can be analyzed by instrumental techniques (Skoog et al. 2018, 196–302) that quickly yield the concentrations of the elements.

It would seem reasonable to presume that the inorganic elements are part of the minerals present in coals. This is not always true. Such elements are said to have an *inorganic affinity*. But elements can also be incorporated in the organic, carbonaceous portion of a coal, on ion-exchange sites, or on coordination sites. They are said to have *organic affinity*. A good correlation between the concentrations of the element in a series of coals and the ash yields from those coals is a signal that the element has an inorganic affinity. A weak or negative correlation means that it is likely the element has an organic affinity. Inevitably there will be cases in which an element is incorporated partially with minerals and partially in organic forms.

This information helps in coal processing. To recover an element with a good inorganic affinity, coal-cleaning operations will provide a "reject" stream having a high concentration of that element. The cleaned coal could be the raw material for chemicals or carbon products, and what once might have been refuse could be treated for recovery of valuable inorganic elements. The recovery of an element with a high organic affinity, without burning the coal first, presents potential challenges. Elements on ion-exchange sites can be removed just as a water softener is regenerated. For elements held on coordination sites, the coal could be leached with an even stronger coordinating agent.

In low-rank coals, organic sulfur may contribute half or more of the total sulfur, but in bituminous coals, the majority of the sulfur is commonly pyritic sulfur. Some organic sulfur may be a remnant of the original plant material. Much of the organic sulfur in coals results from bacterial attack on sulfur-containing compounds to produce hydrogen sulfide. Wet, anaerobic environments favor this process. Hydrogen sulfide then reacts with various materials in the coalifying plant matter to form organosulfur compounds. It also reacts readily with peat (Chou 1990).

Hydrogen sulfide reacts with iron compounds in the swamp water to precipitate iron sulfides, which gradually convert to pyrite. A principal source of the sulfur is the sulfate ion, which is second only to chloride among the anions in saline water. Accumulation and diagenesis of organic matter in marine environments, or in environments that experience occasional incursions of saline water, result in coals of higher sulfur content. Bituminous coals in the interior of the United States were deposited from saline water at a time when a large sea effectively split in half the land mass that is now North America. These coals are high in sulfur. Some Iowa coals contain as much as 8% sulfur. This is such a high sulfur content that, a century ago, there were some ideas about mining these coals not as fuels, but as sources of sulfur (Schobert 1987, 42).

Coals contain, at some level of concentration, every known naturally occurring element except the noble gases and short-lived radioactive elements. Elements with concentrations of less than 0.1% (equivalent to 1,000 parts per million) are considered to be trace elements (Speight 1994, 153). Despite their low concentrations, trace elements in coals are of interest or concern for several reasons.

The concentrations of trace elements and how they are incorporated in coals provide useful information in coal geochemistry. Some trace elements occur with more dominant minerals. For example, lead and zinc can occur as their sulfides—galena and sphalerite, respectively—with pyrite. Trace elements can also occur as isolated grains of their own minerals. Still others, such as boron, can be chemically incorporated in the coal macromolecular framework (Swaine 1990, 36). Plants absorb trace elements as they grow. Copper is considered to have come from the original plant material (Bouška 1981, 170). All trace elements give clues that could relate to the geological setting in which organic matter accumulated, the environment of accumulation, and processes in diagenesis and coalification. Some trace elements in coals have valuable commercial applications, as we will discuss in Chapter 16. Another reason for interest in trace elements is what happens to them when the coal is used, particularly those that cause environmental and human-health problems. We will see this again in Chapter 9. Some trace elements leave the system with the gaseous products of combustion, either as vapors or as tiny ash particles entrained in gases (Miller 2017, 599). Others remain behind with the ash that accumulates at the bottom of the combustor. Some do a bit of both. The different kinds of behaviors during combustion require different ways of dealing with ashes that contain these elements.

Ash formation is a complex process. The large number of inorganic constituents means that there is also a large number of possible reactions or transformations as the temperature rises. Carbonates decompose to oxides, which could react with clays to form aluminosilicates not in the original coal. Calcium oxide from decomposition of calcite can capture sulfur trioxide, forming calcium sulfate. Pyrite oxidizes to sulfur oxides, leaving behind iron oxide. Clays with water in their structures lose the water and produce new aluminosilicate phases. In low-rank coals, as the coal structure breaks apart, the carboxylic acid groups that were holding ion-exchangeable elements decompose, setting the ions free to react with minerals. Compounds of chlorine, such as sodium chloride, decompose with volatilization of chloride. The chemical composition and the specific minerals present in the ash that eventually forms depend on the nature and amounts of minerals or other inorganics that were in the coal initially and on the specific reaction conditions that led to ash formation.

Though minerals become transformed as ash forms, the elements in these minerals are mostly retained in the ash. From the minerals common in coals, we can expect that the ash will contain silicon, aluminum, and iron, with smaller amounts of sodium, potassium, calcium, magnesium, titanium, and phosphorus. These elements are also ones that are considered to be the more common inorganic constituents in plants (Francis 1961, 641–642). Sulfur oxides could be expected to escape as gases and not be seen in the ash unless captured and retained in the ash by alkaline components.

How ash behaves during coal utilization processes depends on its composition. Because it can be very useful to know the actual composition of ash, an ash analysis is commonly done in coal laboratories, though not required by the proximate or ultimate analysis. By long-standing tradition, ash analysis reports the composition as if each of the elements were present as one of their common oxides. An analysis report consists of a list of 10 oxides— SiO_2, Al_2O_3, Fe_2O_3, TiO_2, P_2O_5, CaO, MgO, Na_2O, K_2O, and SO_3—along with the percentage of each in the ash sample. Because low-rank coals accumulate ions of sodium, potassium, calcium, and magnesium, ashes from low-rank coals will as a rule tend to have higher concentrations of these four elements than ashes from bituminous coals. Low-rank coal ashes will also have lower concentrations of silicon and aluminum than will bituminous coal ashes.[5] Trace elements are not determined as a routine part of an ash analysis.

Coal scientists and engineers have built up a collection of empirical equations and practical know-how that allows predicting the behavior of ash on the basis of such analyses. Often, these methods give accurate, useful information about forming and handling ash. Most of the oxides listed in the analysis report do not exist as such in the ash, nor in the coal that produced the ash. Most elements, including much of the silicon, actually occur in ash as various aluminosilicate minerals, not as free oxides. A few can't possibly occur as oxides because they are so reactive that they would never survive contact with traces of moisture. Sodium oxide is an example.

Ash is inevitable. If we consume coal, there will be ash somewhere. Ash is a complex mixture of minerals, mostly aluminosilicates. The components of ash can, and often do, experience further reactions or transformations. At high temperatures, some of the minerals in ash can soften or melt. The result is a sticky, semi-solid material that can lead to operating problems in equipment making use of the coal. The ash can melt entirely to form a liquid slag. In a piece of equipment designed to handle and discharge a molten slag, all is well. But inside equipment not intended to cope with slag, this is serious trouble. What ash does inside a combustion or conversion unit, how we get the ash out, and what we do with it afterword are all significant issues in coal utilization. They will come back in later chapters.

Knowing the temperatures at which the given ash softens or melts helps in the design and operation of equipment for using coals. This behavior is measured in tests of *ash fusibility*. In American practice (ASTM 1984b, 297–301), the ash sample is formed into a triangular pyramid—commonly called a "cone"—and heated at a specified rate. The cone is kept under observation

---

[5] This is another example of the consequences of arithmetic. Since the concentrations of elements necessarily must sum to 100%, if the concentrations of some elements, such as sodium and calcium, increase, then the concentrations of others must necessarily drop to keep the sum at 100%.

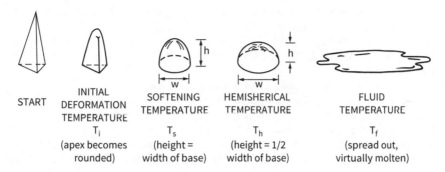

Figure 5.2 The changes observed in the ash fusibility test, from the sample cone at left to complete melting above the fluid temperature, at right.
Artwork by Lindsay Findley, from the author's sketch.

while being heated. The first discernable rounding of the tip is called the *initial deformation temperature*. Continued heating reaches a point at which the former cone is now a "spherical lump" (ASTM 1984b, 297–301) having a height equal to the width at its base, the *softening temperature*. Usually, if only one ash fusibility temperature is reported for an ash, it is the softening temperature. Further heating gets to the *hemispherical temperature*—the height being half the width of the base—and ultimately to the *fluid temperature*, at which point the sample is essentially molten and has spread out in a layer. These transformations are sketched in Figure 5.2. In equipment designed to remove ash as a solid, the operators need to maintain the temperature certainly below the softening temperature and preferably below the initial deformation temperature. But if equipment is designed to remove ash as a molten slag, it is essential to operate above the fluid temperature.

Up to this point we have seen how coals form, how they are classified, and what they are made of. We now begin to look at utilization. We will see how coals are gotten out of the ground, prepared for use, and used in various ways now.

References

ASTM. 1984b. "Standard Test Method for Fusibility of Coal and Coke Ash. D 1857–80." *Annual Book of ASTM Standards. Volume 05.05. Gaseous Fuels; Coal and Coke.* Philadelphia: American Society for Testing and Materials:

ASTM. 1984a. "Standard Test Method for Forms of Sulfur in Coal. D 2492–84." *Annual Book of ASTM Standards. Volume 05.05. Gaseous Fuels; Coal and Coke.* Philadelphia: American Society for Testing and Materials.

Bates, Robert L., and Julia A. Jackson. 1987. *Glossary of Geology.* Alexandria, VA: American Geological Institute.

Berkowitz, Norbert. 1979. *An Introduction to Coal Technology*. New York: Academic Press.

Bouška, Vladimír. 1981. *Geochemistry of Coal*. Amsterdam: Elsevier.

Chou, Chen-Lin. 1990. "Geochemistry of Sulfur in Coal." In: *Geochemistry of Sulfur in Fossil Fuels*, edited by Wilson L. Orr and Curt M. White, 30–52. Washington, DC: American Chemical Society.

Francis, Wilfrid. 1961. *Coal*. London: Edward Arnold.

Gluskoter, Harold J., Neil F. Shimp, and Rodney R. Ruch. 1981. "Coal Analyses, Trace Elements, and Mineral Matter." In: *Chemistry of Coal Utilization. Second Supplementary Volume*, edited by Martin A. Elliott, 369–424. New York: John Wiley & Sons.

Mackowsky, M. T. 1982. "Minerals and Trace Elements Occurring in Coal." In: *Stach's Textbook of Coal Petrology*, edited by E. Stach, M. T. Mackowsky, M. Teichmüller, G. H. Taylor, D. Chandra, and R. Teichmüller, 170. Berlin: GebrüderBorntraeger.

Miller, Bruce G. 2017. *Clean Coal Engineering Technology*. Amsterdam: Butterworth-Heinemann.

Mraw, Stephen C., John P. De Neufville, Howard Freund, Zeinab Baset, Martin L. Gorbaty, and Franklin J. Wright. 1983. "The Science of Mineral Matter in Coal." In: *Coal Science*, edited by Martin L. Gorbaty, John W. Larsen, and Irving Wender, 2:1–63. Orlando: Academic Press.

Schobert, Harold H. 1987. *Coal: The Energy Source of Past and Future*. Washington, DC: American Chemical Society.

Skoog, Douglas A., F. James Holler, and Stanley R. Crouch. 2018. *Principles of Instrumental Analysis*. Boston: Cengage Learning.

Speight, James. 1994. *The Chemistry and Technology of Coal*. New York: Marcel Dekker.

Swaine, Dalway J. 1990. *Trace Elements in Coal*. London: Butterworths.

6
Mining

The earliest exploitation of coal likely involved seams that outcropped on the surface. Pieces could be chipped off and collected using simple tools. When coal readily available at the outcrop was consumed, adventurous people may have followed the seam underground, establishing the first mining operations. Similar to many other technological developments, mining probably began independently in several places of the world and possibly at different times.

Coal mines fall into two broad categories: *underground* and *surface mining* (also called *strip mining* or *open-cut mining*). Choosing between underground and surface mining is based on the depth of the coal seam to be worked. Seams lying near the surface can be extracted by stripping, but which could be prohibitively expensive and disruptive to the surroundings when mining deeplying seams. Each method has its own technical and economic advantages and disadvantages.

Classification of underground mines follows the methods used to reach the coal: shaft, slope, or drift mines. These are illustrated conceptually in Figure 6.1. The depth and pitch of the seam determine the type of access. A *shaft mine* uses a vertical hole dug straight from the surface to the coal seam. A hoist lowers the miners and their equipment to the working level. Tunnels, also called *drifts*, are then driven horizontally into the mine. A *slope mine* provides an access on a slant. A slope can be driven to follow a seam along its pitch or to cut through a hill or mountainside to reach coal underneath. Slope mines provide easier removal of the coal and easier movement of miners and machinery into the mine. Where coal outcrops on a mountainside, a drift mine can be driven straight into the coal. Drift mines are the easiest and cheapest to develop.

Regardless of the type of access, coal has to be removed somehow from the seam. There are three systems: room and pillar mining, pitch mining, and longwall mining. The *room and pillar system* can be used when the seams very little pitch.[1] It involves dividing the coal into a series of blocks. Parallel

---

[1] The angle between the seam and a horizontal plane.

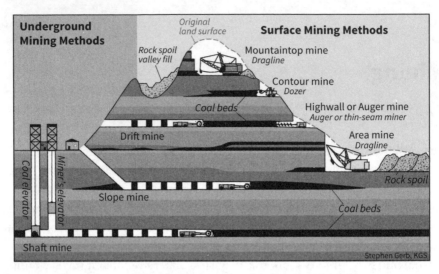

Figure 6.1 The major methods of surface and underground coal mining.
Artwork by Stephen Greb, Kentucky Geological Survey.

tunnels, called *main entries*, are driven into the coal. The main entries connect via a second series of tunnels, called *gangways*, cut at right angles to the main entries. Miners cut the coal into rooms or chambers at intervals along the gangway. A portion of the coal is taken out as the rooms are formed. The rest remains in place as solid pillars to support the mine roof.

The first working leaves roughly 50% of the original seam in the mine. If mining stops at this point, room and pillar mining can be very wasteful. More coal can be recovered by removing the pillars. Removal of the pillars, the second working, is also referred to as the *retreat* or as *robbing the pillars*. It requires additional, careful planning to be done safely. Progressive removal of roof supports makes the second working very hazardous. Robbing the pillars can weaken the roof enough to cause collapse.

Removal of the coal from the seam occurs by continuous mining. A machine creeping along on caterpillar tracks uses cutters to gouge coal from the face of the seam. Coal is automatically loaded onto conveyor belts or into cars. The continuous miner can cut coal so rapidly that the workers attending to roof support, ventilation, and water drainage can't keep pace. The mining machine must be stopped from time to time to allow the rest of the work to catch up. Continuous mining may not be effective with extremely hard coal. Also, the machinery might be too unwieldy to use when the pitch or thickness of the seam vary.

Longwall mining removes coal from one long, continuous face rather than from a number of short faces as occurs in room and pillar or pitch mining.

Mining begins by driving parallel roadways into the seam. The roadways are connected by a tunnel driven at right angles to them; this tunnel forms the face of the seam where the mining operation takes place. In *advance mining*, the tunnel is driven at the front of the seam, and mining operations extend progressively deeper into the coal. The parallel roadways are continuously extended as well. In *retreat mining*, the roadways are driven to the full extent of the seam and then connected by a tunnel. Mining moves from the back of the seam toward the main shafts.

An electrically powered machine moves back and forth across the whole extent of the coal face and removes the coal. A longwall mining machine is illustrated in Figure 6.2. The machine is equipped with heavy revolving chains carrying tungsten-carbide-tipped cutter picks or with a hardened blade somewhat like that of a plow. Such machines can cost hundreds of millions of dollars (Mouawad 2016). The mining machine can cut coal at a rate of up to 5 tonnes per minute. The cut coal moves onto a conveyor belt for removal from the mine. The roof of the mined-out area behind the machine is allowed to collapse.

The coal-cutting machine, the roof supports, and the conveyor are self-advancing. When the machine gets to the end of the exposed seam face, it reverses direction and makes another cut as it travels back to the starting

Figure 6.2 Longwall mining in Poland, showing the hydraulically supported roof, conveyor, and coal cutter.
Courtesy of Milosz Maslanka/Shutterstock.com.

point. Steel canopies held up by hydraulic roof supports provide protection for the miners. Because of the continuous operation, longwall mining is more efficient than room-and-pillar mining. It provides a high rate of production, requires fewer miners, and recovers a higher percentage of coal. The movable hydraulic pillars that support the roof directly over the mining operation eliminate the cost of permanently installed roof supports and allow working at depths at which pillars left by room-and-pillar mining would not support all the weight of the overhead rock. Longwall mining offers a safer system in mines in which the strength of the roof is generally poor.

Worldwide, longwall mining accounts for half the world's total coal production by all methods, although this figure is skewed upward because China still relies heavily on underground mining. In many other coal-producing countries, surface mining dominates. About 50% of the underground production in the United States comes from longwall mines. It also contributed greatly to underground production in European countries. Before the last underground mine in Britain closed at the end of 2015, about 90% of underground production came from longwall mining. It is applicable to various seam thicknesses and pitches and provides a high percentage of extraction of the coal.

Pitch mining is used when the coal seams are steeply inclined. It was often used in anthracite mining in the United States, where in extreme cases anthracite seams might be at 80 degrees to the horizontal. Miners create a gangway along the bottom of the seam. Cuts made at right angles to the gangway pass upward into the seam. Miners climb or crawl up the passageway to reach the seam face. They have to carry or drag their tools and equipment with them. After drilling and blasting, coal is shoveled into a timbered chute. Newly exposed roof must be supported, after which the sides of the chute can be extended. The most strenuous part of a miner's shift can be simply the job of getting from the gangway up to the face of the seam with a complete load of equipment and supplies.

Blasting is the principal method of fracturing and loosening coal. Dynamite, invented in 1867 by the Swedish chemist Alfred Nobel,[2] is made by absorbing nitroglycerin in sawdust, wood pulp, or a chalky material known as kieselguhr.[3] The gases released when dynamite "goes off"

[2] Alfred Nobel, founder of the Nobel Prizes, is otherwise best known as the inventor of dynamite. He was in fact a prolific inventor, holding more than 350 patents. Much of the impetus for the invention of dynamite came from an explosion in the Nobel family's nitroglycerin factory, which killed Alfred's younger brother. Years later, seeing his own prematurely published obituary (it was actually his older brother who had died), which called him a merchant of death, inspired Alfred Nobel to leave much of his considerable fortune to establish the Prizes.

[3] Also called *diatomaceous earth*, this is a soft, low-density sedimentary rock that consists mainly of the silica-rich skeletons of unicellular algae.

generate such high temperatures and pressures that they create a shock wave traveling at supersonic speeds (Glassman 1987, 198). The shock wave fractures the coal.

Blasting in underground mines has to consider that there could be other explosive materials already present in the mine. Methane, discussed in more detail below, can explode at concentrations of about 5–15% in air. Coal dust suspended in air also makes a fine explosive. Both have been responsible for devastating explosions in underground coal mines, with many casualties. The danger of the flame of the original explosion potentially setting off a fire or a secondary explosion led to the development of *permissible dynamite*. Now that other blasting agents are used in addition to dynamite, this class of materials is called *permissible explosives*. They produce only a brief, short flame with little likelihood of setting off a secondary gas or dust explosion. Mining regulations will include requirements for the use of permissible explosives and the classification of these explosives for acceptable use in various mining situations. ANFO, an acronym for ammonium nitrate–fuel oil, has supplanted dynamite. Ammonium nitrate is a very powerful explosive. The ammonium nitrate serves as an oxidizer and the oil provides fuel.

The mine roof must be supported, usually by timbering or roof bolting. *Timbering* involves the use of wood to support the faces of a tunnel. Timbers do not have to support the full weight of all the rocks overhead; the rock or coal forming the sides of the tunnels supports most of this weight. The timbering mostly keeps relatively small, loose pieces of the roof from falling. Soft woods, often preferred as props, have reasonable strength but are elastic enough to bend a bit before they break. Bending props warn miners of an impending *squeeze*, in which the weight of the roof rock suddenly crushes the coal (as well as any miners caught in the way) as the roof settles down to the floor. Timbers made of masonry, concrete, or iron can be used instead of wood, but wood emits audible noises as it is being strained by a possible squeeze. Experienced miners can gauge the imminence of danger from the noise being made by the props.

Roof bolting consists of drilling holes into the roof, inserting steel roof bolts into the holes, and anchoring them in place. Bolts attach the roof of the mine to a self-supporting layer of rock and hold the roof in place. Several thin, weak layers of rock can be bolted together.

In tragic instances the roof supports are overwhelmed by the weight lying above them. In the mid-nineteenth century, a mine in Workington, England, was extended hundreds of meters under the Irish Sea. Eventually the roof gave way, allowing the sea to flood the mine. Thirty-six men and boys died

(E. W. 1873, 92–93). In January 1959, the Knox Mine, just south of Pittston, Pennsylvania, tunneled under the Susquehanna River with only a very little thickness of roof—anecdotal reports were that it might have been only 1 meter—separating the mine from the bottom of the river. Twelve miners died when the river brought about 40 billion liters of icy, sediment-filled water into the mine (Wolensky, Wolensky, and Wolensky 1999). In 2010, 115 miners were rescued from the Wangjialing mine in northern China, trapped for 8 days when a mine flooded (Wong 2010). Some survived by attaching themselves to the walls of the shaft using belts or pieces of clothing so that they would not drown if they fell asleep.

Three problems must be faced in every underground coal mine: dust, gases, and water. A suspension of coal dust in air, ignited by a spark, experiences very rapid combustion. The increase in temperature resulting from heat released during the rapid burning increases the reaction rate, which in turn raises the temperature even more. The entire air–dust suspension burns violently in a fraction of a second, an event observed as an explosion. Coal dust explosions have occurred everywhere that coal is mined underground. The worst, which claimed 1,572 lives, happened in Manchuria in 1942 (Moyer 1969).

Mixtures of methane and air can explode if the concentration of methane is sufficient. In the early days of coal mining, it seemed reasonable to attribute the devastating explosions in underground mines to methane. The actual chain of events leading to such explosions was deduced in 1845, by Michael Faraday. He observed that a methane explosion, which might be relatively localized in a mine, occurs first, raising and igniting a cloud of coal dust, producing a second explosion much more widespread and far more violent than the first. One product of the combustion reaction producing the dust explosion is carbon monoxide (CO), which can be fatal to workers still in the mine or to rescuers entering the mine.

Two approaches can be taken to avoid dust explosions. The first is to minimize production of dust during mining operations and minimize the sparks or flames that could ignite a dust–air suspension. This concern led to the introduction of permissible explosives. The second approach, *dusting*, involves spreading an incombustible material, such as powdered limestone, in the mine. An accidental explosion of methane or the explosion resulting from intentional blasting raises a cloud of dust, but the incombustible stone dust dilutes the concentration of coal particles to the point that their mixture in air will not ignite.

Mine ventilation provides the workers with fresh air and removes gases that may be hazardous. The flow of air occurs through shafts, usually linked in pairs as an intake and a return. Air moves from the intake across each working face and then out through the return. Large fans provide a positive ventilation system. Large mines can require fresh air exchange at rates exceeding 14,000 cubic meters per minute.

The atmosphere in an underground mine can contain several gases of concern to miners. Collectively they are known as *damps*, an Anglicization of the German word *dampf*, meaning vapor. *Firedamp*, primarily methane, is the most troublesome. Coal and the surrounding rocks absorb methane as it forms. Normally, methane desorbs gradually as the mining operations remove coal or rock. Slow seepage of methane from the coal into air in the mine can cause the concentration of methane to reach its explosive range.

Methane, being odorless, tasteless, and colorless, has no properties that make its presence obvious in the mine. Centuries ago, miners used candles or lamps with open flames for illumination. If the methane concentration in the mine reaches a dangerous level, an explosion could be set off by the miners' lamps. As Faraday realized, this creates a suspension of coal dust in the air which immediately causes a secondary, devastating dust explosion. Tens of thousands of miners around the world have lost their lives to methane explosions and, in some places, still do to this day, despite the hazards being recognized for centuries (Mostyn 1677). The invention of a convenient and practical safety lamp was a tremendous boon to miners and, over the years, saved an incalculable number of lives.

In 1812 and again in 1813, devastating explosions occurred in the Felling mine near Newcastle-upon-Tyne. The first took 91 lives. The 1813 victims included nine men, twelve horses, and thirteen boys. Stimulated by such horrific explosions, Humphry Davy developed a safety lamp that used a metallic gauze to prevent the heat of the lamp's flame from igniting a methane-air mixture.

> I surround the candle of the coal-miner with a metal gauze; the flame will not pass through. Confined in its cage, it will not communicate with the gas, and explosions will not take place. (Davy, as quoted in E. W. 1873, 51)

The *Davy lamp*, and evolutionary variations of it, remained standard miners' equipment until eventually replaced by electric lamps.

The safety lamps could also be used as gas detectors. Methane caused the flame to turn blue, and the height to which the flame could be turned while staying blue provided a qualitative estimate of the percentage of methane in the air in that part of the mine. Many mines had a *fire boss*. This person tested for potentially hazardous gases before the full shift of miners was allowed into the mine. The fire boss would wear thick clothing that might be soaked with water. Before the days of the Davy lamp, the fire boss would walk through the mine with a candle attached to the end of a long stick. If there happened to be an accumulation of firedamp, the candle was supposed to ignite it and let it burn out. Most likely there were occasions in which the fire boss himself was ignited, if not completely incinerated. After safety lamps came into use, the fire boss could test for hazardous gases by cautiously observing the color of the lamp's flame when the shield was lifted to allow gases into the lamp. Eventually the job was made redundant by development of gas detectors that could sense the presence of firedamp and of battery-operated electric lights.

Safety practices can't overcome avarice, stupidity, or criminality. Farmington No. 9 mine, West Virginia, November 1968—78 dead. An alarm on a ventilating fan designed to shut off power in the mine if the methane concentration was too high was allegedly disabled by the mine's chief electrician on the previous night (Stewart 2012). Chenzhou, China, February 1995—three explosions in 2 days, 29 dead. Two of the mines were unlicensed and operating illegally; one had installed a water pump in direct violation of safety regulations (Associated Press 1995). The Upper Big Branch mine near Montcoal, West Virginia, April 2010—29 dead, due to methane and coal dust accumulating in a mine with a history of violations for not providing proper ventilation (NRC 2012). The Zasyadko mine, Donetsk, Ukraine, March 2015—24 more dead. The *World Almanac* tallies 141 "notable explosions since 1920," which include 22 coal mine explosions accounting for 4,540 deaths (Janssen 2020, 324–325).

Methane in a seam at high pressure sometimes releases with near-explosive force, blasting ahead of it any coal or rock remaining in the way. Miners call this an ***outburst***. An outburst generates high-velocity, flying fragments of coal or rock, potentially lethal to anyone working nearby. A spectacular outburst in the Morrissey mine, in British Columbia in 1904, took the lives of 14 miners, after an outburst in the same mine a year earlier had killed five.[4] About 2,000 tonnes of coal were blasted loose by the sudden release of an estimated 85,000 cubic meters of methane. Nearly 60 years earlier, enormous quantities of methane were released at the Walker Colliery, Newcastle-on-Tyne. The gas discharge extended some 600 meters and sounding like "the blowing off of an immense high-pressure steam engine" (Moyer 1969, 971). No one was killed, a fact attributed to the use of safety lamps by the miners.

---

[4] George Santayana: "Those who do not remember the past are condemned to repeat it."

Methane also acts a greenhouse gas, an issue to be treated in Chapter 15. Methane emissions from coal mining account for around 9% of global anthropogenic methane emissions (EPA 2021).

Whitedamp, the miners' name for CO, is highly toxic. As little as one-half of a percent of CO in air can be fatal in 2 minutes. Whitedamp is also combustible, potentially contributing to fires or explosions. In the years before the advent of CO detectors, miners tested for the presence of poisonous gases by taking a small, caged bird with them into the mine. Being of much smaller body mass, the bird became unconscious much sooner than a human and therefore served as a warning to the miners to get out of the mine. Often, the birds used were canaries. When they were distressed, they would sing or call frantically. Using canaries as living CO detectors has led to the use of the metaphor "canary in a coal mine" as an expression meaning that something—the metaphoric canary—is providing an advance warning of some impending dangerous situation. A modern example of usage might be a statement to the effect that the worldwide melting of glaciers is a "canary in a coal mine" for global warming.

Blackdamp, also called *chokedamp*, consists mainly of carbon dioxide (CO_2). Blackdamp results from any combustion or oxidation processes involving carbon or its compounds: explosions (deliberate or accidental), coal fires, rotting timbers, and respiration of the workers. Though not poisonous, in high concentrations, blackdamp causes asphyxiation by diluting and displacing the oxygen needed to sustain life.

Stinkdamp, hydrogen sulfide, is notorious as the unforgettable odor of rotten eggs. Though odor is its most obvious characteristic, hydrogen sulfide also is a powerful poison[5] and combustible. *Afterdamp* is the mixture of gases remaining after an explosion or fire has occurred. Afterdamp poses a double problem: it is always deficient in oxygen, possibly causing asphyxiation, and it can contain lethal CO.

Water, often a continual problem in mines, enters by running in from the surface or by seeping in from underground streams. In extreme cases, as much as 20 tonnes of water have to be removed for every tonne of coal dug. Water draining from mines can be highly acidic. Water percolating through a mine is exposed to sulfur compounds in the coal, producing a solution of

---

[5] One measure of the toxicity of gases is the immediately dangerous to life concentration (IDLC). The IDLC value for hydrogen sulfide in air is 100 parts per million. The notorious poison hydrogen cyanide (HCN), used in real life for executions in gas chambers and in numerous detective novels for murders, has an IDLC of 50 parts per million. By this measure, being exposed to hydrogen sulfide is almost as deadly as being in a gas chamber.

sulfuric acid. The acidity originates from conversion of pyrite to sulfuric acid and iron sulfate upon exposure to oxygen in the air and water. In aerated water, ferrous sulfate oxidizes to ferric sulfate, which hydrolyzes to precipitate ferric hydroxide, a yellowish-orange gelatinous mass (Stefanko 1983, 340–341) sometimes called *yellow boy*. In many jurisdictions, environmental regulations require acid mine drainage (AMD) to be neutralized before being discharged to the environment. AMD constitutes a major international environmental problem. Thirty years ago, approximately 19,300 kilometers of streams and rivers, and more than 72,000 hectares of lakes and reservoirs worldwide had been seriously damaged by mine effluents (Kidd 2016). The acidity has a severe effect on local aquatic organisms. It can also be corrosive to anything in the water, such as pumps, pilings, or boat hulls.

The straightforward approach for treating AMD is to react the acid with lime. Mixing AMD with a slurry of calcium hydroxide in water raises the pH to about 9, forming a precipitate that is mainly calcium sulfate. Metal ions that are in solution, such as manganese and iron, will also precipitate. The slurry can be pumped into a clarifier—basically a large tank—that allows the calcium sulfate and metal-ion precipitates to settle as a sludge, with the clear water then discharged.

A coal seam can catch fire at an outcrop, potentially burning for decades. Coal in an active mine can also catch fire, and such an occasion can be very dangerous to the miners. Fires also occur in abandoned mines and in piles of refuse from coal cleaning operations. Fires occur just about every place that coal occurs.

A fresh coal surface exposed to the atmosphere will start to oxidize immediately, even at ambient temperatures. The heat release from slow oxidation in air is barely perceptible. If the heat is not dissipated, the temperature of the oxidizing coal will increase. Because the rates of chemical reactions increase as temperature increases, oxidation will proceed more rapidly, releasing more heat and raising the temperature further, thus kicking off a vicious cycle that ultimately heats the coal to its kindling point. Since no apparent human agency started this fire, the process is commonly referred to as *spontaneous combustion*. Strictly speaking, there is no such thing—something has to start the combustion; it can't be spontaneous. Norbert Berkowitz, an outstanding coal scientist of the twentieth century, suggested the term *autogenous heating* (Berkowitz 1979, 189–192), which is much more accurate but has never caught on in general usage.

A fire once started can spread rapidly. Burning coal produces dangerous afterdamp. The fire could be contained by shutting off access to air. This is difficult, since mines are designed to maintain a steady, vigorous flow of air for the miners. Alternatively, the mine section could be flooded. In the worst case, the mine must be closed entirely, possibly for years.

One of the most remarkable of coal fires is that burning beneath the former town of Centralia, Pennsylvania. The fire appears to have started in late May 1962, and it continues to this day. There are several versions of how the fire first started. The one that seems to have most currency is that an anthracite outcrop was accidentally ignited in an otherwise well-intentioned effort to clean up a rubbish dump—by setting the dump on fire. In 1982, the most re-markable single incident occurred when the earth gave way beneath a 12-year-old boy, seemingly swallowing him: he fortunately survived (Quigley 2007). The situation continued to deteriorate, with decades' worth of legal battles as the residents fought a protracted rear-guard action to save their homes and their town. Today most of what was Centralia has been abandoned and the buildings demolished. The population consists of nine people who forged an agreement to remain in their homes for the duration of their lives, after which their homes, too, will likely be razed. Now, Centralia is a tourist attraction of sorts. No one has a workable solution. There seems to be a sense of resigna-tion to let the fire burn itself out, which is estimated to happen in 2250. An enormous tonnage of anthracite of a quality unsurpassed in the Pennsylvania anthracite fields (Quigley 2007, xvii) and having many potential valuable uses (Chapter 17) is being lost. Complete combustion of 1 tonne of anthracite would produce about 3 tonnes of CO_2,[6] so the Centralia fire is not doing our climate any good, either.

Today, underground mining is a remarkably safe occupation in many countries, though there remain a few unfortunate exceptions. The annual death rate is lower than that for general manufacturing industries. Such was not always the case, and still is not the case in some parts of the world. Even today there are potential long-term consequences for miners, most notably black lung disease.

Miners who successfully escape the perils of work underground can still fall prey to diseases that can take decades to manifest themselves. A chronic shortness of breath among miners came to be known as *miner's asthma*. It results from inhaling dust generated by the mining operation. Two forms of *pneumoconiosis*, the lung disease caused by inhalation of irritants,

---

[6] This is based on the assumption that anthracite, as mined including ash yield and moisture, would be about 80% carbon. One tonne of carbon produces 3.67 tonnes of CO_2.

particularly affect coal miners. **Anthracosis**, commonly known as *black lung disease*, results from inhalation of coal dust. The other, *silicosis*, results from inhaling rock dust.

Black lung disease results from the accumulation of coal dust in small air sacs—the alveoli—in the lungs. Coal dust seems to form sores that eventually become replaced by scar tissue. The process might be triggered by very fine particles of pyrite (Finkelman et al. 2006). Accumulation of scar tissue reduces the ability of the lungs to exchange waste CO_2 for fresh oxygen. Breathing becomes much more labored. A person suffering from black lung disease is readily susceptible to chronic bronchitis and emphysema. In the United States, there had been a slow decline in deaths from black lung disease since about 1990. It appears to be making a resurgence in the coal-producing regions of Appalachia (Valdmanis 2018). In China, deaths from black lung disease far exceed the numbers of miners killed in accidents (Economist 2015).

Silicosis occurs when inhaled particles of silica or siliceous rock are ingested by macrophages, the cells that attempt to engulf foreign objects in the body. Macrophages in the passages of the lungs form fibrous clots that can eventually spread to the openings of the lungs or to the heart. The amount of inhaled silica needed to produce silicosis is usually less than one-fourth the amount coal dust necessary to produce black lung disease.

The particle size of coal or silica dust particularly susceptible to inhalation and collection in the lungs is less than 5 microns. Much of the dust produced during coal mining is in this size range. Several actions mitigate exposure to dust, including water sprays to keep dust down, improving ventilation and dust collection, and requiring the wearing of dust masks. Stone dust used as a precaution against explosions is limestone rather than siliceous rock, to mitigate the unfortunate choice between the immediate danger of a dust explosion and the long-term danger of silicosis. Silicosis is a hazard in many industries other than coal mining. Fewer than 10% of deaths due to silicosis were attributed to coal mining (NIOSH 2008).

Near-surface deposits can be mined by stripping away the soil or rock lying on top of the coal—material called **overburden**—to expose the coal for extraction. The difficulty, which always translates to expense, of operating a surface mine depends in part on the amount of overburden to be removed. An overburden thickness of roughly 200 meters represents the maximum that would be economically or practically acceptable in most situations. The total depth of the mine might be up to about 500 meters. The **stripping ratio** expresses the relative thickness of overburden to the thickness of the coal, also an important factor for the economic operation of a surface mine. The

Figure 6.3 Open pit mining of lignite in Germany using a bucket wheel excavator.
Courtesy of Pixabay.com, Public domain.

tolerable stripping ratio for a specific mine depends on how difficult it will
be to remove the overburden, local environmental and geological conditions,
and the value of the coal expected to be extracted. There is no single value of
the acceptable stripping ratio applicable to all surface coal mines.

Two general methods of surface mining are used: area mining, for level
land, and contour mining, for hilly land. A surface-mining operation is illustrated in Figure 6.3. *Area mining* begins by digging a trench, piling the overburden to one side in a spoil ridge. After the coal has been extracted, a second
trench is dug, parallel to the first one. Overburden from the second trench
is piled as a spoil ridge into the first trench. This process is repeated for suc-
cessive trenches until the lateral extent of the coal deposit has been worked.
Contour mining is used when coal outcrops on a hillside or lies beneath hilly
overburden. Overburden is dumped on the downhill side of the mining operation. This process forms a bench, which consists of the spoil pile of dumped
overburden, the exposed coal seam, and the newly cut high wall of rock or soil
formed by removal of the overburden. Benches are formed along the hillside
as mining proceeds through the coal deposit.

Mining consists of only a few steps. Bulldozers or scrapers clear and level the
area to be mined. Strongly consolidated overburden can be blasted to break
it up for removal. The earth-moving and digging machinery used in surface
mining depend on the size of the operation and the nature of the overburden

and coal to be removed. A *dragline* consists of a bucket mounted on a chain, cable, or rope that passes along the ground to scoop up the overburden. A second cable or rope raises the filled bucket and dumps it onto the spoil pile or into a truck. The biggest have booms more than 130 meters long and are able to hold about 120 cubic meters of overburden in the bucket (Balaban and Bobick 2017). Normally they run around the clock, 7 days a week.

Bucket-wheel excavators consist of a wheel with buckets attached to its circumference. As the bucket wheel, mounted vertically on a boom, rotates, each bucket cuts, removes, and then dumps the coal. The biggest such excavators are among the largest land vehicles on the planet (Atherton 2017). The colossus among bucket-wheel excavators, the German-made Bagger 288, can remove a quarter-million cubic meters of overburden in a day. When the coal has been exposed, it can mine about 240,000 tonnes of lignite per day, loaded onto conveyor belts 3 meters wide. The Bagger 288 itself weighs nearly 12,000 tonnes. When it has to move, it can speed along at a bit more than half a kilometer per hour.

Surface mines require less labor than underground mines. One worker operating a bucket-wheel excavator or dragline can accomplish more than a team of miners working underground. The amount of coal produced per hour worked in surface mines is about triple that produced per hour worked in underground mines. Surface mines generally have a better safety record than underground mines.

The use of surface mining varies widely among countries. Surface mining dominates in the United States and Canada. Less than 10% of China's enormous production of coal comes from surface mines, attributed to the geological conditions in which most Chinese coals are found (Ji et al. 2009). In the United States, the proportion of coal produced by surface mining increased steadily in the last decades of the twentieth century. In 2019, surface mines produced 17 tonnes of coal for every tonne from an underground mine (EIA 2020). The relative tranquility of labor relations in strip mines was also important. The cost per tonne of coal produced by surface mining is lower and the productivity is higher. In favorable circumstances surface mines can recover about 90% of the coal, compared to about 70% from underground mines.

Surface mining, even when practiced by the most conscientious mining companies, is an ugly operation. Underground coal mines do not usually disrupt so wide an area as surface mines. In the United States, strip mining began on a large scale in the Appalachian Mountains in the coal-laden hills of Pennsylvania, West Virginia, and Kentucky. Years ago, unscrupulous or uncaring companies blasted and tore away the overburden, extracted the coal, and then moved to other locations, leaving behind useless pits, spoil piles, and high walls. This devastation of a once-beautiful land likely represents the mental image that many people have today of surface mining.

As mining proceeds, the pits and associated spoil piles eventually cover a large area. The deeper the mine, the wider the area that must be opened to get at the coal. During active mining, the land can no longer be used for the purpose it had prior to mining, whether for agriculture or simply for aesthetic and recreational features. Rainwater carries away mud from the spoil piles. Once-fertile topsoil dislodged by mining can also be carried away by rainwater. The mud then pollutes nearby rivers or streams. Over time, stream or lake beds fill with mud and soil, a process called *aggradation*. In an extreme case, an entire spoil pile can be carried downhill in a devastating mud slide.

Many negative aspects of surface mining can be mitigated through the enactment and enforcement of land reclamation laws. The crucial issue is what happens after the coal has been removed or the mine ceases to operate. Various countries, states, or provinces have passed mined land reclamation laws to require a mining company to restore the area to the condition in which it was before mining began.

Many places with land reclamation laws require a reclamation plan in hand before starting mining. Topsoil must be carefully removed and stored as mining proceeds. Upon completion of mining, reclamation follows. The land can be returned to its original contours by levelling high wall and spoil piles, then topsoil is restored and replanted or reforested. Recontouring, fertilizing, and seeding aim to restore the land to the same topography it had before mining and to restore the plant community that existed before mining. In the best cases, a visitor is unaware that he or she is seeing reclaimed coal mines.

Mined land reclamation is not free. The costs of earth-moving, seeding, planting, fertilizing, and all the ancillary activities have to be borne by somebody, whether the money comes from public or private sources. In the *mountain-topped*[7] areas of West Virginia, it's estimated that it can cost about $6,000 per hectare to reestablish a 30-centimeter thickness of topsoil (Moore 2017).

Comparing the productivity of conventional and continuous mining illustrates a crucial fact almost invariably ignored by politicians of all stripes. Conventional mining could produce about 100–150 tonnes of coal per worker in one work shift; in continuous mining, a worker could produce about 5,000 tonnes in a shift. In countries where the coal industry is in decline, certainly

---

[7] A devastating surface-mining method in which a mountaintop is leveled and the overburden removed in the process is used to fill adjacent valleys.

including the United States, communities in which mining had been a major employer are hard-hit, along with local and regional economies. Populist politicians promise to bring back the lost mining jobs, restoring the former boom times. Unfortunately, they can't. No politician or party, regardless of position on left or right, nor any government or civic agency can restore the lost jobs. They are gone forever. They have vanished from the universe. As political commentator Thomas Friedman puts it, "most lost jobs are outsourced to the past" (Friedman 2006, 276). They have not been destroyed by corporate malevolence, overreaching government regulations, a supposed "war on coal," or by moving across national boundaries to someplace with cheaper labor. The major factor in loss of jobs is the steady improvement in mining productivity, requiring fewer and fewer workers to produce the same amount of coal. The Bagger 288 bucket wheel excavator mines almost a quarter-million tonnes of lignite per day with a crew of three or four (Atherton 2017). Combined with the displacement of coal, in many countries, by renewable energy forms and shale gas (Chapter 14), a resurgence of mining employment seems very unlikely. The situation is further compounded by increasing awareness of global climate change (Chapter 15).

At the 2018 International Pittsburgh Coal Conference in Xuzhou, China, Professor Ge Shirong of the China University of Mining and Technology discussed developments in robotic mining technology, including a mine in China that produces 10 million tonnes of coal per year with a total staff of six people (Ge 2018). All of the underground work is done by robots, suitably equipped with positioning sensors and detectors to indicate whether they are in rock or coal. In the subsequent discussion, an audience member made the important point that the six people involved do not need the work skills that we normally associate with coal mining. Rather, they need expertise in such fields as computer programming and robotics.

The general idea of using control and monitoring systems to operate mining machinery from above ground is not new. It goes back at least to the 1970s (Wood 1980). Now we have a half-century's progress in computers, robotics, sensors, and positioning systems to build on. There is no way to predict how fast robotic mining technology will develop and how widespread it will become. But this is the wave of the future, and it is not good news for traditional mining jobs.

Can't we just figure out a way to use coal without having to mine it first? Two technologies offer options.

Methane, known as firedamp, can pose a serious problem in underground mines. But, methane, when known by a different name—natural gas—is a superb fuel. Even if only for this reason, it can be useful to drain the methane from coal seams before they are mined or extract it without ever mining the coal at all. The term *coalbed methane* (CBM) refers to methane trapped in coal and recovered as a fuel source. Recovering CBM also improves safety in the mine and reduces methane emissions to the atmosphere while at the same time providing a source of excellent and convenient fuel. We will return to CBM in Chapter 11.

Over the years researchers in many countries have explored ways of processing coal right in the seam, to avoid the expense associated with mining coal and transporting it to the place where it will be used. The technology most extensively investigated is *underground coal gasification*, which will also be discussed further in Chapter 11. Other ideas have included injecting coal-eating bacteria underground to digest the coal into some sort of liquid that could be extracted and refined into useful fuels. Decades of efforts have gone into the development of underground processing of coal, but, so far, no technology has emerged that comes remotely close to the scale on which coal is mined and then burned or processed above ground. Currently, China has interest in "chemical mining," using chemical reactions underground to extract products from coal and to evolve away from mechanical mining methods.

Except for CBM and underground gasification, coal has to be brought to the surface by some method. Before we can put coal to work for us, it usually—not always—needs attention to reduce ash yield or sulfur content. We will pursue that issue in the next chapter.

References

Associated Press. 1995. "Gas Explosions Kill 29 Chinese Coal Miners."
Atherton, Kelsey D. 2017. "Charted." *PopSci.com* 3(6): 6–7.
Balaban, Naomi, and James Bobick. 2016. *The Handy Technology Answer Book*. Canton, MI: Visible Ink Press.
Berkowitz, Norbert. 1979. *An Introduction to Coal Technology*. New York: Academic Press.
The Economist. "Shaft of Light." July 18, 2015.
EIA. 2020. "Annual Coal Report." U.S. Energy Information Administration. https//www.eia.gov/coal/annual/index.php.
EPA. 2021. "Frequent Questions About Coal Mine Methane." U.S. Environmental Protection Agency. https://www.epa.gov/cmop/frequent-questions.
"E. W." 1873. *Fuel for Our Fires*, 93. London: Religious Tract Society.
Finkelman, Robert B., Harvey E. Belkin, and Jose A. Centeno. 2006. "Health Impacts of Coal: Should We Be Concerned?" *Geotimes* 51(9): 24–28.
Friedman, Thomas L. 2006. *The World Is Flat*. New York: Farrar, Straus and Giroux.
Ge, Shirong. 2018. "Robotic Mining Technology for Underground Mines." Paper presented at International Pittsburgh Coal Conference, Xuzhou, China, October 16, 2018.

Glassman, Irvin. 1987. *Combustion*. Orlando: Academic Press.

Hananel, S. 2010. "Obama Orders Review of Mines." *Altoona Mirror.* April 18, 2010.

Janssen, Sarah. 2020. *The World Almanac and Book of Facts 2020*, 324–325. New York: World Almanac Books.

Ji, Chang-sheng, Zhao-xue Che, and Qing-hua Chen. 2009. "Surface Coal Mining Practice in China." *Procedia Earth and Planetary Science* 1:76–80.

Kidd, Michael. 2016. "Minerals." In: *Research Handbook on International Law and Natural Resources*, edited by Elisa Morgera and Kati Kulovesi, 327–348. Cheltenham: Edward Elgar.

Moore, Catherine V. 2017. "Turning Appalachia's Mountaintop Coal Mines into Farms." *Yes!* 83: 26–29.

Mostyn, R. 1809. "Of Damps in Mines" (no. 136. p. 890, volume XII (1677)). Reprinted in *The Philosophical Transactions of the Royal Society of London*, from their commencement in 1665, in the year 1800. London: C. R/ Hutton.

Mouawad, Jad. 2016. "Crusader in the Coal Mine." *New York Times*, April 30.

Moyer, Forrest T. 1969. "Hazards of Mining." *Encyclopaedia Britannica*, 14th ed., vol. 5, 971. Chicago: Encyclopaedia Britannica.

NIOSH. 2008. "Most Frequently Recorded Industries on Death Certificate, U.S. Residents Age 15 and Over, Selected States and Years, 1990–1999." National Institute for Occupational Safety and Health. https://wwwn.cdc.gov/eworld/Data/Silicosis_Most_frequently_recorded_industries_on_death_certificate_US_residents_age_15_and_over_selected_states_and_years_1990-1999/176.

NRC. 2012. "April 2010 Upper Big Branch Mine Explosion—29 Lives Lost." Safety Culture Communicator. Case Study 4. Washington, DC: United States Nuclear Regulatory Commission.

Quigley, Joan. 2007. *The Day the Earth Caved In*. New York: Random House.

Stefanko, Robert. 1983. *Coal Mining Technology: Theory and Practice*. New York: American Institute of Mining, Metallurgical, and Petroleum Engineers.

Stewart, Bonnie E. 2012. *No. 9: The 1968 Farmington Mine Disaster*. Morgantown: West Virginia University Press.

Valdmanis, Richard. 2018. "A Tenth of U.S. Veteran Coal Miners Have Black Lung Disease: NIOSH." Reuters. https://www.reuters.com/article/us-usa-coal-blacklung/a-tenth-of-us-veteran-coal-miners-have-black-lung-disease-niosh-idUSKBN1K92W1.

Wolensky, Robert P., Kenneth C. Wolensky, and Nicole H. Wolensky. 1999. *The Knox Mine Disaster*. Harrisburg: Pennsylvania Historical and Museum Commission.

Wong, G. 2010. "115 Rescued from Coal Mine in China." *Altoona Mirror*, April 6.

Wood, Peter A. 1980. "Less-Conventional Underground Mining." International Energy Agency Report. No. ICTIS/TR12.

7
Preparation

Various operations, collectively belonging to the practice known as *coal prep-aration*, crush or grind coals to a specified size and reduce moisture, ash yield, and sulfur content. These operations prepare coals for use coal utilization processes and ensure a reasonably consistent quality product.

Coals as they come from the ground are referred to as *run-of-mine* coals. In many, but not all cases, run-of-mine coals are treated by a variety of operations before they are ultimately used. Anything done to improve the quality of a coal is a benefit to the user. Processes and operations used to improve coal quality collectively comprise the field of *coal beneficiation*. Much of the focus of beneficiation is on reducing the amount of undesirable material associated with the coal. Since minerals or rocks have no heating value, they contribute nothing to the production of energy nor to other products made from coals, but their ash adds the burden of having to be collected and disposed of. As we reduce mineral matter, calorific value necessarily goes up.

Many coal beneficiation processes work by taking advantage of the differences between physical or surface properties of coals and their associated mineral matter or rock. Employing these processes constitutes the field of coal preparation. The term *coal cleaning* is sometimes used to refer to processes for removing sulfur, minerals, and rock. Coal cleaning processes are divided into those relying on advantages of physical differences between coals and their impurities (i.e., physical cleaning) and those that utilize chemical reactions, or chemical cleaning. *Coal washing* is a form of coal cleaning that involves wet systems, meaning the use of water that may often contain dissolved or suspended compounds. Both in the coal literature and in practical coal business these terms are often used interchangeably.

Coal beneficiation converts run-of-mine coals into products that meet the specifications of the customer. Equally important, beneficiation provides a product of consistent quality. It is possible to design coal utilization equipment to burn or convert to burn almost any particular coal—but not to burn every possible coal. A power plant boiler will be designed and built for effective and efficient use of a coal having established specifications, such as calorific value and ash yield. Good engineering will allow a unit to function well with a coal supply having small deviations from the coal specifications but not for large swings in the properties of the coal being burned. The coal supplier needs to be confident that the coal being sold is and will continue to be a product of consistent quality. The coal user needs to be comfortable that the coal being purchased will be of that same quality not just day-to-day but also over long periods of time.

Every kilogram of water that enters a boiler consumes 2.26 megajoules of energy merely to be evaporated. This high heat of vaporization means that some of the energy released from burning the coal is used simply in driving off water. Reducing total moisture by 1% raises the overall efficiency of production of heat by at least a tenth of a percent (Luckie and Leonard 1991). In most coal combustion systems, energy lost simply to vaporize water will never be efficiently recovered. Excess moisture can lead to handling problems and can greatly limit coal mills and other preparation systems (Nowling 2016). Reducing moisture also enhances calorific values, especially for low-rank coals, which have high moisture contents as mined. Less water associated with coal means lower costs for transportation. A higher calorific value indicates a greater amount of energy per tonne or that fewer tonnes need to be shipped to provide a given amount of energy. During winter, coals that are particularly wet can freeze during transportation, causing problems in unloading. One percent moisture in a coal is equivalent to 4% ash yield (Luckie and Leonard 1991) in terms of the effects associated with each on process efficiencies.

Dewatering refers to processes for removing water from coals. "The most elusive constituent of coal to be measured in the laboratory is moisture" (Hessley, Reasoner, and Riley 1986, 187). The elusiveness derives in large part from the fact that water can occur in coals in at least three ways: *free water*, also called *surface water*, is found on the surfaces of coal particles and in the small spaces between particles. *Inherent water* is physically adsorbed in the pores of coal particles. A third form, particularly in low-rank coals, is chemically sorbed (i.e., *chemisorbed*) onto the macromolecular framework of the coal. Much of the free water can be removed by mechanical processes. It's not so easy with inherent water, which may require mild heating.[1]

As a good general rule, purely mechanical process steps running at or near ambient conditions are less costly than process steps requiring significant energy inputs. For dewatering, mechanical removal of free water is less costly than thermal drying for removal of inherent water. Spending the money for

[1] Heating the coal to several hundred degrees Celsius, which is not part of any commercial coal-drying operation, liberates two more kinds of moisture: some water will be released from the thermal breakdown of the organic structures in the carbonaceous framework. Some will also be released from the thermal dehydration of minerals, such as clays.

an energy-intensive process such as thermal dewatering requires vigilance to make sure that the water just driven off is not later reabsorbed.

Run-of-mine coals must be crushed or ground before they can be beneficiated. Pieces of run-of-mine coal must be reduced in size to match the requirements of the equipment in which it might be used. The need for size reduction raises questions of how hard coal is, how readily it breaks apart, and how hard the pieces are to grind. Hardness and friability—the susceptibility to breaking— also are important properties affecting transportation and handling of coal prior to its use. Depending on their intended uses, beneficiated coals might also be further reduced in size and possibly separated into fractions corresponding to relatively narrow ranges of sizes.

Friability expresses the resistance of coals to abrasion (van Krevelen 1993, 470–477), or the tendency of pieces of coal to break. Friability tests are remarkably low-tech, though certainly standardized. The common approach relies on the *drop-shatter test*. As its name implies, it involves allowing a sample to fall a specified distance onto a steel plate. The ratio of average particle size of the sample after the test to average particle size of the original sample gives an indication of size stability (ASTM 1984a, 265–269). Size stability relates to friability: the higher the size stability, the lower the friability.

Grindability indicates how easy or difficult it will be to grind coal to a specified distribution of particle sizes in some type of grinding equipment. The *Hardgrove grindability index* (HGI; ASTM 1984b, 253–259), named in honor of its developer, is one of many tests used to assess grindability. Ralph Hardgrove, working at the Babcock and Wilcox Corporation facility in Ohio in the early 1930s, wanted to assess the relative hardnesses of coals for preparing feed for pulverized-coal–fired boilers. In this test, a sample of coal of specified particle size range is put into a bowl in which eight steel balls run in a circular pathway. A ring, placed on top of the grinding balls, puts a specified weight on the balls. The apparatus is operated through 50 revolutions. At the end of the test the value of HGI is determined from the weight of particles smaller than 74 microns. Bituminous coals have the highest HGI, indicating the easiest grindability. A high index signals relatively easy grindability. The maximum HGI is reached at 90% carbon (van Krevelen 1993, 470–477), consistent with a minimum in hardness at this point. Anthracites and lignites are more difficult to grind. HGI values serves as a useful guide for calculating the sizes of grinding equipment needed to process coals.

Run-of-mine coal usually is in large lumps. Fine grinding might be desired to reduce particle size to below 75 microns. Size reduction is best performed in several stages, used in series (Austin 1991). No single piece of equipment could reduce a lump to 75 microns in a single operation. The product from size reduction contains material of a range of particle sizes. Size separation is used to remove particles larger than wanted. Depending on the process in which the coal will be utilized, it may also be helpful to remove particles of too small a size, the *fines*. The array of combustion or conversion processes in subsequent chapters are made more effective by appropriate sizing of the coal being used.

Vibrating screens of woven wire cloth can be manufactured with different-sized openings, down to about 0.15 millimeter. A series of screens having different sizes of apertures can be used to remove the coarse material, then to prepare several size fractions, and finally to collect the fines. An alternative is the liquid–solid cyclone, usually called a *hydrocyclone*. Cyclones rely on the effects of the centrifugal forces set up in a swirling flow of water. They operate with a slurry of coal in water, taking advantage of centrifugal force to push the slurry toward the interior wall of the cyclone. Coarse particles are thrown to the wall and discharged from the bottom. Fine particles are not in the cyclone long enough to travel all the way to the wall so are discharged with the overflow. Hydrocyclones can also be used in series.

Whether, and how extensively, a particular coal is cleaned before use depends on the market specifications—usually calorific value, ash, and sulfur—that have to be met to sell that coal competitively. Removal of mineral impurities usually relies on the difference in specific gravity between coal and the minerals.[2] The specific gravity of most coals ranges from about 1.2 to 1.5. Specific gravities of minerals found associated with coals are much higher. For example, kaolinite and quartz each have specific gravities of 2.6, and pyrite has a specific gravity of 5. If a crushed coal is placed in a fluid having a specific gravity intermediate between those of the coal and the minerals, the coal will float and the minerals will sink. Separation of coal from the minerals can then be effected by using various mechanical systems.

How extensively a particular coal is cleaned depends on the characteristics of that coal and on economics. The characteristic of most importance is the way in which the minerals are incorporated in or with the coal. Minerals deposited in coals after they had formed usually concentrate either along

---

[2] *Specific gravity* is the ratio of the density of a substance to the density of a reference material. Commonly, the reference substance used for specific gravities of liquids and solids is water. In the metric system, the density of water is 1 gram per cubic centimeter. Since dividing by 1 does not change the numerical value, specific gravity and density are numerically equal. The difference arises from the fact that specific gravity has no unit (i.e., is a dimensionless number), whereas a reported value of density must always be accompanied by a unit, such as grams per cubic centimeter or kilograms per cubic meter.

bedding planes or in major cracks and fissures in the coal. Their removal is fairly easy. So, too, is removal of rocks and minerals inadvertently included with the coal during mining. In other cases, minerals were incorporated either in the original plant material or mixed with the accumulated organic material in early stages of coal formation. In these cases, removal of minerals by simple physical processes may become virtually impossible because it is not feasible to grind the coal fine enough to liberate such tiny grains of minerals. And low-rank coals contain some inorganic constituents chemically bound to the coal matrix itself. No amount of grinding could liberate these cations.

Success of any cleaning process depends on the size distribution of the coal fed to the process—the raison d'être for size separation. As a rule, the smaller the particle size, the more minerals are liberated from the coal and the better the separation. The extent of size reduction has to be tempered by how much grinding is practical and economic and by the size requirements of the equipment in which the coal will be utilized. Economically, coal beneficiation must strike a balance between the cost of coal cleaning and either the increased revenues resulting from the greater marketability of the cleaned coal or the cost savings from processing cleaned coal.

Success in cleaning a particular coal can't be predicted from its conventional analysis report (i.e., proximate, ultimate, and calorific values). Individual coals have to be evaluated in the laboratory, a common method being the *float-sink test*. This test involves shaking the coal sample with liquids of various specific gravities. Usually two fractions—the *floats* and the *sinks*—are formed. This simple test is illustrated in Figure 7.1. Compared with the original coal, carbonaceous material will be enriched in the floats, and the proportion of minerals

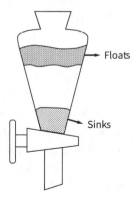

Figure 7.1 A simple laboratory-scale float-sink test provides preliminary information about the washability of coals.
Artwork by Lindsay Findley, from the author's sketch.

will be higher in the sinks. The amounts of floats and sinks as a percentage of the original coal, as well as the ash yields of the floats and sinks, are measured. Sometimes a third split, the *middlings*, forms. Middlings have about the same specific gravity as the liquid medium used for the test, tending to remain in suspension rather than to float or sink. The data from these tests allow the coal preparation engineer to determine the liquid specific gravity that would be required to achieve a desired ash level in the floats and can predict the yields of floats, sinks, and middlings.

In principle, a continuous gradation runs from pure coal, as might be expressed as a 0% ash yield, through coal of increasing mineral content, to rock of decreasing carbon content, and finally to "pure" rock that would be 100% ash. In practice, a perfect separation between coal and minerals cannot be achieved in any reasonable coal cleaning process. An optimum practical level of cleaning for a particular coal depends on its **washability**, as might be determined from float-sink testing; from the ash value desired, or at least willing to be tolerated, in subsequent utilization of the coal; and from the economic tradeoffs associated with cleaning to various degrees of ash levels. Depending on the particle size of the coal being treated, cleaning processes are divided into the categories of coarse-coal and fine-coal cleaning. The size division between these categories usually lies at about 10 millimeter particles (Schobert 1987, 126–132).

Coarse-coal cleaning principally employs jigs, heavy-media separators, concentrating tables, or cyclones. Coarse-coal cleaning equipment all achieve a separation between coal and mineral matter based on differences in their specific gravities. Each represents a different mechanical tactic for doing this. Jigs rely on a pulsating flow of water to cause a bed of coal to stratify according to specific gravity. Cleaned coal rises to the top and washes out of the jig. Particles of high specific gravity leave as a waste stream. Jigs are relatively simple devices, designed and built to achieve large production rates, and economical to operate. Heavy-media separators use a liquid of specific gravity intermediate between the coal and mineral impurities, commonly a suspension of magnetite in water. Coal feeds continuously into a vessel having a recirculating magnetite suspension. Cleaned coal exits from the top and waste from the bottom. Shaking and swirling of the water make a concentrating table a large, mechanically operated analogue of the gold miner's pan from years ago. The primary distinction is that the miner panning for gold wanted to retain the high specific gravity gold and wash away lower gravity, worthless rock. Cyclones using a dense medium have a better recovery of clean coal than hydrocyclones, which rely on water as the fluid.

Differences between the surface properties of coal and minerals are exploited in two techniques for fine-coal cleaning: froth flotation and oil agglomeration. The way in which water molecules interact with the surface of a solid depends on its chemical nature. *Hydrophobic* surfaces show little interaction with water molecules. Other surfaces readily attract water molecules and are said to be *hydrophilic*. Surfaces of substances consisting largely of carbon and hydrogen atoms are hydrophobic. Hydrocarbon substances, including bituminous coals and anthracites, fall into this category. Surfaces that contain abundant oxygen atoms, such as low-rank coals and clays, will be hydrophilic.

When air is blown through a slurry of finely ground coal in water, air bubbles attach to hydrophobic surfaces of the coal particles. As the air bubbles float to the top of the container, the attached coal particles float right along with them. Surfaces of hydrophilic mineral particles, such as clays, will be well-covered by water molecules and much less likely to have air bubbles attach to them. The mineral particles sink. This difference in behavior forms the basis of *froth flotation*. Skimming off the froth removes the coal; the mineral-rich residue can be removed from the bottom of the unit. Froth flotation is especially useful for cleaning fine coal. Separation of coals from minerals is never complete. Not all minerals are liberated by grinding, some minerals may be hydrophobic and thus float, and some coal particles may be hydrophilic, especially coals having oxygen atoms chemically incorporated on their surfaces.

Oil agglomeration also depends on the nature of the coal surface, in this case on the ability of a hydrophobic surface to be wetted by a nonpolar liquid, such as diesel fuel. A suspension of coal in water is stirred with an oil. Oil preferentially coats hydrophobic coal surfaces but does not interact with hydrophilic mineral surfaces. Oil-coated coal particles agglomerate so that the comparatively large agglomerates are easily removed by screening. Like flotation, oil agglomeration works best on finely ground coal. Flotation gives a higher yield of cleaned coal and generally provides more sulfur removal, but agglomeration provides a greater reduction of ash yield. Oil agglomeration also dewaters coal because the oil adsorbed on the surface displaces moisture. Oil agglomeration could be run in combination with a flotation unit to achieve reductions of sulfur, ash, and moisture.

Modern coal preparation plants use several cleaning operations in parallel (Noble and Luttrell 2016). Each cleaning operation is most effective, and most efficient, only for a defined range of particle sizes. Current practice in the United States uses dense medium vessels or cyclones for particles greater than 1 millimeter, water-based separations for 0.15–1 millimeter particles, and froth flotation for particles smaller than 0.15 millimeter.

For at least a half-century, sulfur removal has been a prime target of coal cleaning because of regulatory requirements limiting sulfur oxide emissions from power plants. Neither organic sulfur nor fine inclusions of pyrite in pores[3] are amenable to removal by physical cleaning. Many chemical and biological cleaning methods have been tried for removing organic sulfur. Much ingenious chemistry has been explored. Some processes have shown potential in the laboratory, but so far none has been commercialized. Only two even made it as far as pilot plant trials (Kawatra and Eisele 2001, 320–344).

The lack of commercial chemical cleaning processes is due to equipment technology and process economics. As a rule, it is very hard to find any chemical process that can be cheaper than a relatively simple physical separation. Physical cleaning processes use equipment that is comparatively easy to build and maintain, translating to low costs for both the original investment and for the operating and maintenance costs. Physical cleaning processes operate at or near ambient temperature and pressure. Added chemicals are either not required or used only in small amounts. In comparison, some proposed chemical cleaning processes require high temperatures and sometimes high pressures. An example is the Gravimelt Process (Anastasi et al. 1990), in which coals are reacted with molten sodium hydroxide at around 400°C for several hours. As described, the process involves six separate operations using a total of 160 pieces of equipment. It does a splendid job of reducing sulfur content and ash yield—down to less than 1% each—from bituminous coals. But it's not easy. It's not likely to be cheap.

Chemical cleaning processes result in greater initial investment for the equipment and higher operating and maintenance expenses. Chemical processes consume larger quantities of chemicals, or use more expensive ones, than physical processes do. Provision has to be made for limiting chemical exposure of workers and for safe, acceptable disposal of reaction products. Since most of the sulfur in most coals is pyritic, removing as much pyrite as possible by relatively inexpensive physical methods and then coping with the rest of the sulfur after the coal is burned seems to be a more economical approach than the alternative: setting up expensive chemical processes to remove organic sulfur and any residual pyritic sulfur not removed by physical processes.

---

[3] Strictly speaking, tiny pyrite inclusions in pores or coalified plant cells could be removed by physical processes, provided that the coal could be ground fine enough to liberate them from the coal matrix. This requires grinding the coal to a particle size smaller than that of the pyrite inclusions, a task that would be both difficult and expensive. Also, such finely divided coal could be hazardous to handle because of the increased likelihood of dust explosions.

Coals that have been cleaned usually must be dewatered. Dewatering removes water picked up during handling or processing (e.g., water acquired during physical cleaning processes). Further drying reduces moisture content in the coal below its value in an as-mined condition.

Mechanical drying processes are easier and less costly but are less effective in dewatering. The simplest process is sedimentation. A suspension of coal particles in water is allowed to settle until the solids can be collected and the water poured or pumped off for reuse. Rapid rotation in a centrifuge sets up forces that throw wet coal against a screen. Water passes through the screen while coal is held back. A suction applied to the other side draws off the water. Centrifugation for dewatering coal is not much different from the operation of the spin cycle of a domestic clothes washing machine. In thermal drying, hot gas, usually from combustion, directly contacts the wet coal in a rotating cylinder that allows the wet coal to tumble through the hot gas (Luckie and Leonard 1991), much like a domestic clothes dryer. Hot gases evaporate water from the coal. Thermal drying can be expensive, up to 10 times as costly as mechanical dewatering (Parekh and Matoney 1991), because of the high heat of vaporization of water.

Coal cleaning does not destroy the unwanted components of coals; instead, it separates coals into at least two products. One is the product we want—a coal having lower sulfur content, ash yield, or both, and a higher calorific value than the as-mined material, ready to be used. There is another product, too, a mineral-rich material that often has little or no commercial value. A wide variety of terms—culm, bone, bony coal, refuse, slack, duff, or slag—has been coined in different regions or countries to refer to the products of coal cleaning that have no commercial demand. The question of what to do with these residues must be confronted.

Years ago, industry often favored the straightforward approach of simply piling coal cleaning waste on the ground. This practice creates eyesores that can at times be deadly. Though only a small percentage of the coal is rejected as waste, the enormous tonnage of mined coal leads to waste piles that grow to sizable proportions over the years. The worst tragedy associated with coal cleaning waste occurred in Wales. In October 1966, steady rains saturated a slag heap near the Merthyr Vale colliery to a point at which the wet material suddenly slipped down the mountainside above the town of Aberfan, destroying the school, several houses, and a farm, killing 116 school children and 28 adults (Blakemore 2020).

Refuse piles have accumulated over the course of decades. They are remarkably devoid of plant life. It might be reasonable to expect that seeds from surrounding native vegetation would be blown or carried onto the piles and that, year by year, the piles would be colonized by plants. This does not happen, for a seemingly simple reason: the piles are black. During the summer season the black surfaces of the piles absorb enough solar energy to reach temperatures up to 75°C (Schramm 1966). Temperatures this hot are lethal to newly sprouted seedlings attempting to push upward through the surface of the pile.

Returning the waste to the mine as fill provides a solution to dealing with the discarded material from coal beneficiation. One alternative is to use the refuse as a building material. In Donetsk, Ukraine, a neighborhood where miners live had houses that were built of slag mixed with cement (Greene 2014).

Another option is to burn the waste. Why go to the trouble and expense of cleaning coal only to burn the discarded material anyway? Most boilers used in coal-fired power stations often require cleaned coal, in part to help meet air-quality standards. Since cleaning never provides perfect separation into mineral-free coal and coal-free minerals, these wastes have a measurable heating value—albeit not nearly so high as the original coal—so have potential fuel applications.

In *fluidized-bed combustors*, a bed of inert material, such as sand, is suspended in a column of air blown upward through it. The suspended bed looks like a fluid and in some ways acts like one. Fuel burns in the hot bed. Fluidized-bed combustors can be designed to handle a remarkably wide range of fuels, including coal beneficiation residues. They operate at comparatively low temperatures for combustion systems, about 750–1,000°C. At these temperatures, finely ground limestone or dolomite can be added to the bed to react with and capture sulfur oxides formed during combustion. Capturing sulfur oxides right in the bed minimizes or even eliminates the need for a separate unit to capture the sulfur oxides before they can be emitted to the environment.

Some jurisdictions provide economic incentives, such as tax rebates or reductions in tax rate to encourage the "re-mining" of refuse banks or slag heaps as a source of fuel. Doing so helps get rid of these blights on the landscape, makes another form of fuel available, may reduce the need for additional new mining, and may eventually help make the area amenable to land reclamation.

Almost invariably coal must be stored for some time between mining and its use. Some mining companies may store excess production to have a ready supply on hand for sale when demand increases. Mine-mouth plants provide for enough storage to manage demand if irregularities or short-term disruptions in mine production should occur. Prudent customers store a reserve supply of coal to guard against disruptions in the supply chain. Some facilities store huge quantities of coal—possibly a 90-day supply—to guard against the effects of labor disruptions in the mining or transportation industries or to be able to take advantage of favorable coal prices on the spot market.[4] Stockpiling coal becomes attractive if increases are expected in coal prices or transportation costs.

How hard could stockpiling coal be? Find a convenient spot, dump it on the ground, and pile it up with bulldozers. Or, if there's not so much coal, pour it into bins or bunkers. That should do it, except—as in so much in life, the devil is in the details, and in this case there are a lot of details.

As coal dries, it may break apart into smaller pieces. Physical degradation of coal pieces during storage—known as decrepitation, *slacking*, or spalling—mainly affects low-rank coals. Slacking can be serious for lignites and subbituminous coals, though of less concern for bituminous coals. Anthracites can be stored with no slacking problem. Some low-rank coals have very high moisture contents as mined, exceeding the normal equilibrium moisture of the coal. On exposure to air, drying begins immediately. Extensive moisture loss causes shrinkage of the pieces of coal. Shrinkage sets up mechanical stresses inside the coal particles, which can build to a point at which they fracture the coal. In extreme cases, slacking can be very extensive in a matter of days.

Formation of small particles can result in losses during transportation, handling, or processing simply because the smaller particles are more susceptible to being blown out of rail cars or trucks by wind. And, as particle size decreases, the total surface area of a given mass of coal increases.[5] Increased surface area provides increased chances that the coal will react with whatever medium surrounds it—in this case, oxygen in the air. Such reaction causes reductions of calorific value and reduction or loss of caking properties. It may also lead to autogenous heating. There will be more instances of the important connections among particle size, surface area, and reactivity in upcoming chapters.

Loosely packed piles, exposed to weather, support the easy movement of air through the pile, providing plenty of oxygen to the coal. Such cases can

---

[4] A *spot market* is a system for buying commodities for immediate delivery, often with the commodity—such as coal—and the payment changing hands within 2 days. In this context, buying on the spot market would contrast with establishing a long-term contract, for months or years, between a coal consumer and a coal supplier.

[5] For example, given an equal weight of watermelons and cherries, the many cherries collectively have far more total surface area than the few watermelons.

witness a 40% reduction of the original calorific value in 6 months' storage. The coal is on its way to being worthless. Well-consolidated stockpiles, in which tight packing of coal particles minimizes air flow, might have calorific-value losses of about 1% per year, even over the course of several years. Loss of caking properties from a coal intended for metallurgical coke production is calamitous—the coal has become useless. Mild oxidation can do this.

A serious problem of coal storage involves the heating, and sometimes ignition, of the coal. In the worst cases, this leads to the stockpile burning. We saw this issue in Chapter 6, in the context of mine fires. Though almost universally known as "spontaneous combustion," the more accurate adjective is "autogenous"—self-produced (Berkowitz 1979, 191).

The likelihood of autogenous heating is difficult to predict. Whether a particular coal will experience autogenous heating depends on its rank and type, its composition, the ease of its reactivity toward oxygen, its particle size, and its porosity. As a rule, lower-rank coals are more susceptible, possibly because of a proportionately higher amount of reactive liptinite and vitrinite macerals (Mastalerz et al. 2011). Autogenous heating also depends on how readily air and moisture can access the interior of the coal pile and how easily heat can be conducted out of it. Prevailing temperature, air currents and humidity also have roles.

Oxygen from air is adsorbed onto the coal surface. This process is *exothermic*, releasing energy called the *heat of adsorption* (Hoffman 2000). The exothermic reactions of the adsorbed oxygen with carbon atoms on the coal surface begin the oxidation processes. Because these processes occur on the surface, particle size has a strong influence. Smaller particles have a larger surface area to volume ratio, consequently exposing more of the carbon to atmospheric oxygen.

Volatile matter content is strongly associated with self-heating. A higher volatile matter content suggests a greater amount of light hydrocarbons—such as methane—that can be liberated as a coal begins to heat autogenously. If these highly combustible gases accumulate in the pile, they could add to the danger of fire. Oxygen content is also associated with increased autogenous heating potential. Lower-rank coals, typically having higher volatile matter and higher oxygen contents, are much more likely to suffer autogenous combustion.

An additional heat comes from wetting. Analogous to oxygen adsorption, when a liquid wets the surface of a solid, heat is also liberated, in this case known as the *heat of wetting*. Heat liberation occurs because the atoms or molecules on the surface of a solid are in a higher energy state than those in the interior. When a new surface is formed, such as by molecules of water, the atoms or molecules formerly on the solid surface relax to a lower energy state; as they do, the energy released by this transition appears as heat. This also contributes to autogenous heating.

Pyrite can act as a catalyst for the oxidation reaction of the coal (Nowling 2016). When exposed to moist air, pyrite itself can oxidize, providing additional heat to the coal pile. Humphry Davy mentioned that seams of pyrite-rich coal along the southeast coast of Scotland experienced "spontaneous" combustion (Davy 1980, 131). Slag piles containing a high proportion of pyrite mixed with combustible material can smolder continuously because of the heat liberated from oxidation of pyrite. Though pyrite has often been suggested as a leading cause of autogenous heating, low-rank coals, particularly prone to autogenous heating, commonly have low values of pyritic sulfur.

Trouble really starts when autogenous heat, regardless of what caused it in the first place, cannot escape from the coal pile as it is being liberated. Heat confined in the pile must necessarily raise the temperature. Raising the temperature increases the rate at which reactions take place. We have already seen this dependence of reaction rate on temperature in Chapter 6, in connection with coal-dust explosions and fires in coal seams. The dependence of reaction rate on temperature is a fundamental rule of chemistry, first developed by the brilliant Swedish scientist Svante Arrhenius (Cantor 2020, 91–108). Increasing the reaction rate increases the rate at which heat liberated from the reactions further raises the temperature of the pile. That further temperature rise increases the reaction rate still more. This sequence becomes a vicious cycle that eventually raises the coal to its ignition temperature, resulting in its catching fire. Ignition temperature depends on the particle size and rank of coal, the amount of oxygen in the air immediately surrounding the coal, and the velocity of air currents moving oxygen into and conducting heat out of the pile.

Loss of coal certainly is an economic waste, but if a fire breaks out, property damage, injury, or loss of life can occur. Autogenous heating and ignition of coal in the hold of a ship, for example, puts the ship and its crew in great danger. A fire caused by autogenous heating in the coal bunkers of the battleship USS Maine in Havana harbor in 1898 is strongly suspected to have been the root cause of the explosion and sinking of the ship (Nowling 2016). The explosion was originally attributed to a Spanish mine. The sinking touched off the Spanish–American war, in which the American battle cry was "Remember the Maine!" Well, yes, but don't forget about the pyrite. Likely a fire caused by autogenous heating of the bituminous coal in the ship's coal bunkers eventually reached the magazine, touching off the munitions.

Limiting access of air and moisture to the interior of a coal pile minimizes autogenous heating. As the stockpile is being built, great care needs to be taken to compact the coal as the best preventative measure. Temperature monitors in the pile sense the first warning signs of trouble, places in the pile where the coal is heating. For coal stored in a bin or bunker, the first approach to combating the problem is to seal tightly all possible air access points, restricting the availability of oxygen. If this doesn't help, or if the coal is not in a contained volume, the hot coal has to be dug out—not a pleasant job—and allowed to cool. It might be possible to stop autogenous heating by spraying the hot coal with water, but this could be counterproductive if heat of wetting is the source of the problem.

The common approach to stockpiling is to build piles in the open air. Good stockpiles are shown in Figure 7.2. In all likelihood they have been carefully blended to ensure consistent coal quality going into the plant and carefully compacted to ensure minimal changes in storage. An alternative approach is to enclose the stored coal inside domes (Pyper 2017). A plant near Menkeqing, in northern China, utilizes three domes, each of which is capable of storing 60,000 tonnes. Run-of-mine coal goes into Dome One when it arrives at the facility. Coal is taken out of this dome for cleaning and size reduction. The prepared coal goes to the other two domes, from which it can be taken for loading onto trains. The overall footprint for coal storage with this arrangement is smaller than open-air stockpiles. Coal dust is less likely to get into the environment.

Figure 7.2 Well-built stockpiles of coal outside a power plant.
Courtesy of Rudmer Zwerver/Shutterstock.com.

Not all coals undergo beneficiation or preparation treatment. For those that do, the operations that we have just seen offer a better product to the user. If nothing else, coal preparation can provide a cleaned coal of consistent quality, allowing smoother plant operations. Reduced ash yield means less ash to be collected, handled, and disposed of. This benefits equipment operation, where ash is a nuisance. Reduced sulfur content means less cost for subsequent sulfur-capture units. Probably it would be possible to design and operate coal utilization equipment on crushed run-of-mine coal, but coal preparation makes life cleaner, smoother, and easier. It provides a vital link in the chain of events from coal in the ground to coal being burned or converted to valuable products.

Coal preparation plants may have to deal with thousands or tens of thousands of tonnes of coal per day. They are often loud, wet, and sometimes dirty. Yet the reasons that they work—specific gravities, hydrophobicities, surface tensions—all depend on the compositions and structures of the coals and their minerals. Now that the coals have been mined and prepared, in the next six chapters we will see the current technologies for using coals. We will start with the big one: burning coal to produce steam for electricity generation.

References

Anastasi, J. L., E. M. Barrish, W. B. Coleman, W. D. Hart, J. F. Jones, L. Ledgerwood, L. C. McClanathan, R. A. Meyers, C. C. Shih, and W. B. Turner. 1990. "Molten Caustic Leaching (Gravimelt Process) Integrated Test Circuit Operation Results." In: *Processing and Utilization of High-Sulfur Coals III*, edited by R. Markuszewski and T. D. Wheelock, 371–78. Amsterdam: Elsevier.

ASTM. 1984b. "Standard Test Method Grindability of Coal by the Hardgrove Machine Method. D 409-85." *Annual Book of ASTM Standards. Volume 05.05. Gaseous Fuels; Coal and Coke.* Philadelphia: American Society for Testing and Materials.

ASTM. 1984a. "Standard Method of Drop Shatter Test for Coal. D 440-49." *Annual Book of ASTM Standards. Volume 05.05. Gaseous Fuels; Coal and Coke.* Philadelphia: American Society for Testing and Materials.

Austin, Leonard G. 1991. "Size Reduction." In: *Coal Preparation*, edited by Joseph W. Leonard and Byron C. Hardinge, 187–220. Littleton, CO: Society for Mining, Metallurgy, and Exploration.

Berkowitz, Norbert. 1979. *An Introduction to Coal Technology*. New York: Academic Press.

Blakemore, Erin. 2020. "How the 1966 Aberfan Mine Disaster Became Elizabeth II's Biggest Regret." History. https://www.history.com/news/elizabeth-ii-aberfan-mine-disaster-wales.

Cantor, Brian. 2020. *The Equations of Materials*, 91–108. Oxford: Oxford University Press.

Davy, H. 1980. *On Geology: The 1805 Lectures for the General Audience*. Madison: University of Wisconsin Press.

Greene, David. 2014. *Midnight in Siberia*. New York: W. W. Norton.

Hessley, Rita K., John W. Reasoner, and John T. Riley. 1986. *Coal Science*. New York: John Wiley & Sons.

Hoffman, Wesley P. 2000. "Chemi-Sorption Processes on Carbons." In: *Sciences of Carbon Materials*, edited by Harry Marsh and Francisco Rodríguez-Reinoso, 437–483. Alicante: Universidad de Alicante.

Kawatra, S. Komar, and Timothy C. Eisele. 2001. *Coal Desulfurization*. New York: Taylor and Francis.

Luckie, Peter T., and Joseph W. Leonard. 1991. "Dewatering. Part 2: Thermal Dewatering." In: *Coal Preparation*, edited by Joseph W. Leonard and Byron C. Hardinge, 581–604. Littleton, CO: Society for Mining, Metallurgy, and Exploration.

Mastalerz, Maria, Agnieszka Drobniak, James C. Hower, and Jennifer M. K. O'Keefe. 2011. "Spontaneous Combustion and Coal Petrology." In: *Coal and Peat Fires: A Global Perspective*, edited by Glenn B. Stracher, Anupma Prakash, and Ellina V. Sokol, 48–62. Amsterdam: Elsevier.

Noble, Aaron, and Gerald H. Luttrell. 2016. "Classification in Coal Preparation." *World Coal.* 25(6): 21–26.

Nowling, Una. 2016. "Who Moved my Btus? The Pitfalls of Extended Coal Storage." *Power.* 160(12): 42–46.

Parekh, B. K., and Joseph P. Matoney. 1991. "Dewatering. Part 1: Mechanical Dewatering." In: *Coal Preparation*, edited by Joseph W. Leonard and Byron C. Hardinge, 499–580. Littleton, CO: Society for Mining, Metallurgy, and Exploration.

Pyper, Rebecca L. 2017. "Massive Storage Multiplied." *World Coal* 26 (2): 29–31.

Schobert, Harold H. 1987. *Coal: The Energy Source of Past and Future.* Washington, DC: American Chemical Society.

Schramm, Jacob R. 1966. "Plant Colonization Studies on Black Wastes from Anthracite Mining in Pennsylvania." *Transactions of the American Philosophical Society* 56 (Part I): 1–194.

van Krevelen, Dirk W. 1993. *Coal: Typology - Physics - Chemistry - Constitution.* Amsterdam: Elsevier.

8
Electricity

The use of coal in electricity generation is the most significant impact of coal on daily life. Coal's place in the electricity sector is declining because of increasing contributions made by natural gas and renewable energy and because of concerns about reducing carbon dioxide (CO_2) emissions. Even so, coals remain an important fuel for electricity generation worldwide, with about 40% of world electricity generation coming from plants burning coal. By a great margin most of the coals produced in the world are used for electricity generation. Without coals, the electricity systems of many major industrialized countries would be severely crippled.

The capacity of electricity-generating stations, commonly called "power plants," is usually expressed in units of megawatts (MW). The capacity expressed in this way represents the maximum output of a plant. It does not always reflect the amount of electricity generation because plants are shut down from time to time, either for planned maintenance or occasionally due to some unforeseen event such as severe weather.

Nowadays, a mid-sized plant might generate about 400 MW of electricity and consume 10,000 tonnes of coal a day. Large plants have outputs exceeding 1,000 MW. One megawatt will power 650 average homes (Eskom 2015; UtiliPoint 2012). The 400 MW plant in this example would provide electricity to 1.3 million people.[1] A single, large railroad car of coal—about 100 tonnes— supplies enough coal either to heat a modest house during the winters for 10 years or to operate a typical modern power plant for 15 minutes. Such plants provide electricity to feed into regional or national power grids. Via the grid, their electricity might reach customers hundreds or thousands of kilometers away.

Electricity is not produced from coal. Rather, heat from burning coal is used to boil water and heat water to produce steam. The steam is fed into a turbine

---

[1] This is based on assuming five people in 650 homes. An important word here is *homes*. This estimate for the number of people served by 400 MW would be lower if electricity produced in that plant were consumed in a mixture of domestic, industrial, and commercial enterprises.

directly connected to a generator; the generator makes the electricity. Why steam? Why coal?

In 1831, Michael Faraday demonstrated conclusively that an electric current in one circuit can induce a current in a different separate circuit. He showed that the effect could be achieved by moving a magnet in and out of a coil of wire. The American scientist Joseph Henry independently observed similar phenomena.[2] Their work is the basis of *electromagnetic induction*— the ability of an existing current or magnetic field to generate a new current or field. A magnetic field can generate an electric current in a coil of wire, or an electric current can produce a magnetic field, provided that either the magnet or the wire is moving relative to the other.

By the mid-nineteenth century, many engineers had contributed to the development of practical devices for continuous generation of electricity. Early generators operated by rotating a coil of wire inside a magnetic field. Reliable generation of electricity then came down to the question of how to turn the generator, and keep it turning, as reliably and cheaply as possible. In the waning decades of the nineteenth century, the prime mover of choice was a reciprocating steam engine. Converting the motion of a reciprocating engine to rotary motion, as would be needed by a generator, is not difficult. The problem had already been solved in steam locomotives and in ship propulsion. The reciprocating steam engine is big, heavy, slow, and not very efficient at converting the energy of coal into the work that it was intended to do. Development of the steam turbine by the Anglo-Irish engineer Charles Parsons provided a solution.

Any turbine is designed to convert the energy in a moving fluid—such as water, wind, or steam—to rotary mechanical motion. The fluid passing through the turbine, the *working fluid*, should be one that can be generated easily on demand, around the clock, without restrictions of local geography or weather; that is inexpensive and readily available; and that is reasonably safe to work with and safe in the environment. Water, in the form of steam, satisfies these criteria better than any alternative. Coupling a steam turbine to a generator answers the question of how to turn the generator, and keep it turning, as reliably as possible. Then a new question arises: How to generate steam as reliably and cheaply as possible?

---

[2] Joseph Henry, like Faraday, contributed to several scientific fields. He organized a system of reporting weather that eventually became the US Weather Bureau (now known as the National Weather Service), and was the first director of the Smithsonian Institution, in Washington, D.C.

Any heat source can be used to produce steam: a fuel, the concentrated rays of the sun, or heat produced by nuclear reactions. Coal dominated these choices for many decades and remains important in many parts of the world. Coal has two advantages: among the likely competitors, it was until recently the cheapest in terms of cost per megajoule of heat energy released. It is widely distributed in the world. Coal can be stockpiled to ensure against disruptions in the supply chain. We know how to build and operate coal-fired power plants. A competent engineering firm can design and construct a plant that will burn the coal specified by the owner and that will operate at the planned level of generation. The accumulated body of engineering know-how provides a powerful inducement. A new coal-fired plant might cost about $3,500 per kilowatt of capacity (EIA 2020). The 400 MW plant mentioned earlier would cost about $1.4 billion dollars. Investors might feel it more prudent to invest those dollars into generating capacity for which there are dozens of successfully operating examples (i.e., coal firing) than to venture off into other technologies where there is greater risk.

Modern electric-generating plants require prodigious amounts of steam. The amount of steam required often exceeds 500,000 kilograms per hour. Making steam on this scale depends crucially on two factors: the rate at which heat can be produced and the rate at which heat can be transferred into water or steam. The rate at which heat can be produced relates to the rate at which the coal can be burned. The rate at which heat can be transferred depends on available surface area across which heat will be transferred.

A steam-generating system consists of the *furnace*, where heat is liberated by burning coal, and the *boiler*, where water absorbs heat to generate steam. In modern systems, the furnace and the boiler are physically the same piece of equipment. Because of that, the term "boiler" often means both the place where coal is burned and where steam is generated. Boilers used for large-scale production of steam represent design approaches to increasing the amount of surface for transferring heat while keeping the overall size within reason. The solution puts water or steam inside of tubes with the very hot combustion gases flowing outside of them. This design, a *water-tube boiler*, represents the standard approach used in modern power plants, almost universally adopted for supplying steam to turbines for generating electricity. Such a boiler is shown in Figure 8.1.

Boilers are configured to enclose as large a volume as possible while presenting a relatively small surface to the outside world. The large volume gives ample space for combustion to occur, while the small surface helps minimize heat loss from the boiler to its surroundings. The boiler, usually a big

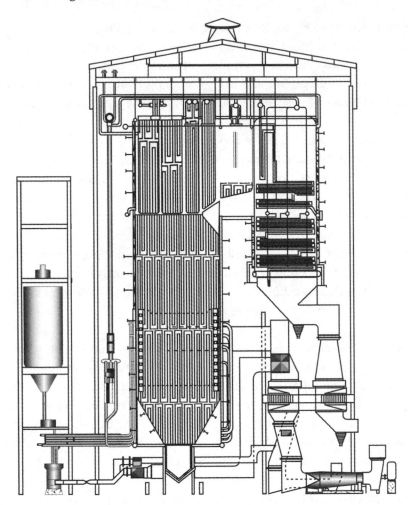

Figure 8.1 A modern water-tube boiler, showing the immense amount of piping involved in the water and steam tubes.

Courtesy of Taylor and Francis, from author's earlier publications.

rectangular box, could be 10 to 20 stories tall, depending on the size of the plant (i.e., its generating capacity) and on the rank of coal used. Figure 8.2 shows an idea of the crucial parts of the plant.

To consider how to get the fastest release of heat from coal, as a model (recall George Box!) coal can be thought of as being pure carbon. Taking that simple model, combustion can be represented by the equation for the carbon–oxygen reaction:

$$C + O_2 \rightarrow CO_2.$$

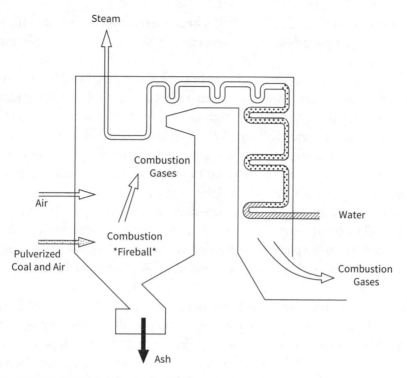

Figure 8.2 Simplified sketch of pulverized coal combustion to generate steam. Ash leaving the boiler is bottom ash. The steam will go on to the turbines (Figure 8.3). The combustion gases will be treated as shown in Chapter 9.
Artwork by Lindsay Findley, from the author's sketch.

Because coals are solids and oxygen is a gas, the carbon–oxygen reaction can occur only where the oxygen molecules have access to carbon atoms on the surface of the coal particles.

As a coal particle begins to heat in the combustion environment, volatiles will be driven off into the surrounding gas phase. This process starts as soon as a coal particle is introduced to a flame. Some of the volatiles are highly combustible hydrocarbon compounds coming from the breakdown of the coal's macromolecular framework. They ignite and burn very rapidly, on a time scale of milliseconds. As gaseous species, the volatiles intimately mix with air, so that the combustion of the volatiles takes place in a single phase, a *homogeneous reaction*.

Fixed carbon and minerals remain as a solid particle of ***char***.[3] To attain complete combustion, the char must ignite and burn. Char burn-out is much

---

[3] We will see in Chapter 10 that the situation is a bit more complicated. Some bituminous coals soften and go through a fluid phase, commonly but incorrectly called "melting." The solid that eventually forms is called a "coke." For now we needn't worry about that.

slower than the burning of the volatiles because reaction is limited to the surface of the char particles, a *heterogeneous reaction*. Char and oxygen are in different phases.

Eventually, some 95–99% of the carbonaceous material will have burned, leaving behind the ash. The entire lifetime of a coal particle, from its injection into the burner, through volatiles combustion and char burn-out, is very short, on the order of 1 second (Tillman 1991, 305). The combustion rate will be limited by the particle surface area that can be accessed by oxygen molecules. The rate at which heat can be released to raise steam will be governed by the surface area of the coal. To increase heat release, the surface area available for reaction must be increased. This can be done by pulverizing coal to extremely fine particle sizes, less than 74 microns. As particle size is decreased, the total surface area of all the particles in a given mass of coal (not the surface area of an individual particle) increases. Increased surface area means increased reaction rates.

Using finely pulverized coal introduces a mechanical issue. Coal burned as large pieces can be supported on a grate that holds the burning coal away from the floor of the furnace and allows air to circulate through burning coal. This can't be done with finely pulverized coal. A suspension of pulverized coal in air is blown into the boiler, an approach is called *suspension firing* or, often, *pulverized-coal* firing.

The pulverized-coal system doesn't need mechanical support for the coal while it burns. Not needing a grate reduces restrictions on equipment size. Burning lumps of coal on a grate is limited to plants producing less than about 10 MW. In the United States, pulverized-coal firing for electric power generation was first used around 1920, by the Milwaukee Electric Railway and Light Company. This seems to have been the first application of pulverized-coal firing in the world. Since then, it has become the principal method of firing in large coal-burning power stations.

Coal will have been crushed before it reaches the boiler. Final grinding to very fine particles takes place in pulverizers directly adjacent to the burners. At such fine particle sizes the coal–air mixture can be handled almost like a gas. The air used to blow the coal into the burners, called *primary air*, provides a fraction of the total amount of oxygen needed for complete combustion. The rest of the necessary oxygen, *secondary air*, is introduced around the burners to ensure that most of the combustion occurs near the tips of the burners.

Pulverized-coal units operate on about 20% more air—called *excess air*—than is needed for complete combustion. This results in some 80–90% of the chemical energy in the coal, expressed by its calorific value, being captured as heat energy in steam entering the turbines. This figure is known as the *boiler efficiency*.

Complete burning of the coal requires thorough mixing of coal and air to make sure of an abundance of oxygen molecules in the vicinity of the particle surfaces. A very turbulent flow of the coal–air suspension facilitates such mixing. The turbulent swirling needed for complete mixing of coal and air can be created by the arrangement of burners in the boiler. The temperature of the turbulent combustion "fireball" reaches about 1,500–1,650° C (Tillman 1991, 305). Most boilers using pulverized-coal firing have been designed to remove the ash as a solid, rather than as molten slag, requiring that the rate of heat release and the temperature be controlled to prevent melting.

A boiler, once designed, cannot fire any random coal. Every large boiler is a one-off design tailored to the coal specifications supplied by the purchaser. Some small deviations from the exact coal specifications have to be anticipated and accommodated in the design. The more the coal being used deviates from the original design specifications, the less effective and efficient will be the operation. Extreme deviations might make it impossible to operate the boiler at all.

The fact that these installations are one-off designs has significant implications for cost. Every boiler is designed around a particular set of coal specifications. The great variability among coals results in a corresponding great variability in specific boiler designs. Boilers must be erected on site. In comparison, the components of solar panels and wind turbines can be manufactured to standard designs and transported to the point of use. Solar and wind installations are modular, so can be expanded easily when needed. This gives an economic advantage to these renewable energy forms.

Many electricity-generating plants operate as mine-mouth plants. This helps keep the coal quality from varying excessively. It also minimizes transportation costs. Stockpiles of incoming coal can be built using blending to create a coal supply that represents an average of quality of coal from the mine, helping to ensure that the blended coal has properties close to the specifications for which the boiler was designed. (As shown in Figure 7.2.) When the electricity company has a long-term contract with a coal supplier—or even owns the mine itself—a detailed strategy can be established for mapping coal quality within the mine. This helps to plan blending strategies, if needed. It also provides an early warning of long-term changes in coal quality that might be expected as mining progresses.

A coal-fired electricity-generating plant converts the thermal energy in a coal into useful electric energy needed by consumers. The thermal energy is in the chemical bonds of the coal piled up in the yard, and the electrical energy is supplied to an electricity distribution system. *Efficiency* in general terms represents the ratio of the amount of output of the desired product to the amount of material or energy put into the process. For electricity plants, energy engineers sometimes speak of the *coal pile-to-busbar efficiency*.[4]

Overall plant efficiency is determined from the amount of useful electric energy leaving the plant and being fed to the distribution system relative to the amount of chemical energy in the coal fed into the front of the plant. Efficiency of operation can be enhanced by increasing the temperature and pressure of the steam. Less coal is required to produce a given amount of electricity, accompanied by reduced CO_2 emissions. When the steam temperature exceeds 374°C and the pressure goes above 220 bars, steam enters the supercritical region[5] in which it is a fluid that is neither a liquid nor a gas (Maize 2018). Boilers operating above the critical point achieve significant gains in efficiency. Steam temperature in a supercritical plant is about 590°C. Getting the temperature to 650–760°C enters a realm of operation called *ultrasupercritical* (Maize 2018).

A conventional plant, operating below the critical point, might achieve efficiencies of about 33%. Ultrasupercritical plants can reach 45% (Yeager 2016). This change represents an enormous improvement, with reduced coal consumption and reduced emissions. China is in the forefront of ultrasupercritical power plants. A coming generation of plants might go even further, to about 700°C and 360 bars (Yeager 2016).

The importance of efficiency lies in the fact that, ultimately, somebody has to buy the coal that is burned to make the steam that is used to turn the turbine and generator. In many countries that somebody is the consumer paying the monthly electric bill. The higher the efficiency, the more electricity can be generated from burning a given amount of coal. From the other perspective, the higher the efficiency, the lower the amount of coal to be purchased to produce a given amount of electricity. A 40% efficiency of a very modern, state-of-the-art plant might seem low but has resulted from a century's worth

---

[4] A busbar is the device that collects or concentrates the generated electricity for feeding into a distribution system.

[5] When a liquid, kept in equilibrium with its vapor, is heated, the density of the liquid decreases and the density of the vapor increases. It is possible to reach a temperature and pressure, called the *critical point*, at which the densities of the liquid and vapor are equal, which means there is no longer a distinction between liquid and vapor. The conditions at which this occurs are called the *critical temperature* and *critical pressure*.

Above the critical point the substance is neither a gas nor a vapor, but simply a fluid. Often, a substance at temperatures and pressures above the critical values is referred to as a *supercritical fluid*. When used in this context, the term "supercritical" has absolutely nothing to do with a nuclear reactor becoming supercritical. That is a totally different phenomenon, unrelated to coal technology.

of steady, evolutionary improvements in the engineering of plant design and operation. Ninety years ago, plant efficiencies of less than 20% were common.

A plant efficiency of about 35% leads to another question: Where does the other 65% go? The generator loses about 1%. Turbine efficiency could be about 45%. About 10% of the energy goes straight up the stack in the heat in the hot gaseous combustion products. Steam moving through the plant, from boiler to turbine and from turbine to condenser, inevitably loses heat. Warming the cooling water in the condenser (discussed below) absorbs more heat. Fans provide air for combustion, pumps circulate water through condensers and cooling towers, and emission-control equipment must be operated. All of these operations consume still more energy. This hypothetical plant, reasonably efficient by today's standards, throws away the equivalent of almost 2 MW of wasted thermal energy for every 1 MW of electrical energy put into the distribution system. Two-thirds of all the energy extracted from the ground by coal miners is thrown away as hot air and hot water. By the time the electricity actually reaches the consumer, more losses will have been encountered in the wires and transformers. Consequently, only about 30% of the thermal energy in the coal reaches the consumer as electrical energy at an outlet in the wall.

Conversion of thermal energy in coal to electric energy can be expressed as the *heat rate*, the amount of heat from coal required to generate 1 kilowatt-hour (MJ/kWh) of electricity. The amount of coal needed to achieve a given heat rate depends on the calorific value of the coal. Using a high-volatile bituminous coal with a calorific value of 29 megajoules per kilogram (as-received), the heat rate of a modern, well-maintained power plant is about 0.3 kilograms of coal burned per kilowatt-hour of electricity generated. This is one-half the typical heat rate obtained in plants before the middle of the last century. In the first very small plants, at the end of the nineteenth century, the heat rate was about 4 kilograms per kilowatt-hour (Klein 2008).

The steady, quiet work of combustion engineers in cutting the heat rate in half over the past 80 years has a direct impact on the coal industry. Suppose that population growth and increased industrial activity would have doubled the demand for electricity during those years. This is the same time period over which the amount of coal needed to generate a kilowatt-hour of electricity was cut in half. Arithmetic is inexorable: $2 \times \frac{1}{2} = 1$. In other words, consumer and industrial demand for electricity doubles, but coal production does not change.[6] This is not a result of malevolent government conspiracies

---

[6] This simple example does not take into account the displacement of coal from electricity generation markets by natural gas and such renewable energy sources as solar and wind. Including those factors means that coal demand does not stay steady—it declines.

or the machinations of environmentalists—it's because we have gotten a lot smarter about how to burn coal.

A plant having a heat rate of 0.3 kilograms per kilowatt-hour generates enough electricity from 1 kilogram of bituminous coal to operate three 100-watt light bulbs for about 10 hours or to operate a laptop computer nonstop for about 5 days. Since a great many of us around the world consume electricity, quite often operating more than one device at a time, and we do so continuously for hours or days on end, collectively we have an enormous demand for electricity. In 2019, before the economic dislocations of the global COVID-19 pandemic, world electricity demand was about 24,000 terawatt-hours (IEA 2019). This is why we build large pulverized-coal–fired electricity plants that consume tens of thousands of tonnes of coal per day.

Ultrasupercritical plants offer the promise of increasing efficiency to 45% or lowering the heat rate to about 0.28 kilograms per kilowatt-hour. These changes seem very small, but when enough plants of improved efficiency are online, seemingly small changes will be applied against hundreds of millions of tonnes of coal. One hundred million tonnes of coal, burned in plants having heat rates of 0.28 instead of 0.30 kilograms per kilowatt-hour will generate an additional 37 billion kilowatt-hours of electricity, roughly equivalent to the annual electric power production of such countries as Denmark or Serbia (Wikipedia 2020). Alternatively, electricity now produced from 100 million tonnes of coal could be generated from 88.5 million tonnes in plants having the lower heat rate. If the coal contained 70% carbon as-fired (dry basis), the lesser tonnage needed to be burned would reduce CO_2 emissions by about 30 megatonnes.

Steam passes through the turbine at high pressure and high temperature. Steam pressure and temperature drop as it drives the turbine. Low-pressure, low-temperature steam leaving the turbine must be condensed back to water and recycled to the boiler. Relatively cold water from a natural source is brought into the condenser to absorb the heat given off when steam condenses back to liquid water.[7] Many plants have been built close to natural water sources to take advantage of a ready source of cooling water. Heat lost

---

[7] Everyone cooking or making coffee sees that it takes heat to boil water—to convert water into steam. Even when the water has been heated to its normal boiling point, 100°C, an additional input of heat is required to change the water to 100°C steam. This additional heat is called the *heat of vaporization*. When steam condenses back to water, the additional heat is released as the *heat of condensation*. On a practical level, this is why very large condensers are needed at power plants—to deal with the heat of condensation. On a personal level, this is why being burned by steam is a far more serious injury than being burned by hot water, even if both are at 100°C. The heat released by the condensation of steam to liquid is also absorbed into the affected part of the body, injuring the skin and tissues much more than if the burn were caused only by hot water.

from the steam increases the temperature of the cooling water as it passes through the condenser. The heated water cannot be discharged immediately back to the environment. Doing so might upset the local environment by raising the temperature of the natural water source, facilitating an unnatural growth of aquatic plants or animals. This problem is sometimes known as *thermal pollution*. Heated water leaving the condenser outlet has to be cooled back to near-ambient temperature before it can be discharged. This is done in a *cooling tower*. The "back end" of a power plant is shown in Figure 8.3.

Ash forms initially in the turbulent fireball of the burners. Ideally, as particles of ash form, they should drop to the bottom of the boiler and accumulate there as *bottom ash*, collected in hoppers. Some ash particles are so small that they become entrained in the rapid flow of combustion gases and are carried through the plant. Years ago, these fine particles, *fly ash*, would depart with the flue gases, eventually to come down somewhere else. Now, most industrialized nations have requirements for capturing and disposing fly ash before it becomes an environmental problem. We will revisit this in Chapter 9.

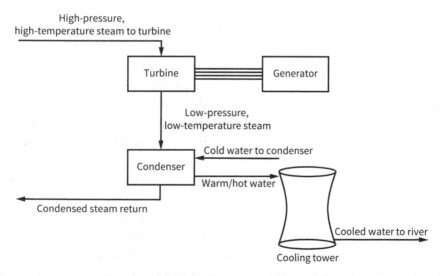

Figure 8.3 Simplified sketch of steam and water handling in a power plant. Steam from the boiler (Figure 8.2) operates the turbines, is then converted back to water in the condenser, and the water will be recycled to the boiler. Cooling water used in the condenser will become hot and must be sprayed into a cooling tower before being discharged into the environment.
Artwork by Lindsay Findley, from the author's sketch.

Ash can also accumulate inside the boiler, sometimes causing serious operating problems. The behavior of the ash must be an integral consideration in designing and operating combustion systems. The inorganic, non-combustible part of the coal can be at least as important as—even more important than—the behavior of the combustible, carbonaceous part.

Slagging results from the formation of molten ash inside the boiler. Ash disposal mechanisms designed to handle a relatively powdery ash that has never been molten may not be able to cope with large chunks of resolidified slag. Although slag is very hot, it acts as a thermal insulator and reduces the amount of heat transferred from the combustion gases to the water or steam. Large deposits of slag in the boiler can loosen and fall off because of partial melting. Extreme deposits can be the size of compact cars. Such a mass hurtling down through the boiler can cause catastrophic damage. Boiler repairs and electricity purchased to make up for production lost while the boiler is shut down can be very costly.

Fouling results from the formation of solid or semi-solid ash deposits on boiler tubes. Fouling deposits also act as insulators. In severe cases, they can clog the passageways for gases, upsetting the combustion process and forcing the boiler to be run at a lower rate of combustion. Fouling is of special concern in low-rank coal combustion. It probably is the most serious problem affecting the use of these coals.

Boiler-tube fouling during combustion of low-rank coals begins with the formation of a layer rich in sodium sulfate on the tubes. Ash particles striking the sodium sulfate layer begin to accumulate on the front of the tubes, facing the oncoming gases. A layer of ash particles accumulates on the sodium sulfate layer. As the layer thickens, it becomes a better and better thermal insulator. In addition to reducing heat transfer into the water or steam, this also has the effect that each newly added sublayer of ash particles cools more and more slowly. Eventually, the accumulation of such particles becomes thick enough so that the outermost particles cool so slowly that they have time to undergo a partial melting. Up to this point, the ash deposit is small and poorly consolidated. Once partially molten material forms, further accumulation of ash particles results in a continuous, relatively hard solid that can grow to be very large. In extreme cases, deposits can grow so rapidly that a clean boiler can be forced to shut down after only a few days of operation.

Sodium content of low-rank coals indicates their propensity to foul. In low-rank coals, most or all of the sodium occurs bonded to the organic acid groups in the coal structure. As the coal is heated, these acid groups decompose, freeing sodium ions that had been bonded to them. At about the same time, clay minerals in the coal are being dehydrated by the heat, opening and expanding their structures. Sodium ions can penetrate the opened structure and form new sodium aluminosilicate compounds having low melting points (Falcone and Schobert 1986). Some sodium–aluminum–silicon oxide compositions melt around 750°C (Kracek 1963). Such low-melting compounds form the matrix of the solid mass of the deposit. Ash fusibility temperatures vary widely, but even initial deformation temperatures are commonly 1,100°C and upward, so the formation of such low-melting-point phases can have a significant impact on ash behavior.

A pulverized-coal boiler with the emission-control devices that will be discussed in the next chapter represents mature, well-tested technology. Nevertheless, combustion engineers continue pursue new ways of generating heat or electricity from coals for many reasons: to develop lower-cost systems and reduce the investment required for a new plant; to achieve more efficient and cheaper control of emissions, particularly CO_2; to improve flexibility in operation to allow efficient response as demand for electricity changes; and to develop systems capable of burning lower-quality coals or capable of accommodating coals of variable quality as premium-quality coals become less plentiful.

We introduced fluidized beds briefly in Chapter 7 for burning the refuse from coal cleaning. The *combustor* is a vertical shell with a plate on the bottom. Small-diameter holes in the plate distribute the upward flow of air through the bed of particles, hence its name—*distributor plate*. The bed initially rests on the distributor plate. As air passes through the plate, the bed becomes fluidized. In some cases, the bed consists entirely of coal, pulverized to particle size less than 1 or 2 millimeters. Alternatively, the bed can include an inert material such as sand, or a chemically reactive material such as limestone. Limestone will capture the oxides of sulfur directly in the bed before they can escape. Water tubes immersed in the bed and additional tubes mounted in the space above the bed capture the heat released from the burning coal to generate steam. Particles of ash or partially burned coal that are sufficiently fluidized to be swept from the bed are captured in a cyclone.

Fluidized-bed combustion typically proceeds at temperatures substantially below those of pulverized-coal–fired units.[8] Ash fusibility is less important because the bed temperature is lower than the lowest ash fusion temperatures

---

[8] Typically about 750–1,000°C in a fluidized-bed combustor, much lower than the 1,500–1,650°C in the hottest part of the combustion region of a pulverized-coal unit.

of most coals. Caking coals present no problem in a bed of sand because the coal particles are kept far enough apart to avoid agglomeration. The fluidized bed can be designed to handle coals from lignites to anthracites exhibiting any degree of caking property and any ash fusibility behavior. Temperatures are too low to cause the reactions of nitrogen with oxygen.

Because the bed temperature is below the temperature at which calcium sulfate would decompose, sulfur oxides can be captured and retained in the bed. Enough sulfur capture can usually be achieved to meet air quality regulations without the need for additional emission control equipment downstream of the combustor. Some low-rank coals may contain enough calcium and magnesium in their ashes to supply adequate sulfur capturing capacity without the addition of limestone.

Many of the attributes of fluidized beds make them attractive for use in small industrial installations, either for electricity generation or for heat. Their ability to tolerate variations in coal quality provides an advantage for the user not wanting to enter into a long-term coal supply contract. Meeting sulfur emission regulations by adding inexpensive limestone to the bed instead of buying and maintaining a separate sulfur capture system provides another advantage. Fluidized-bed systems experience less ash deposition than pulverized-coal–fired units, translating into reduced maintenance.

The air velocity can be increased to a value at which bubbles can be observed forming in the bed. A combustor that operates this way is sometimes referred to as a *bubbling fluidized bed combustor* (BFBC). Bubbles add to the turbulence and agitation in the bed. Pushing air velocity even higher gets to a point at which the bed particles become entrained in the air flow, such that the air–particle suspension exits the top of the combustor. The mixture of air and particles passes into an adjacent cyclone, effectively separating the particles from the gases. The solid particles are recycled back into the combustor. A system operating in this way is called a *circulating fluidized bed combustor* (CFBC).

Any time that more hardware or more steps are added to any process, the complexity and cost of the process will increase. There needs to be a good reason to choose the more complex CFBC, knowing that the simpler BFBC units certainly work well. The CFBC concept allows for designing and building larger units of higher capacity than can be done with a BFBC (Miller and Miller 2008). More than 3,000 CFBC plants are in operation in China (Maize 2018), producing about 90,000 MW of power.[9] One of the latest developments

[9] It seems remarkable that, by 2000, China was building and installing more electricity generation capacity *per year* than the *total* of the installed capacity in Britain.

combines supercritical operation with CFBC combustion. The first such plant was built in Sichuan Province nearly 20 years ago. These installations all rely on large CFBC units. The CFBC plant in Tufanbeyli, Turkey, is successfully burning lignite of about 5 megajoules per kilogram calorific value, 40–52% moisture, and 20–31% ash yield (Patel 2016), a very poor-quality coal.

Combined-cycle plants generate electricity in two steps. Most common is a *combined-cycle gas turbine* (CCGT) plant. The first step uses a turbine that operates on hot gases produced by burning a gaseous fuel, either natural gas or a synthetic gas produced from coal. Hot gases enter the turbine at about 1,100°C and exit at about 500°C. The gases exiting the turbine are about the same temperature at which flue gases in a conventional pulverized-coal–fired power plant generate steam. The steam system in a conventional plant operates at temperatures between about 565° and 30°C. By adjusting operating conditions, a combined-cycle plant can take in heat at 1,100°C and eject heat at 30°C. Operating over such a wide temperature range increases the efficiency of the plant.

The maximum possible thermal efficiency for a conventional steam plant is about 64%, while a combined-cycle plant provides about 77%, based on the temperature differences in the system. They represent the maximum possible efficiency that could be achieved *if* there were no other factors in play. Unfortunately, there are plenty of other factors, such as friction in mechanical equipment, heat losses, and energy needed to operate ancillary equipment. The actual efficiency for coal in to electricity out is never as high as the ideal efficiency.

Where natural gas is abundant, it is the simplest and easiest approach to firing the gas turbine. Since large quantities of coal will be consumed in any case, an option is to convert some of the coal to a synthetic gaseous fuel. Another tactic uses fluidized-bed combustors operated at high pressure to generate hot gas for the gas turbine. The Osaki Power Station in Hiroshima, Japan, ran for nearly 20 years this way (Komatsu et al. 2001), but has since been supplanted by a plant using coal gasification. Regardless of the gaseous fuel, hot gases exiting the gas turbine provide the heat needed to make the steam for the steam turbine cycle.

Combined-cycle electricity generation using gas made from coal is done in installations called **integrated gasification combined cycle** (IGCC) plants. The combustion components of such a plant are similar to those in a combined-cycle natural gas plant. In addition to the benefits of higher efficiencies, in an IGCC plant the gas coming from the coal can be cleaned of potential pollutants before being burned. Cleaning the gas at this stage is easier than cleaning the much higher volumes of gas coming from a pulverized-coal–fired boiler. The necessary equipment can be smaller, resulting in lower investment and operating costs. Effective removal of sulfur oxides means that such plants could handle high-sulfur coals, which are not as expensive as cleaner coals (Schon and Small 2006). The first tests of an IGCC plant employing gas produced from coal were conducted in Luren, Germany, in 1972. Droplets of coaltar in the gas caused some problems, but the combined-cycle system itself worked well. In the United States, an IGCC plant at Southern California Edison's Cool Water station near Daggett, California, ran from 1977 to 2015 with good results. The Cool Water plant was retired in 2015, likely to be replaced by a solar facility. By 2013, there were six commercial-scale IGCC plants operating in the world, with a total generating capacity of about 1,700 MW (Miller 2017).

Any present-day technology for using coals to produce electricity also makes an abundant amount of heat. This is why many power plants have cooling towers. A useful strategy for increasing the overall efficiency of a coal-fired installation captures and makes use of what would otherwise be wasted heat. Facilities that produce both electricity and useful heat are called *combined heat and power* (CHP) plants, or *cogeneration plants*. Conceptually, heat exported to off-site users displaces the coal or other fuel that would have been used to generate such heat at the site where it is needed—the *thermal host*. Thermal hosts include, as examples, commercial buildings, universities, petroleum refineries, and hospitals. Overall efficiency is improved by generating electricity and using some of the available heat that would otherwise have been lost. Overall efficiency could rise to the 80% range (Balaban and Bobick 2016).

District heating uses the extra heat to generate hot water that is then piped to a town for residential or business heating. Besides increasing the overall efficiency of the power station, district heating centralizes delivery of fuel and control or collection of wastes. Smaller power plants may work with a nearby industry or with an industrial park to supply high-temperature heat for generating steam or for process heat. A remarkable example was the now-defunct town of Pyramiden, a Russian mining settlement on the Norwegian island of Svalbard (Andreassen, Bjerk, and Bjornar 2010). A coal-fired power plant provided electricity to the town and the mine. Excess heated cooling water was piped to the town, heating the houses, providing each apartment with hot water, and heating the community swimming pool. Heated water also warmed barns for livestock that produced milk, meat, and eggs, as well as manure to fertilize soil in a heated greenhouse.

Conventional ways of using coals as energy sources to generate electricity work well. Today's pulverized-coal–fired plants rely on a remarkable array of hardware: crushers, pulverizers, conveyors, boilers, pumps, turbines, generators, emission-control equipment, cooling towers, etc. Approaches such as fluidized-bed combustion will not shorten the list by much, if at all. This lengthy catalog of "stuff" needed to build and run a plant translates into substantial capital investment costs and continuing operating and maintenance costs.

The ultimate way of producing electricity from coal lies in eliminating combustion. Obtaining energy directly from coal requires that carbon in the coal must be converted to CO_2 and that the liberated energy be used as heat energy, or converted to another form of energy, such as electricity.

The most efficient process is one in which the carbon, oxygen, and CO_2 are all at the same temperature so that no energy is wasted to heat the CO_2 or to expand the volume of the gaseous products. Such a process represents a *reversible system*. In a truly reversible system, the properties of the system—notably temperature and pressure—never vary from those of the surroundings by more than an infinitesimal amount. A reversible system is an ideal that can never be attained in practice, but we can come close. An *electrical cell* represents a practical, nearly reversible system that keeps reactants and products (i.e., coal, oxygen, and CO_2) at the same temperature. Since the main reason for burning carbon in coal is to raise steam to make electricity, the prospect of generating electricity directly from coal using an electrochemical process offers interesting possibilities.

In 1923, Gilbert Lewis and Merle Randall speculated on this prospect in their now-classic text, *Thermodynamics* (Lewis and Randall 1923). They called attention to the possibility of a perfectly efficient process involving coal and oxygen producing CO_2 and said that "This might be done . . . if we could devise a galvanic [electrochemical] cell with reversible electrodes of carbon and oxygen. But hitherto all such attempts to obtain 'electricity direct from coal' have failed."

A *fuel cell* is an electrochemical device, which, like the more familiar batteries, makes electricity from the reactions of chemicals. In a fuel cell the reactants are kept separate from each other and brought into contact with the electrodes only when electricity is needed. A fresh supply of reactants can be made available continuously, so a fuel cell never "dies."

Direct carbon fuel cells (DCFC) have been developed to use coal directly as the fuel. In a DCFC the chemistry remains the same as in a combustion unit: carbon reacts with oxygen to produce CO_2, liberating energy. Then what is gained? First, DCFCs provide a great increase in efficiency of converting the thermal energy in coal to electric energy. Conversion is done in one step: coal in, electricity out. In a pulverized-coal plant, coal burns to liberate heat, heat generates steam, steam turns a turbine, and the turbine turns a generator. The smaller number of steps provides an immediate efficiency gain. Whenever more steps are added to any process, the overall efficiency of converting the input to the output decreases. Second, DCFCs operate at much lower temperatures than combustion processes, reducing heat losses compared to a combustion system. DCFC efficiencies are now in the 55% range (Zarzycki, Kacprzak, and Bis 2018).

DCFCs inevitably produce CO_2 because the key chemical reaction is just the same in a DCFC as it is in a combustor. Each kilogram of carbon (not coal) fed to the cell produces 3.7 kilograms of CO_2. But the higher efficiency of the DCFC relative to a combustion-based process means that less coal needs to be consumed to produce the same amount of electricity. Reduced coal consumption means reduced CO_2 emissions while still meeting consumer demand for electricity. CO_2 produced in a DCFC comes off as essentially pure CO_2. In conventional combustion, CO_2 is diluted with large quantities of nitrogen, some unconsumed oxygen, and a few other gaseous products. To capture CO_2 for reuse or for sequestration, it is much easier to deal with a gas stream that is pure CO_2 rather with a dilute stream in which CO_2 is only one of many components. DCFCs operating on coal offer the potential of cleaner, more efficient electricity generation, and a process better able to be coupled with CO_2-capture technologies.

This chapter has focused primarily on making electricity. There is another side to the story, the question of how to deal with all the other products of the plant; namely, the products of combustion of the coal. That's the subject of the next chapter.

References

Andreassen, Elin, Hein B. Bjerck, and Bjornar Olsen. 2010. *Persistent Memories: Pyramiden—A Soviet Mining Town in the High Arctic*. Trondheim: Tapir Academic Press.

Balaban, Naomi, and James Bobick. 2016. *The Handy Technology Answer Book*. Canton, MI: Visible Ink Press.

EIA. 2020. *Cost and Performance Characteristics of New Generating Technologies, Annual Energy Outlook 2020*. U.S. Energy Information Administration: Washington.

Eskom. 2015. "What Is a Megawatt?" Eskom Corporate Affairs. www.eskom.co.za.

Falcone, Sharon K., and Harold H. Schobert. 1986. "Mineral Transformations during Ashing of Selected Low-Rank Coals." In: *Mineral Matter and Ash in Coal*, edited by Karl S. Vorres, 114–127. Washington, DC: American Chemical Society.

IEA. 2019. "World Energy Outlook 2019." International Energy Agency. https://www.iea.org/reports/world-energy-outlook-2019/electricity.

Klein, Maury. 2008. *The Power Makers*. New York: Bloomsbury Press.

Komatsu, Hideaki, Masakatsu Maeda, and Masaru Muramatsu. 2001. "A Large Capacity Pressurized Fluidized Bed Boiler Combined Cycle Power Plant." *Hitachi Review*, 50: 105–109.

Kracek, Frank C. 1963. "Melting and Transformation Temperatures of Mineral and Related Substances." *U.S. Geological Survey Bulletin*. No. 1144-D.

Lewis, Gilbert N., and Merle Randall. 1923. *Thermodynamics and the Free Energy of Chemical Substances*. New York: McGraw-Hill.

Maize, Kennedy. 2018. "Advanced Coal Technologies Improve Emissions and Efficiency." *Power* 162(11): 42–43.

Miller, Bruce G. 2017. *Clean Coal Engineering Technology*, 467–506. Amsterdam: Butterworth-Heinemann.

Miller, Bruce G., and Sharon F. Miller. 2008. "Fluidized-Bed Firing Systems." In: *Combustion Engineering Issues for Solid Fuel Systems*, edited by Bruce G. Miller and David A. Tillman, 275–340. Amsterdam: Elsevier.

Patel, Sonal. 2016. "The Coal Refuse Dilemma: Burning Coal for Environmental Benefits." *Power* 160(7): 56–58.

Schon, Samuel C., and Arthur A. Small. 2006. "Climate Change and the Potential of Coal Gasification." *Geotimes* 51(9): 20–23.

Tillman, David A. 1991. *The Combustion of Solid Fuels and Wastes*. San Diego: Academic Press.

UtiliPoint. 2012. "What is a Megawatt?" http://www.utilipoint.com/2003/06/what-is-a-megawatt/.

Wikipedia. 2020. "List of Countries by Electricity Production." Wikipedia. https://en.wikipedia.org/wiki/List_of_countries_by_electricity_production.

Yeager, Robert. 2016. "Pushing the Ultra Envelope: Advanced Power Technologies Are Mainstream in China." *Power* 160(11): 50–51.

Zarzycki, Robert, Andrzej Kacprzak, and Zbignew Bis. 2018. "The Use of Direct Carbon Fuel Cells in Compact Energy Systems for the Generation of Electricity, Heat and Cold." *Energies* 11:3061–3071.

9
Environment

Almost everything discussed in the previous chapter is devoted to one cause: to boil water. We can use electricity because generators are turned by turbines running on high-temperature, high-pressure steam. The steam is generated in boilers for which coal is a major source of heat. The products of combustion of coals can be determined as if coals were a simple mixture of four elements—carbon, hydrogen, sulfur, and nitrogen—and mineral matter. The combustion process involves the reactions of these elements with oxygen from air and the transformations of the mineral matter.

Complete combustion of carbon forms carbon dioxide (CO_2). Hydrogen in the coal framework burns to water, which will exist as steam at the temperatures of combustion. Organic and pyritic sulfur will produce sulfur oxides. Sulfur has two stable oxides, the dioxide, SO_2, and the trioxide, SO_3. At high temperatures, sulfatic sulfur decomposes to SO_3. It's reasonable to expect that sulfur might produce both oxides, although usually SO_2 greatly predominates. The proportions of the di- and trioxides can vary depending on the combustion temperature and amount of excess air in the combustion system. To avoid having to use such expressions as "sulfur dioxide and/or trioxide," the mixture of the two is referred to as SO_x, pronounced "socks."

Nitrogen is more complicated because there are seven oxides of nitrogen (Miller 2017, 129). Those that are reasonably stable as gases are nitrous oxide, N_2O; nitric oxide, NO; nitrogen dioxide, NO_2; and dinitrogen tetroxide, N_2O_4. Nitric oxide is the dominant contributor. As with sulfur oxides, these compounds are lumped together as NO_x, pronounced "knocks." In addition, there are two distinct sources of nitrogen in a combustion system. First, coals contain some amount of nitrogen bonded to the coal structure, typically in the range of 1–2% by moisture-and-ash-free (maf) basis. The air needed to sustain combustion contains 79% nitrogen as the remarkably stable N_2 molecule. At modest temperatures and pressures N_2 does not take part in chemical reactions. At the high temperatures of combustion, some nitrogen from the air will react with oxygen to produce more NO_x. To distinguish the two sources, they are referred to as *fuel NO_x*, from nitrogen atoms bonded to the coal structure, and as *thermal NO_x*, from the high-temperature reaction of N_2 and O_2 molecules in the combustor.

Some inorganic components of coal can contribute in a small way to these emissions. Carbonates break down to liberate CO_2. Similarly, sulfate minerals will decompose to SO_x (Frost et al. 2005). By far the greatest weight of the inorganic components contribute to formation of ash—as bottom ash, fly ash, and, though we hope not, ash in fouling and slagging deposits. Fly ash is the product of concern. Concentrations of inorganic elements in coals vary from one coal to another, even within one seam of the same coal. The extent of the various reactions which produce ash also varies depending on the temperatures in the boiler and on the times over which these reactions can take place. Consequently, the composition of ash is also variable. Following the examples of sulfur and nitrogen, ash is sometimes facetiously referred to as RO_x (pronounced "rocks").

Many decades ago, standard practice allowed combustion products to pass out of the boiler, up the stack, and out into the environment. Gradually it became evident that each product of coal combustion contributes to one or more environmental problems. In response, most nations issued, and continue to issue, regulations governing emission of combustion products into the environment. These regulations stimulated development of approaches to remove potential pollutants from coal before it is burned, as we've seen in Chapter 7, or to burn the coal but capture combustion products before they can be released. These approaches add to the complexity of operation of an electricity-generating plant, add to initial investment and continued operating costs, and therefore ultimately add to the cost of electricity to the consumers. However, these approaches are also doing an increasingly good job of reducing pollution and protecting the environment, as we will now see.

SO_x and NO_x both play a role in the environmental problem commonly known as *acid rain*. A more inclusive term is *acid deposition*, the fallout of acidic compounds in solid, liquid, or gaseous forms. Wet deposition, or *acid precipitation*, results from airborne acids incorporated in snow, sleet, and hail as well as rain. Dry deposition comes from particles of acidic compounds in smoke or dust. Dry deposition sticks to the ground as well as to surfaces. Rain dissolves or dislodges these particles, making stormwater runoff more acidic. Dry deposition accounts for just about half of acid deposition.

In a completely pristine unpolluted location, where anything coming down from the sky must be pure, "pure" rain is already mildly acidic. Chemically pure water has a pH of 7, indicating that it is neutral (i.e., neither acidic nor basic). CO_2 is mildly soluble in water. In the atmosphere, CO_2 dissolves in rain, producing a solution of carbonic acid that has a pH of about 5.6. When sulfur and nitrogen oxides in the atmosphere become converted to sulfuric and nitric acids and dissolve in rainwater, the pH is lowered drastically, in extreme cases to about pH 3. The pH scale is logarithmic, which means that a change of two pH units (e.g., from 5 to 3) represents a hundred-fold (10^2) increase in acidity. Acid deposition with a pH of 3 is 10,000 (10^4) times more acidic than chemically pure water. Highly acidic deposition accumulated for decades in such industrial regions as northeastern United States and neighboring parts of Canada, the north of England, and in Europe.

Acid deposition corrodes building materials such as limestone, marble, or metals. Irreplaceable monuments, building facades, and statues, some that have survived for millennia, have been ruined by acid rain in a period of a few decades. Acid deposition acidifies bodies of natural water, destroying their aquatic life. It also destroys terrestrial plant life. Inhaling "acid mist" irritates the respiratory tract and exacerbates existing respiratory problems, such as emphysema.

In the mid-1800s, the British chemist Robert Angus Smith recognized that sulfuric acid in the air of heavily industrialized Manchester was the likely cause of the rusting of metals and fading of dyed fabrics. His book, *Air and Rain: The Beginnings of a Chemical Climatology*, qualifies him to be considered the "father" of acid deposition chemistry. Unfortunately, the book promptly vanished into obscurity, taking with it the concept of acid deposition.

All gases, regardless of composition, are mutually soluble because a gas, in contrast to liquids and solids, consists of molecules that do not experience strong intermolecular forces.[1] Gas molecules move freely through a volume that is very large compared to the volume occupied by the molecules. The free molecular motion allows gases to mix freely. Gaseous SO_x and NO_x mix completely with the atmosphere and can be transported in the air hundreds or thousands of kilometers from their source. They can move from one region of a country to another, across national borders, and even to different continents. By the mid-1990s, satellite monitoring had shown a region of highly acidic precipitation nearly circling the globe, from Britain to Central Asia.

Both organic and pyritic sulfur oxidize primarily to SO_2. The eventual fate of sulfur depends on the relative amounts of pyritic and organic sulfur, on the specific conditions of combustion, and even on prevailing weather. Low-rank coals produce an ash containing some alkaline components, such as calcium oxide. Alkaline ash can react with acidic SO_x to capture and retain sulfur. Up

---

[1] The molecules of a gas do experience very weak attractive forces when they come very close together. Under most normal conditions these forces are negligible.

to about 15% of the sulfur originally in the coal could be retained in the ash (Maloney et al. 1978). Without emission control equipment the rest would escape to the atmosphere as SO_x.

Worldwide, about two-thirds of sulfur emissions to the atmosphere are anthropogenic (Dahiya and Myllyvirta 2020). For some time the leading source of anthropogenic sulfur emissions has been fuel combustion, particularly coal, accounting for about one-fourth of the sulfur oxides in the atmosphere. Recently, decreasing coal use coupled with steady improvements in SO_x capture has led to fertilizers and pesticides from agriculture becoming a significant anthropogenic source (Hinckley et al. 2020). Producing such metals as copper and nickel from their ores also generates large amounts of SO_x. Natural sources include sulfur released by plankton in the oceans, decaying vegetation, and volcanoes. In the atmosphere, a portion of SO_2 is converted to sulfur trioxide, and then to sulfuric acid or sulfate salts. Sulfur compounds in the atmosphere return to land or water in three ways: being adsorbed on surfaces or by reacting with them, being washed from the air in precipitation sulfurous acid coming from dissolution of SO_2 in water, and being washed from air as sulfuric acid or sulfate salts.

Two ways exist for dealing with sulfur emissions: either reduce the amount of sulfur going into the boiler in the first place (i.e., burn less sulfur) or capture the SO_x before it can leave the plant to get into the environment. Taking steps to reduce the amount of sulfur burned constitute *pre-combustion strategies*. Actions to capture the SO_x that forms before it can escape to the environment represent *post-combustion strategies*. In some cases, a company will do both.

Coal preparation reduces the amount of sulfur going into the boiler. Commercial processes reduce only the pyritic sulfur. Even so, in many coals pyritic sulfur constitutes the majority of the total sulfur, so coal preparation still results in a significant drop in sulfur content. This is good from the standpoint of reducing sulfur emissions, but it comes at a cost. Coal cleaning to reduce the sulfur content could add, say, $3–4 per tonne to the cost of the coal. For a plant that burns 10,000 tonnes of coal per day, coal cleaning could represent an extra cost of about $30,000 per day to the electric company, and, ultimately, to us in our electric bills.

A second way of burning less sulfur involves fuel switching. The plant could burn coals having lower sulfur contents than the coal ordinarily used. Even blending the usual fuel supply with lower-sulfur coals could have an effect. It may not be easy to find coals with low sulfur content within a convenient distance of the plant. Companies that own large supplies of low-sulfur coal know what they have and are likely to charge accordingly. If supplies of coals having acceptably low sulfur contents are available only at long distances from a particular power plant, transportation costs might become significant. Such costs would have to be a factor to be balanced against the costs of preparation for more local coals or against the costs of installing and operating SO_x emissions control equipment.

Rather than switching to a coal of lower sulfur content, the coal used by the plant could be co-fired with an entirely different fuel. Options include natural gas or various forms of biomass. Neither contains sulfur, so the blend will have a sulfur content lower than that of the coal itself. This approach will require some modifications to the fuel handling system, burners, and boiler, for which somebody has to pay. Co-firing with biomass was successfully conducted on large scale at the Drax power station in the north of Yorkshire, England. This is a 3,900 megawatt (MW) plant, a very large installation. Between 2010 and 2016, the Drax plant operated successfully with co-firing. After 2016, it was converted to operate completely on biomass. In the United States, plants in the James Rogers Energy Complex in North Carolina have been converted to co-fire natural gas (Downey 2018).

Capturing SO_x before it can escape, a practice known as *flue-gas desulfurization* (FGD), occurs in units called *scrubbers*. Scrubbers take advantage of the fact that SO_x dissolves in water to produce acids, which will react with alkaline compounds such as hydroxides or carbonates. The first flue-gas desulfurization system, in 1932 at the Battersea station in London, used a spray of water from the Thames to dissolve the SO_x. The resulting solution of sulfurous and sulfuric acids was pumped back into the river. Flue-gas desulfurization became common only because of increasing environmental awareness and accompanying regulations, beginning in the 1970s. Scrubbers can remove more than 90% of the SO_x from flue gases.

A scrubber uses a slurry of lime (calcium hydroxide) or limestone (calcium carbonate) to react with the SO_x. The products, calcium sulfite and calcium sulfate, precipitate as a sludge. In such scrubbers, also called *non-regenerative scrubbers*, the active agent—lime or limestone—is used only once and then becomes incorporated into the sludge. Scrubbers do not destroy pollution. The scrubber transfers the environmental problem from one of air pollution, SO_x in the flue gas stream, to one of potential land or water pollution, scrubber sludge. Disposal of the sludge, not emissions of gaseous SO_x, becomes the problem. Many plants impound sludge in lagoons. Over time, more and more land area must be devoted to sludge lagoons. An alternative approach returns the sludge to the mine as fill. In either case, water-soluble compounds can leach from the sludge and contaminate groundwater. At best, the leaching will produce hard water.[2] At worst, leaching could introduce harmful trace elements such as arsenic and selenium into the groundwater.

Calcium sulfate in sludge is usually in its hydrated form, gypsum, $CaSO_4 \cdot 2H_2O$. Income from sale of byproduct gypsum could offset some costs of operating the scrubber. This material can be applied in the manufacture of the wallboard used in building construction. Roughly half of all the wallboard manufactured in the United States is made with gypsum that comes from power plants. Gypsum has other uses, such as in cement manufacture and in soil treatment. Gypsum added to poor soils improves their structure[3] by adding calcium, as well as enhancing the exchange of water and air. Gypsum also helps to neutralize acidic soils.

Regenerative scrubbers recycle the active reagent, often producing a sulfur-containing material as a byproduct. Recycling the sulfur capture agent eliminates the recurring expense of having to buy it, except for small amounts needed to replace inevitable losses. Depending on the specific process, the byproduct could be elemental sulfur, liquefied SO_2, or sulfuric acid. The byproduct can be sold for revenue to offset some of the operating expenses. An attractive target would be a byproduct useful for making sulfuric acid, one of the most important of industrial chemicals.

Scrubbers work very well. In the late 1930s, before scrubbers were used in the United States, an estimated 2,500 tonnes of sulfurous acid fell in the Chicago area per day. In the last quarter of the twentieth century, during a time when coal use in the United States had roughly doubled so that sulfur emissions might have been expected to double as well, SO_x emissions actually dropped by 30%. But, again, scrubbers do not destroy pollution. Sulfur has only been converted from very dilute SO_x in the combustion gases to much more concentrated scrubber sludge.

Scrubbers are not the only reason that SO_x emissions from electricity generation have fallen. Steady improvements are being made in the heat rate of boilers, meaning that less coal is required to generate the same amount of electricity. Even with no other factors being considered, less coal means less sulfur going into the plant. In addition, the percentage of total electricity generated

---

[2] Hard water contains dissolved salts of divalent ions such as Ca^{+2} and Mg^{+2}. It is undesirable because these ions react with soaps to produce a "soap scum." In equipment that heats or boils water, carbonates and sulfates of these ions form deposits or "scale" that restrict water flow and heat transfer. If Fe^{+2} is one of the ions, it will leave orange-yellow scale on kitchen or bathroom fixtures. About the only good news is that hard water is not considered a health hazard.

[3] In soil science, the word "structure" has a meaning very different from when we talk about molecular structure. Soil structure is described by the way that the individual particles of clay, silt, and sand pack together into aggregates. In a soil that has a good structure, air and water circulate through the soil, whereas a bad structure characterizes a soil with poor circulation of air and water.

in coal-fired plants has been decreasing, with more electricity coming from sulfur-free fuels such as natural gas.

Scrubbers use a lot of water. In a plant of 1,000 MW generating capacity, water consumption in the scrubbers is about 4,000 liters per minute. This issue will be of increasing concern in those countries or regions already facing water supply problems. The flue-gas desulfurization system does not operate itself; it uses some of the electricity produced in the plant, an amount that otherwise could be sold to consumers. The electricity that could have been sold but must be used inside the plant instead is called a *parasitic loss*.

Two alternatives to wet scrubbers exist. One uses alkaline compounds injected into the flue gas system in a powdered form via a pneumatic system. This approach is known as a *dry scrubber*. Though many compounds are alkaline enough to react with SO_x, economic considerations would limit the choice to sodium or calcium compounds. After the injected material has absorbed SO_x, the "spent" material can be collected in dry form without having to deal with a sludge. In best cases, dry scrubbers can reduce SO_x emissions up to 95% (Larson 2017).

A hybrid version of wet and dry scrubbers uses a slurry of a sorbent such as lime, but in which the solid concentration is much higher than in a conventional wet scrubber. Injecting the slurry into the hot flue gases promptly evaporates the water, leaving the sorbent as a fine powder. In these hybrid scrubbers the product is again a dry, powdery material that can be collected for disposal.

These approaches for reducing sulfur emissions have been considered in light of currently conventional pulverized-coal–fired combustion systems. A more radical alternative is to change the combustion system itself. One way uses fluidized-bed combustion, in which a sulfur-capture agent such as limestone is added to the bed to keep most of the sulfur from getting into the gas. This alternative might be a choice for new construction, but it would be a difficult and costly retrofit to an existing plant having a conventional pulverized-coal combustion system.

Regardless of the strategy adopted to deal with SO_x emissions, somebody— most likely the consumer of electricity—has to pay. Running a coal cleaning plant requires energy, which must be purchased from someplace. If coal cleaning were co-located with an electricity-generating plant, this energy would come most conveniently from the generating plant itself. This parasitic loss decreases the net efficiency of converting thermal energy in the coal to electrical energy leaving the plant to be sold to consumers. A scrubber system reduces the efficiency of conversion of thermal energy in the coal to electrical energy by about 1%. Scrubbers represent a capital investment equivalent to roughly one-third of the cost of the entire plant. Once purchased and installed, non-regenerative scrubbers have a continuous cost for lime or limestone. All scrubbers have costs for maintenance and for salaries of the workers who operate or maintain them. These costs have to be recouped somehow, often via the inevitable monthly electric bill sent to the consumer. The parasitic consumption of electricity represents lost revenue to the power company.

Emissions trading provides an additional approach to dealing with SO_x. In such a system, each facility that produces SO_x operates under a permit specifying the maximum level of SO_x that it can legally release. As the goal, each company should achieve, and not exceed, its allowed level of emissions. Exceeding the maximum can result in fines or other penalties. Those facilities that perform better, with SO_x emissions below the permitted level, receive pollution credits. Each credit conveys the right to emit a specified quantity of SO_x. Credits can be sold to another organization that is unable to meet its emission allowance. This creates a financial incentive to achieve significant reductions of emissions because extra credits sold to companies not meeting targets represent income for plants doing a good job in reducing emissions. Buying these credits allows companies that cannot meet SO_x standards to continue operation, but it's hoped that few companies would be willing to tolerate for long the continuous expense of having to buy pollution credits. This creates a financial incentive to work toward reducing emissions. Credits have been bought and sold in private transactions and in public auctions.

The United States experience with SO_x emissions trading began in 1990. The Environmental Protection Agency created a nationwide market for SO_2, limiting the total amount that could be emitted nationally and giving allowances to power plants based on past SO_2 emissions. Power companies had permission to trade their SO_x allowances. In 15 years, this *cap-and-trade system* reduced SO_x emissions by about 35% although, in the same period of time, electricity generation was up by 25%.

Since 1990, the emissions of SO_2 in North America and Europe have dropped by some 70–80% (Poetzscher 2020), reflected in a global decrease up to about 2000. The rapid pace of industrialization and electrification in China caused global emissions to rise until 2006. From then on, sulfur emissions began again to decrease, as Chinese pollution control efforts took effect. In the early 1960s, about 60% of global SO_2 emissions came from the developed countries. Today, about 80% of these emissions come from the developing countries because sulfur emission controls have been deployed so successfully in the developed nations (Zhong et al. 2020).

Selecting the optimum choices involves a tradeoff between how much damage we are willing to accept from acid deposition and how much extra we are willing to pay on electric bills or taxes. Scientists or engineers cannot calculate that answer. Chemists can improve our understanding of the reactions involved in SO_x and NO_x formation, capture, and transport in the atmosphere. Ecologists can study the effects of acid deposition on water, soil, and ecosystems. Engineers can design systems to clean coal, control NO_x and SO_x formation during combustion, and capture these materials from the flue gas. Economists can calculate the costs to a utility of installing and using these systems or switching to low sulfur coal, as well as the additional cost of electricity to the consumer. But, ultimately, decisions must be made by society as a whole, in the voting booth, lobbying elected officials or members of regulatory agencies, and even testifying before such bodies. What level of damage is acceptable? What costs are acceptable?

Even if it were possible to reduce the nitrogen contents of coals to 0.000% (it isn't), there would still be NO_x emissions from a combustion system. This is because of the formation of thermal NO_x at the temperatures of the coal–air flame in the boiler. In contrast, if the sulfur content of a coal could be reduced to zero, there would be no SO_x emissions to worry about. In 2017, 60% of all anthropogenic NO_x emissions were from three sectors: energy generation, industry, and transportation (McDuffie et al. 2020). About half of the emissions from energy generation and industry was attributed to coal combustion.

Nitrogen molecules react with oxygen to form nitric oxide (NO). In the presence of additional air, NO quickly converts to nitrogen dioxide. Nitrogen dioxide is brown; in badly polluted air it gives the sky a ghastly brown color. It dissolves in water droplets in the air producing nitric acid, a component of acid deposition. Worse, nitrogen dioxide is toxic, "one of the most insidious gases" (Budavari et al. 1989). Inhalation causes inflammation of the lungs, leading to accumulation of fluid that could cause death.

At high altitudes NO_x reacts with *ozone*, the form of oxygen with three oxygen atoms per molecule (O_3). This contributes to the reduction of the ozone layer. Some 10–30 kilometers above Earth's surface, the atmosphere contains increased concentrations of ozone. The ozone concentration in the so-called ozone layer is small, about 15 parts per million (0.0015%). However, it is crucial for human health. High-altitude ozone intercepts some of the ultraviolet light that is part of solar radiation. Problems attributed to ultraviolet exposure include an increased susceptibility to skin cancer and cataracts. Each 1% reduction in ozone concentration increases the amount of ultraviolet reaching the surface by about 2%. Because NO_x reacts with ozone, an increase in high-altitude NO_x could contribute to reduced ozone concentrations, which would lead to more exposure to ultraviolet.

As with SO_x, human activities are not the only sources of NO_x. Soil, forest fires, and lightning both produce NO_x. Microbes in the soil break down nitrogen-containing compounds to release NO_x. This process is especially significant in areas that use high dosages of nitrogen fertilizers (Almaraz et al. 2018). Lightning bolts heat the surrounding air to about 28,000°C (NOAA 2018), easily driving the formation of thermal NO_x. Natural emissions account for about 40% of the NO_x released to the atmosphere. The most notorious role of nitrogen oxides in air pollution, their contribution to smog formation, is mainly a problem of vehicle exhaust emissions rather than related to coal-fired power plants.

NO_x is harder to control than SO_x. First, no easy, cheap way exists to remove nitrogen from coal. Pyritic sulfur is relatively easily separated because of the difference in specific gravities between pyrite and the carbonaceous material. Nitrogen, on the other hand, is chemically incorporated in the molecular framework of coals. Reducing nitrogen in a pre-combustion coal-cleaning process would require chemical processing, likely much more expensive than a physical process based on specific gravity differences. Second, scrubbers rely on the fact that calcium sulfate is not soluble in water. An analogous system for "flue gas denitrogenation" could not function in the same way because the nitrate salts of all elements are soluble in water and would not precipitate. Third, thermal NO_x formation is almost inevitable. Not only that, the higher the combustion temperature, which is needed for increased efficiencies in supercritical and ultrasupercritical power plants, the more thermal NO_x will be formed.

Changes to the way the combustion process is operated can reduce thermal NO_x formation. Low-NO_x burners control the coal–air mixture and the temperatures attained during combustion. Reaction of nitrogen with oxygen becomes appreciable only at temperatures greater than 1,480°C. The reverse reaction, nitric oxide decomposing to nitrogen and oxygen, can take place if the gases cool slowly from this temperature. Using lower flame temperatures, combined with giving any nitric oxide that does form time to decompose, are the keys to reducing thermal NO_x. Achieving this relies on a tactic of *staged combustion*. In staged combustion, coal burns in a fuel-rich flame produced by delaying the mixing of secondary air. In this first stage, NO_x converts back to nitrogen. In the second stage, more air is made available to complete the combustion, particularly char burn-out. Having only low amounts of excess air also helps. Temperature does not affect the formation of fuel NO_x significantly, but this form often represents only a small fraction of the total NO_x.

A second tactic, *re-burning*, also aims to reduce formation of thermal NO_x. In this case, coal burns in a lower region of a boiler, where combustion in this region provides some 80–90% of the heat needed to generate the steam. The re-burn region, higher in the boiler, uses an addition of natural gas. Methane in the gas reacts with NO_x, decomposing it to nitrogen. Co-firing, in which coal is combusted simultaneously with a fuel of near-zero sulfur content, such as biomass, has been mentioned as a way to reduce SO_x formation. NO_x emissions are also reduced by co-firing. Combinations of these tactics can reduce NO_x emissions even further.

For about the past 40 years the reaction of ammonia with NO_x has been deployed to convert NO_x to harmless nitrogen and water vapor. Other nitrogen-containing compounds, such as urea (H_2NCONH_2), can be used instead of ammonia. The reactions occur on the surface of a catalyst, usually employed at temperatures of 300–400°C. The technology, *selective catalytic reduction* (SCR), is now used around the world. It can remove up to about 95% of NO_x (Miller 2017, 526–530).

In operation, ammonia is adsorbed on the catalyst. NO_x reacts with the adsorbed ammonia. Activated carbon (Chapter 17) can also be used as the catalyst support. This option could be attractive because activated carbons can be made from coals, often by reaction with steam. There is no shortage of coal or steam at a power plant. Perhaps a clever process design could provide a way of making the SCR catalyst support right on site.

Most industrialized nations have made, and continue to make, significant strides in controlling SO_x and NO_x emissions. The greatest threats from acid precipitation will come from developing or underdeveloped nations as they become significant emitters of SO_x and NO_x. Many such countries lack the legislative or regulatory structures needed to control emissions. Impoverished countries lack the financial resources to install modern—but expensive—emission control equipment. These countries have rapidly growing populations; more people account for more energy consumption, even at a constant level of per capita energy use. But, in addition, the people of such countries not unreasonably aspire to improved standards of living, certainly including an electricity supply that is available "24/7." Ratcheting up the standard of living increases energy consumption. In the absence of regulatory structures and the ability to fund pollution control equipment, increased energy consumption brings with it increased emissions to the environment.

Dealing with ash in any manner requires coping with the huge amount that is produced. A medium-sized electric plant might consume 10,000 tonnes of coal per day. If that coal has an ash yield of 10% as fired in the boiler, then the plant produces 1,000 tonnes of ash every day, ash that must be collected and used or disposed of in environmentally acceptable ways.

The various components of ash have greater densities than the carbonaceous part of the coal. It would be reasonable to expect that, as coal burns in a pulverized-coal flame, particles of ash should drop to the bottom of the boiler, where they can be collected for removal. The bottom ash—the larger, relatively coarse particles of ash—represents some 15–40% of the total amount of ash (Miller 2017, 257). The highly turbulent flame results in very small ash particles being carried upward and out of the boiler with the hot gaseous products of combustion. If these particles—fly ash—escape into the environment, they contribute to the category of pollutants also termed *particulate matter*. This term refers generally to solid particles and liquid droplets found in the air. Particulate matter includes not just fly ash, but also soot, smoke, and dirt or dust particles. Electricity generation using all fuels, not just coal, contributes about 15% of all the particulate matter in the air (Weagle et al. 2018).

Bottom ash collects in an ash hopper. From there, it can be removed and sent to a landfill, put back into the mine, or used as fill material in mined land reclamation. In these applications, it is concerning that rain water or groundwater could leach various compounds from the ash that could be health hazards if they made their way into the water supply. Far better is to use ash in economically beneficial ways, either using the ash itself or using it as a raw material for recovery of valuable byproducts. Doing so provides an additional revenue stream for the company or utility that is producing the ash and, at the same time, reduces the volume of ash in the environment. Bottom ash can be used as a filler in concrete and in road construction.

Fly ash particles will contain some of the trace elements found in coals. Compounds of some trace elements, such as lead, mercury, and arsenic, are hazardous to health. In a badly polluted environment, fly ash particles might adsorb sulfuric or nitric acids, which, as a minimum, will irritate tissues in the respiratory tract and lung. Fly ash may contain some carbon as soot from incomplete combustion or partially combusted char particles. When combustion is very inefficient, the emissions could include large molecules of highly aromatic compounds, some of which are known or suspected carcinogens, as well as small amounts of highly toxic dioxins,[4] which are persistent environmental pollutants.

---

[4] Dioxin itself is a cyclic molecule containing two atoms of oxygen and four of carbon. Its many derivatives are collectively known as *dioxins*. The worst of the family contains two aromatic rings attached to the parent dioxin structure and two chlorine atoms attached to each aromatic ring: 2,3,7,8-tetrachlorodibenzo-*p*-dioxin (TCDD). Dioxins are produced during the combustion of fuels, the incineration of wastes, and in such natural processes as forest fires and volcanoes. They are carcinogens and disrupt hormonal activity. Stored in fat cells, dioxins can persist in the body for a decade.

Fly ash not captured by control devices will be carried out of the stack and released to the environment. These particles vary in size from about 0.01 micron to 10 microns in diameter. Fly ash can be a problem in several ways. Perhaps the least of these is the deposition of a layer of ash particles on anything outdoors, or even indoors. This can be a nuisance and certainly an aesthetic issue. Particles of about 0.01 to 1 micron diameter can be trapped in the respiratory system. Smaller particles represent only a small fraction of fly ash as a percentage of mass, but because each particle is so minute, the number of particles in this size range is enormous. Such particles have large surface areas in proportion to their diameters,[5] often with many tiny cracks and crevices. These features make these particles very good at absorbing other pollutants. When such pollutant-laden particles enter the lungs, the absorbed pollutants are released to the surrounding tissue at higher concentrations than if they had simply been inhaled directly from the atmosphere. This release of pollutants straight into the tissues irritates and damages lungs and exacerbates respiratory problems such as emphysema.

The greatest hazards come from particles smaller than a diameter of 2.5 microns, having the symbol $PM_{2.5}$. About 30 of the largest particles classified as $PM_{2.5}$ would collectively make up the thickness of a typical human hair. $PM_{2.5}$ is of particular concern because these extremely small particles can penetrate most deeply into the tiny air passages in the lungs. Relative to larger particles, they pose the greatest problem to health.

The control method used in most power plants, an *electrostatic precipitator* (ESP), passes the combustion gases between electrodes. Fly ash particles acquire an electrical charge from the high-voltage electric field. Their electrical charge causes them to be attracted to one of the electrodes. From time to time the collected ash particles can be removed by rapping on the electrodes to dislodge the particles into a hopper. The ESP was invented by the American physical chemist Frederick Cottrell, who graduated from high school at 16 and from the University of California at Berkeley at 19, and who was alleged to read textbooks as rapidly as if they were novels.

Electrostatic precipitators came into use in 1923, only a few years after the introduction of pulverized-coal firing. Performance requirements have become increasingly stringent over the years as air quality standards have become tighter. To achieve a stack discharge that appears clear to the eye, the particles in the gas must be at a level no greater than 200–400 grams per cubic meter of gas. However, the gas entering the stack has particulate

---

[5] As an example, assuming a spherical ash particle, a particle with a diameter of 2 microns would have a surface area of 12.57 square microns.

concentrations several hundred times greater than this level. Achieving a completely clear discharge requires the precipitator to operate at efficiencies of 99.0–99.7%. Getting almost any kind of equipment to operate at 99% efficiency is a major engineering feat. The design and operating conditions of an ESP must be matched to the amount and properties of the fly ash being collected, particularly its electrical properties.

Fabric bags can trap ash particles, approximating the workings of a gigantic vacuum cleaner. Filter bags are used in multiples inside a structure called a *baghouse*. A baghouse offers collection efficiencies that are at least as good as those of a precipitator, possibly up to 99.99% (Miller 2017, 444). Like an ESP, a baghouse has to be cleaned periodically. There are various ways of doing this. Reversing the direction of air flow through the baghouse blows collected ash particles off the fabric. Or the fabric bags can be mechanically shaken to dislodge ash.

Effective collection of fly ash to control emissions leads to the question of how best to dispose of it in an acceptable manner. The amount of fly ash generated each year is about 350 million tonnes, from the 11 countries producing most of it[6] (Dwivedi and Jian 2014). In the United States, coal ash represents the second-largest amount of waste material, with only household refuse produced in greater quantity (Ritter 2016). Much of the ash is discarded in various ways, such as returning it to the mine as fill or dumping it in waste ponds. Almost 1.5 billion tonnes of coal ash are stockpiled in the United States alone (Ritter 2016). The need for a better idea was dramatically illustrated in December 2008, when a dike in a waste storage pond collapsed at the Kingston power plant in Tennessee. The collapse of the dike released nearly 4 billion liters of a water slurry containing 4 million cubic meters of ash (Overton 2016). Homes were flooded, and the nearby Emory River was polluted. It took 7 years and a billion dollars to clean up the mess.

Fly ash commercial applications could provide a useful, potentially valuable material. If appropriate processes could be developed to sell ash for conversion into useful byproducts, the revenue could offset environmental control and compliance costs, rather than incurring a cost for disposal. By subsidizing ash recycling, China is able to recycle about 60% of its ash. In the United States, the proportion is about half.

---

[6] In descending order, India, China, the United States, Germany, the United Kingdom, Australia, Canada, France, Denmark, Italy, and Netherlands. India and China together account for about 55% of the world total.

Fly ash often has *pozzolanic properties*, meaning that it can react with lime and water at normal temperatures to produce a cement-like product. About half of the concrete made in the United States contains at least some fly ash. In addition to concrete, fly ash can also be used in bricks and plaster. Meeting restrictions on NO_x emissions might result in limitations on combustion, particularly affecting the extent of char burnout. This leaves unburned carbon in the ash. Specifications on the percentage of allowable unburned carbon in fly ash to be used in cement vary in different countries and vary depending on the intended use (Federal Highway Administration 2017). An allowable upper limit of 6% seems fairly common. Ash of higher carbon contents is not suitable for concrete and may not be usable for other potential uses.

Numerous other uses exist. Fly ash mixed with soil can act as a stabilizer, improving its load-bearing capacity. Blended with aluminum alloys, fly ash helps to form metal-matrix composites, which have improved strength properties and hardnesses relative to the original alloy but lower densities. The metal-matrix composites are harder and stronger but lighter. Fly ash is often alkaline and has the capacity to absorb moisture. These characteristics can make ash an inexpensive soil amendment to improve the properties of acidic soils for plant growth.

We've seen that coals contain many elements in trace amounts. When a particular coal is burned, the various trace elements will be partitioned; some will tend to distribute about evenly in the bottom ash and fly ash, some be enriched in the fly ash and correspondingly depleted in the bottom ash, and a few are so volatile that they are emitted with the flue gases in the vapor phase. This is of concern for two reasons. Some of these trace elements can cause human health problems. If they are leached out of ash and enter the water supply, this represents a potential problem. Even worse, they can enter the human body directly by inhalation of vapors or fine particulate matter. On the other hand, some trace elements have valuable uses. Germanium and gallium are examples, both having applications in electronics. Coal ash could be treated as a source of such elements. This topic we will defer to Chapter 16.

In modern electricity-generating plants with state-of-the-art emission control equipment, and with pre-combustion coal cleaning, there is little reason to be concerned about potential health effects of trace element emissions. Very serious problems can ensue, but generally from use of domestic or small industrial combustors with few or no emission controls or from the burning of coals of unusually high concentrations of trace elements (Ding et al. 2001).

At least 20 trace elements have been flagged as being potential health hazards. For the past 30 years the United States has maintained a list of nearly 200 substances defined to be *hazardous air pollutants*, often known as HAPs. The HAPs include 11 trace elements, the most notorious being arsenic, mercury, and lead.[7] At least 9 more trace elements are regulated by other official standards or legislation.[8]

The International Energy Agency put the inorganic elements in coals into three classes, depending on how they partition in combustion (Miller 2017, 424). Class I elements are those that are distributed about equally between bottom ash and fly ash. These elements are relatively non-volatile and include aluminum, iron, and silicon. The Class II elements are enriched in fly ash. These include some of the more volatile elements such as potassium and sodium. Elements in Class II tend to be more volatile than those of Class I. The elements in Class III—mercury, chlorine, and fluorine—are so volatile that they are emitted in the vapor phase. They are not even enriched in fly ash, but rather remain as vapors.

Some 40–50 years ago, when there was great concern about SO_x emissions from power plants and their role in acid precipitation, the comment might be made that sulfur is the component of coal that everybody loves to hate. The mantle has now passed to mercury.

Mercury is a highly volatile element whose toxic effects have been known for a long time. Mercury emitted to the environment is gradually converted to the organometallic compound methylmercury (CH_3Hg) (Nordberg and Cherian 2005). Anaerobic bacteria present in aquatic sediments metabolize mercury compounds, such as mercuric sulfate, into methylmercury. Mercury works its way up the food chain into fish, then to birds, and eventually to us. Ingestion of mercury compounds can lead to neurological problems, beginning with tremors, memory loss, and slurring or stuttering of speech.

Mercury occurs in coals at concentrations of less than 1 part per million. In most coals the concentrations are lower than 0.2 ppm (Valković 1983, 157). The literature suggests that most or all of the mercury is present as a sulfide. This so-called inorganic mercury is found with sulfide minerals: pyrite, the zinc sulfide mineral sphalerite, or sometimes as its own sulfide, cinnabar. Mercury can show both inorganic and organic associations. In some coals, mercury shows some affinity for associating partially with organic portion of the coal. When coals are

---

[7] The remaining eight are antimony, beryllium, cadmium, chromium, cobalt, manganese, nickel, and selenium.

[8] These additional elements include barium, boron, chlorine, fluorine, molybdenum, radon, thorium, uranium, and vanadium.

burned, mercury is considered to be fully emitted in the vapor phase, though a portion might be retained in the fly ash[9] (Swaine 1990, 134).

The portion of mercury that is associated with sulfide minerals, particularly pyrite, can be reduced during coal cleaning and never enter the combustion system in the first place. Mercury that happens to be in or on fly ash particles can be captured by particulate control equipment downstream of the boiler. The major problem is that most or all of the mercury will be in the vapor phase in the flue gases. A method is needed for mercury capture. Activated carbons can do that. The preferred approach for using activated carbon involves injection of powdered activated carbon[10] into the flue gases downstream of the scrubber but ahead of the precipitator or baghouse.

In the atmosphere, mercury occurs dominantly as gaseous elemental mercury (GEM). About 90% of the mercury in plants is in this form (Shotyk 2017). Mercury emitted from US coal-burning utilities represents about half of the total anthropogenic mercury emissions in the United States (EPA 2018). This is also about 1% of total worldwide anthropogenic mercury (EPA 2019). Coal combustion in all countries is responsible for about one-fifth of mercury emissions.

In the late 2000s, activated carbon producers began to commercialize mercury capture technologies based on their products (Miller 2017, 573–576). For about 5 years, activated carbon alone seemed to be on the way to being the premier technology for mercury capture. Then it was observed that activated carbon treated with bromine compounds offers improved performance (Reisch 2015). Bromine compounds oxidize mercury in flue gas, converting it to mercury bromides, which are then more easily captured. An additional benefit is that activated carbons treated with bromine can reduce the total need for carbon by about 80%, reducing cost and making material handling easier. Brominated activated carbons are more expensive than untreated carbons but, in at least some applications,[11] do a better job so overall are more cost-effective than the untreated carbons. Activated carbons made from coals, including anthracites, could be an attractive option for this application (Song et al. 2020). As appropriate for the specific combustion application, bromine can also be added upstream in the system.

[9] The reports that some mercury is retained *in* the fly ash don't rule out the possibility that vapor-phase mercury may have condensed *onto* fly ash particles as the flue gases cooled after leaving the boiler. The normal boiling point of elemental mercury is 356°C.
[10] The major forms of commercial activated carbons are *powdered* and *granular*. Granular and powdered activated carbons differ in terms of particle size. Various associations or societies have slightly different standards for discriminating between the two forms. As one example, material of which 95–100% passes a sieve with 0.177 mm openings is classified as powdered activated carbon, larger material as granular.
[11] Such as in plants firing low-rank coals of the western US.

Carbon not consumed in the combustion reaction can be emitted with other particulates. These carbon emissions, as particles of soot, are referred to as *black carbon.* The black carbon is a pollution problem because rain does not wash it out of the atmosphere. Consequently, it can be transported for long distances, traveling far from its original source. Coal combustion in modern power plant boilers is not a significant source of black carbon, possibly not more than 10% of total black carbon emissions. Black carbon can be produced in older and smaller boilers that do not do as efficient a job in completely burning the coal. Home heating systems are really small and inefficient coal combustion systems that sometimes are quite good at producing black carbon.

Countries with large numbers of coal-fired power plants that are not well equipped with emission control equipment often experience hazy skies. The haze is mainly due to fine particles of sulfate compounds. The sulfates originate from SO_x, which becomes oxidized to sulfates (SO_4^{-2}) in the atmosphere. It's suspected that black carbon, emitted with the SO_x, might act as a catalyst for this oxidation reaction, making the haze worse.

The overall layout of a modern coal-fired electricity-generating plant might look like Figure 9.1. The combination of scrubber and ESP or baghouse, together an appropriate NO_x control strategy, would remove close to 99% of potentially harmful pollutants.

Thanks to continued steady progress in both fundamental science and engineering development, the clean-up of "socks, knocks, and rocks" has been

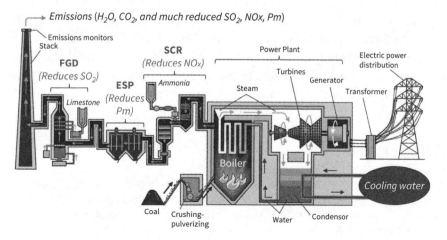

Figure 9.1 Overall layout of a power plant, from coal in to electricity out. Most of the NO_x, SO_x, and RO_x are captured in the equipment shown to the left of the boiler. Courtesy of Kentucky Geological Survey.

highly effective and continues to get better. Substantial strides are being made in technology aimed at the large-scale utilization of coal with very low emissions from the plant. In the time from the passage of the Clean Air Act in the United States (1970) through the year 2012, US gross domestic product grew by more than 200%. In the same time period, the emissions of NO_x, SO_x, and particulate matter decreased by an average of 72%, even with this substantial economic growth. Scrubbers, low-NO_x burners, and units for capturing ash can make it possible to create electricity-generating plants or synthetic fuel plants with very low emissions compared to today's technology. None of this technology is free. Building coal plants to have minimal impact on the environment adds to both the initial investment and to continuing operating costs. What is needed is the collective will to pay those costs in the form of an increased electric bill and the insistence that elected or appointed officials work to make sure that environmental issues are addressed with the best available technology. But, given that will, what else would we need to worry about? CO_2.

An overwhelming consensus in the scientific community agrees that CO_2 emissions and global climate change are directly linked and constitute an increasingly serious problem for humanity. Consensus also exists that human activities, including the burning of fuels, contribute significantly to the steady increase in atmospheric concentrations of CO_2. These points are one of the major issues that moves us away from today's heavy reliance on fossil fuels in general and on coal in particular. High-volatile bituminous coal produces 1.2 times more CO_2 per megajoule of heat released than does an equal weight of petroleum; it produces 1.9 times more CO_2 per megajoule than an equal weight of natural gas.[12] On the basis of CO_2 emissions per amount of heat energy released, coal inevitably comes off worst in a comparison among fossil fuels. Pressure to reduce or eliminate fossil fuel combustion is already most intensely applied to coal. For the coal industry, and for the ways that coal will be used in the future, global climate change is unquestionably a major "game-changer." This issue is important enough to merit its own chapter, Chapter 15.

References

Almaraz, M., E. Bai, C. Wang, J. Trousdell, S. Conley, et al. 2018. "Agriculture Is a Major Source of NO_x Pollution in California." *Science Advances* 4: doi: 10.1126/sciadv.aao3477.
Budavari, Susan, Maryadele J. O'Neil, Ana Smith, and Patricia E. Heckelman. 1989. *The Merck Index*. Rahway, NJ: Merck.

[12] These calculations are based on a Kentucky high-volatile B coal of 80.6% moisture-and-ash-free; a Mexican fuel oil of 84% carbon; and a natural gas of 97% methane content, methane being 75% carbon.

Dahiya, Sunil, and Lauri Myllyvirta. 2020. *Global SO₂ Emission Hotspot Database.* Delhi: Center for Research on Energy and Clean Air, and Greenpeace India.

Ding, Zhenhua, Baoshan Zheng, Jiangping Long, H. E. Belkin, Robert B. Finkelman, Chaogang Chen, Daixing Zhou, and Yunshu Zhou. 2001. "Geological and Geochemical Characteristics of High Arsenic Coals from Endemic Arsenosis Areas in Southwestern Guizhou Province, China." *Applied Geochemistry* 16: 1353–1360.

Downey, John. 2018. "Duke Energy Wrapping Up $65M Gas Co-Firing Project for Its Cliffside Coal Units." *Charlotte Business Journal*, November 19, 2018.

Dwivedi, Akash, and Manish K. Jain. 2014. "Fly Ash: Waste Management and Overview: A Review." *Recent Research in Science and Technology* 6: 30–35.

EPA. 2018. "Mercury Emissions." U.S. Environmental Protection Agency Environmental Topics. https://cfpub.epa.gov/roe/indicator.cfm?i=14.

EPA. 2019. "Mercury Emissions: The Global Context." U.S. Environmental Protection Agency Environmental Topics. https://www.epa.gov/international-cooperation/mercury-emissi ons-global-context.

Federal Highway Administration. 2017. "Fly Ash Facts for Highway Engineers." https://www. fhwa.dot.gov/pavement/recycling/fach03.cfm.

Frost, R. L., M. L. Weier, and W. Martens. 2005. "Thermal Decomposition of Jarosites of Potassium, Sodium and Lead." *Journal of Thermal Analysis and Calorimetry* 82: 115–118.

Hinckley, Eve-Lyn S., John T. Crawford, Habibollah Fakhraei, and Charles T. Driscoll. 2020. "A Shift in Sulfur-Cycle Manipulation from Atmospheric Emissions to Agricultural Additions." *Nature Geoscience* 13: 597–604.

Larson, Aaron. 2017. "Improved Emission Controls and State-of-the-Art Ash Handling Extend Gallatin's Life." *Power* 161(10): 24–26.

Maloney, K. L., P. K. Engel, and S. S. Cherry. 1978. "Sulfur Retention in Coal Ash." U.S. Environmental Protection Agency Report. No. EPA-600/7-78-153b.

McDuffie, Erin E., Steven J. Smith, Patrick O'Rourke, Kushal Tibrewal, Chandra Venkataraman, Eloise A. Marais, Bo Zheng, Monica Crippa, Michael Brauer, and Randall V. Martin. 2020. "A Global Anthropogenic Emission Inventory of Atmospheric Pollutants from Sector- and Fuel-Specific Sources (1970–2017): An Application of the Community Emissions Data System (CEDS)." *Earth System Science Data.* doi.org/10.5194/essd-2020-103.

Miller, Bruce G. 2017. *Clean Coal Engineering Technology.* Amsterdam: Butterworth-Heinemann.

NOAA. 2018. "How Hot is Lightning?" National Oceanic and Atmospheric Administration. https://www.weather.gov/safety/lightning-temperature.

Nordberg, Monica; and M. George Cherian. 2005. "Biological Responses of Elements." In: *Essentials of Medical Geology*, edited by Olle Selinus, Brian J. Alloway, José A. Centeno, Robert B. Finkelman, Ron Fuge, Ulf Lindh, and Pauline Smedley, 179–200. Amsterdam: Elsevier.

Overton, Thomas W. 2016. "Coal Ash Hits the Big Time." *Power* 160(3): 18.

Poetzscher, James. 2020. "Learn More about Atmospheric Gases." Greenhouse Maps. https:// greenhousemaps.com/learn/.

Reisch, Marc S. 2015. "Bromine Bails Out Big Power Plants." *Chemical and Engineering News* 93(11): 17–19.

Ritter, Stephen K. 2016. "A New Life for Coal Ash." *Chemical and Engineering News* 94(7): 10–14.

Shotyk, William. 2017. "Arctic Plants Take Up Mercury Vapour." *Nature* 547: 167–168.

Song, Guanrong; R. Deng, Z. Yao, H. Chen, Carlos Romero, Thomas Lowe, J. Gregory Driscoll, Boyd Kreglow, Harold H. Schobert, and Jonas Baltrusaitis. 2020. "Anthracite Coal-Based Activated Carbon for Elemental Hg Adsorption in Simulated Flue Gas: Preparation and Evaluation." *Fuel* 275: 117921.

Swaine, Dalway J. 1990. *Trace Elements in Coal.* London: Butterworths: London.

Valković, Vlado. 1983. *Trace Elements in Coal*. Boca Raton, FL: CRC Press.

Weagle, C. L., G. Sinder, C. Li, A. van Donkelaar, S. Philip et al. 2018. "Global Sources of Fine Particulate Matter: Interpretation of PM$_{2.5}$ Chemical Composition Observed by SPARTAN Using a Global Chemical Transport Model." *Environmental Science and Technology* 52: 11670–11681.

Zhong, Qirui, Huizhong Shen, Xiao Yun, Yilin Chen, Yu'ang Ren, Haoran Xu, Guofeng Shen, Wei Du, Jing Meng, Wei Li, Jianmin Ma, and Shu Tao. 2020. "Global Sulfur Dioxide Emissions and the Driving Forces." *Environmental Science and Technology* 54: 6508–6517.

10
Coke

By far the most important use of coal is to use the heat from its combustion to raise steam for electricity generation. Second is the use of coal in metallurgy, most notably in producing coke to be used in smelting of iron ores. Along with its alloys, the steels, iron is one of the most versatile materials available to engineers. It is relatively easy to form into shapes, strong and durable in use, somewhat resistant to corrosion. Some metals can be extracted from their ores more easily but lack some of the desirable engineering properties of iron. Others are superior in some applications but are more difficult to obtain from their ores.

Extraordinary changes swept through Western society from about 1760 to 1840. Machinery was increasingly substituted for hand-work performed by humans. Heavy work once done by humans or draft animals was replaced by more powerful and faster steam engines, increasingly centralized into factories or mills, rather than being done on a small scale in homes. New processes were developed for producing iron and steel. These great changes have come to be known as the Industrial Revolution. What made the Industrial Revolution possible was inexpensive, large-scale production of iron. That was in turn made possible by the availability of large deposits of high-quality coal, first in Great Britain and soon after in Belgium, Germany, and the United States.

Iron, the fourth-most abundant element on Earth, occurs in chemical combination in many minerals. The important ores of iron are its oxides or carbonates: hematite, magnetite, limonite, taconite, and siderite. To obtain metallic iron, the ore must be *smelted* by being heated with a material that will react with and remove the chemically combined oxygen to liberate the metal. Such materials act as reducing agents. Various forms of carbon are the cheapest and easiest reducing agents to use in producing iron.

The utility of carbon as a reducing agent for making iron comes from a curious quirk of the stability of chemical compounds toward heat. As the temperature is raised, oxides of metals become increasingly less stable with respect to their constituent elements. But carbon monoxide becomes increasingly stable with increasing temperature. The implication of the difference in stability with increasing temperature is that any metal oxide could be reduced with carbon provided that the oxide-plus-carbon mixture could be heated hot enough to pass the temperature at which carbon monoxide becomes more stable than the metal oxide. In practice, not all metal oxides can be reduced this way, because the metal reacts with carbon to form stable carbides. Titanium is an example.

Using charcoal to smelt iron ore had been established by about 1000 BCE. Even in the seventeenth century, Britain's increasing industrialization made serious demands on available supplies of wood for making charcoal. Making enough charcoal to sustain a growing iron industry consumed prodigious quantities of wood, up to 3 hectares of trees per tonne of iron produced. But wood was also vitally needed for ship-building. Cutting forests for wood to feed the growing industries of ship-building and iron-making became so severe that the amount of available wood sharply constrained the expansion of the iron industry. Even in the 1600s, production of iron—miniscule amounts by today's standards—could not be sustained on charcoal. A modern industrial society based on charcoal-smelted iron would be impossible. Though iron ore was abundant, iron-making could not expand until something was found to replace charcoal.

Iron-makers turned to bituminous coals as a substitute. Direct use of bituminous coals to smelt iron ore gives a poor grade of iron. Heating coals in an iron-making furnace begins to break apart their molecular frameworks, just as in the volatile matter test. As the structure breaks apart, sulfur forms hydrogen sulfide, which reacts with iron to precipitate iron sulfide at the boundaries between individual grains of iron that form as the molten iron solidifies. When the iron piece is reheated to be worked into a useful shape, it can break or tear along grain boundaries that have been weakened by accumulated iron sulfide. This converts a potentially useful item into scrap.

The shortage of charcoal also affected the brewing industry. Brewers were looking for an alternative heat source to dry hops and malt. A similar problem occurred: hydrogen sulfide contaminated the hops and malt, ruining the flavor of the beer. Heating wood without letting it burn completely produces a better fuel—charcoal. What not the same with coal? The problem might be solved by heating coal just enough to drive off the undesirable gases and then using the residual solid as fuel. Heating some bituminous coals in the absence of air produces a hard, porous solid of high carbon content, a material that we now call *coke*. Iron-makers began to experiment with coke as a possible fuel and reducing agent for their furnaces.

One was Abraham Darby, who established a brass foundry in Bristol. Being able to use iron instead of brass appealed to metalworkers because ironware could be much cheaper, so would sell more widely than comparable articles of brass. Darby moved his operations to the Shropshire coal fields near the town of Coalbrookdale. His prior connection with the brewing industry may have been what led him to consider the possibility of using coke in iron-making. Darby first operated his iron furnace on "charked coal" in January 1709. That year, the furnace produced about 80 tonnes of iron.

Compared to charcoal smelting, using coke cost less and used less fuel per tonne of iron produced. Even so, decades passed before coke smelting became fully accepted. In the early 1700s, iron made with coke was brittle and hard to shape, partly because coke needs a greater oxygen supply to burn fiercely than does charcoal. Coke is more difficult to ignite and to get to burn vigorously compared to coal or charcoal. The iron furnaces of the time provided an insufficient amount of air. Successful use of coke in an iron furnace depends on supplying a veritable blast of air to the lower sections of the furnace. Darby and his son developed a taller furnace with the necessary powerful blast of air. By the mid-eighteenth century, coke had become the standard fuel for iron smelting. Darby's development of the practical blast furnace represents a beginning step of the Industrial Revolution.

A *coke* is any carbon-rich solid produced at high temperature and that passed through a fluid state in its conversion from raw material to product. Carbonaceous solids that do not pass through an intermediate fluid state are called *chars*. These definitions apply regardless of the raw material used. Cokes made from coals for use in blast furnaces are often called **metallurgical cokes** (or "met coke") to distinguish them from other kinds of cokes, such as petroleum cokes. For our purposes we can use the word "coke" to mean metallurgical cokes. An example of coke is shown in Figure 10.1.

All coals leave a solid carbonaceous residue when heated in the absence of air. In the laboratory, the residual solid represents the fixed carbon and ash. For some coals, the residual solid seems to have fused or melted and then resolidified into a single piece. Coals that show this apparent fusing behavior are called *caking coals*. They occur in the bituminous rank range, caking being particularly pronounced with medium-volatile bituminous coals. Caking is not a true melting, which is a completely reversible process.

Metallurgical cokes must fulfill three roles when in use. Coke supplies heat for melting the metal and slag and for driving **endothermic** processes inside the furnace. It is the source of the primary agent, carbon monoxide, for reducing iron oxide. And, it has to have the mechanical strength to support the great weight of the ore-coke-limestone mixture—the **burden**—above it in the furnace while giving the bed permeability to let gases flow upward and

Figure 10.1 Metallurgical coke.
Courtesy of Wojmac/Shutterstock.com.

metal and slag to travel downward. Within the family of caking coals lies a subset of coals that produce a solid product having physical properties and chemical constitution ideal for use as coke in metallurgical uses. They are the *coking coals*.

Caking behavior can be assessed using a rather low-tech measurement, the *free swelling index* (FSI). It involves heating a measured quantity of coal under standardized conditions,[1] using a crucible of standardized size and shape. Caking coals form a single, solid carbonaceous piece usually called a *button*. The value of FSI can be assigned by comparing the size and shape of the button with a standardized chart. Figure 10.2 shows examples of low, medium, and high FSI results. FSI provides a good, relatively simple screening test for caking or coking behavior. Values range from 1 to 9; generally, a good coking coal has an FSI greater than 4. Coals that do not form a coherent button have an FSI of 0.

The best coking coals are low- and medium-volatile bituminous rank. They should have a low ash yield. A large amount of mineral matter in the coal interferes with the formation of coke by diluting the fluid that forms as the coal softens. In the blast furnace, the ash melts along with a limestone flux and any mineral impurities in the ore, forming slag. As the ash value of the coke

---

[1] To 820°C for 2.5 minutes.

Figure 10.2 Examples of the coke buttons produced in the free swelling index (FSI) test. An FSI of 1 is a coal that agglomerates without swelling, FSI of 4 is a moderately swelling coal, and FSI of 9 is a highly swelling coal.
Artwork by Lindsay Findley, from the author's sketch.

increases, more heat must be used—which really means wasted—for melting it, and larger amounts of slag have to be handled. Good coking coals should contain little sulfur and phosphorus to minimize these elements getting into the iron.

Coking behavior depends on how a coal softens into the fluid state, on the rate at which volatiles evolve while the coal is fluid, and on how the fluid mass eventually resolidifies into coke. Volatile matter content, ash yield, the size distribution of coal particles charged to the oven, and bulk density come into play as well (Mackowsky 1982). FSI by itself does not provide a complete characterization. Further evaluation involves measuring the fluidity of the plastic mass and the change in its volume that occurs as the coal passes through this stage.

A standard way of measuring fluidity is the *Gieseler test*. This device used in the test, called a *plastometer*, employs a stirrer immersed in a mass of the coal to be tested, pulverized to extremely small particle size.[2] How fast the stirrer can turn depends on the torque applied to the stirrer and how easily the sample permits the stirrer to turn (i.e., the fluidity of whatever liquid may form). The torque applied to the stirrer shaft is kept constant so that the speed of revolution of the stirrer indicates fluidity. As the test sample is heated, a temperature will be reached at which it becomes fluid enough to allow the stirrer to begin to turn. A dial attached to the shaft shows how fast the stirrer is turning. Because the torque on the stirrer remains constant, how fast it turns and how fast the dial spins depend on how fluid the sample is. Numerical fluidity data are expressed by noting the number of dial divisions passed per minute. A fluidity curve has three characteristic points. The temperature at which fluidity first becomes 20 dial divisions per minute (ddpm) determines the **softening temperature**, T_s. Continued heating causes fluidity to increase until it reaches a maximum; the temperature at which this occurs is the **temperature of maximum fluidity**, T_m. Heating beyond T_m causes the sample to

---

[2] Less than 425 microns.

become less and less fluid until the fluidity has decreased back to 20 ddpm: the *resolidification temperature*, T_r. The difference between T_s and T_r indicates the temperature region in which the fused mass displays fluid behavior, called the *plastic range*.

In many countries most of the good coking coals have already been mined and made into coke. Finding coals that have all of the desirable properties for producing good coke is becoming increasingly difficult. Medium- and low-volatile bituminous coals that make the best coking coals bring a premium price relative to ones burned for making steam in electric plants (so-called *steam coals* or *thermal coals*). Today, coke plants often blend several coals to obtain a coke of appropriate composition and properties. Regardless of how many coals comprise the blend, and regardless of the properties of the individual coals, blends have to be evaluated carefully and thoroughly in test ovens before they can be accepted for use in commercial-scale coke ovens.

Why is it that caking behavior is restricted to bituminous coals, and especially to the medium- and low-volatile bituminous coals? What causes a caking coal to turn from a solid into a fluid state, to remain fluid for a time as temperature increases, but then eventually to turn back into a solid as temperature goes up further?

A temperature around 350°C marks the beginning of the extensive breakdown of the macromolecular framework of the coal. Chapter 4 introduced the idea that most coals, other than anthracites, can be thought of as similar to crosslinked polymers,[3] also indicated by the "squiggles" in the Hirsch diagrams of coal structure. When coals begin thermal breakdown, crosslinks in the macromolecular structure of the coal come apart. The structure starts to fragment into smaller molecules. Some fragments escape as gases or vapors. Larger molecular products remain in a fluid state even at these elevated temperatures. Since the extent of crosslinking decreases as rank increases, medium- and low-volatile bituminous coals have the fewest crosslinks of any of the various ranks. Many caking coals show Gieseler softening temperatures not far above a point at which thermal decomposition could be expected to begin. Caking coals derive their characteristic behavior in part from the relatively low number of crosslinks.

The chemical bonds in the molecular frameworks of coals are formed by the sharing of a pair of electrons between the two atoms in the bond. When such bonds break, each fragment produced by bond-breaking now has one

---

[3] In the anthracites, there might be electronic interactions, not crosslinks, between the very large aromatic ring systems. In this respect, anthracites are the proverbial different breed of cat.

electron that is unpaired. Chemical species with unpaired electrons are called *radicals*. Radicals react readily and quickly to form products with new, stable, electron-pair covalent bonds. A carbon radical could be stabilized with hydrogen, as C· + ·H → C:H. A fragment of the coal macromolecule that has been "capped" with hydrogen may be small enough to be in the fluid state, at least for a while.

The only available hydrogen is the hydrogen atoms incorporated in coals. To illustrate, we use the simple molecule 1,2,3,4-tetrahydronaphthalene, also called tetralin. It has four more hydrogen atoms than naphthalene. Naphthalene is fully aromatic. Tetralin contains one ring of carbon atoms fully saturated with hydrogen and a second, aromatic ring. It is the exemplar of the class of compounds called *hydroaromatics*. Hydroaromatic structures included in a larger molecule readily give up the "extra" hydrogen atoms to form the corresponding aromatic substance. Tetralin can give up, or donate, four hydrogen atoms to form naphthalene, a process that we can represent as tetralin → naphthalene + 4 H·. Here are the hydrogen atoms needed to react with carbon radicals.

Hydroaromatic units in the coal macromolecule provide hydrogen to stabilize radicals that come from breakdown of the coal structure. One part of the structure gives up hydrogen, which means that it is becoming relatively carbon-rich, while the stabilized products are gaining hydrogen (i.e., becoming relatively hydrogen-rich). We've seen this in the context of coalification, and we will see it again. The combined characteristics of few crosslinks but lots of hydroaromatic structures converge in the medium- and low-volatile bituminous coals.

T_s usually occurs not far above the temperature region in which thermal decomposition of most coals begins. Some crosslinks have started to break; radical fragments have been stabilized with H·, thus providing the beginnings of a fluid material. As temperature increases beyond T_s, fluidity increases also. As temperature goes up, more and more bonds are being broken and smaller, more fluid products are formed.

With increasing temperature, fluidity eventually hits a maximum, at T_m. The steadily ascending curve of increasing fluidity changes slope and now begins to decrease. On the "uphill" side, from T_s to T_m, bonds are breaking, radicals are forming, and the radicals are being stabilized by H·. The needed H· comes from hydroaromatic regions within the coal. But, sooner or later, we use up all the hydroaromatic hydrogen. There is not much more H· available.

With very limited H·, the dominant reactions will now likely involve carbon radicals reacting with each other, C· + ·C → C:C, a process sometimes called *radical recombination*. Carbon–carbon bonds start forming, and these will provide the framework for a highly carbon-rich solid. More and more carbon bonds lead to bigger and bigger molecular structures, harder to keep in the fluid state. How can we eke out the last bits of H· in this stage of the process? They could come from dehydrogenative polymerization (Figure 4.2), where aromatic structures coalesce into ever-larger ones, tearing some of the H· away from the edges of the aromatic rings. As this happens, the carbon-rich product will have large and growing aromatic systems. Although the time is much shorter and the temperature much higher, the reactions are not much different from those accompanying anthracite formation.

We observe the fluidity to decrease even as temperature continues to rise. Eventually, bigger and bigger aromatic-carbon-rich structures lead to a fluid too viscous to stir any longer—T_r. A final push of heating leads to a solid of very high carbon content, sometimes called *semicoke*. Heating semicoke to nominally 1,000°C causes structural rearrangements that lead to metallurgical coke. Similar chemical processes in coke formation and at the high-rank end of coalification should reasonably lead to products of similar composition. Metallurgical coke has about 96% carbon and 0.9% hydrogen; a meta-anthracite about 95% carbon and 0.6% hydrogen, on a moisture-and-ash-free (maf) basis (Newman et al. 1967).

Suppose there was a coal of unusually high hydrogen content, high enough that there would be a nearly continuous supply of H· in the coal that could be used for capping radicals. In such a case, perhaps the system would not run out of hydrogen, and T_r would not happen. Behavior akin to this has been observed by Shaoqing Wang and his colleagues at China University of Mining and Technology (Wang et al. 2015). A coal sample from the Mingshan mine in southern China, with 6.4% hydrogen (maf) goes through a maximum fluidity but never resolidifies in the Gieseler test. Even using coals of more typical hydrogen content, say ≈5% hydrogen (maf), if we made the effort to ensure adequate H· by supplying extra hydrogen ourselves, and continued to supply hydrogen, we should be able to avoid T_m and convert most of the coal into liquid. The idea certainly works: it is the fundamental basis of the so-called *direct liquefaction of coal* (Chapter 13).

During heating, vapors of volatiles cause it to swell. As the viscous fluid begins to resolidify, tracks left by volatiles passing through the fluid do not refill with more fluid but remain as pores in the eventual solid coke. A porous structure facilitates reactivity of the coke when it is used in the blast furnace. Above T_m, dehydrogenative polymerization reactions form ever-larger aromatic structures in the liquid. Beyond T_r, such molecules will try to align into vertical stacks, which help form strong, rigid solid structures. A wide plastic range provides ample opportunity for evolution of vapors and for the growth and alignment of large aromatic molecules. Both processes result in desirable coke properties.

Over the course of the past century the amount of coke required to produce one tonne of iron—called the *coke rate*—decreased from about 1.1 tonnes of coke per tonne of iron to 0.35 in the very best cases (Wozek 2013). Making a tonne of iron today requires only one-third the amount of coke (and thus one-third the amount of coking coal) than the industry practice of a century ago. This is due to continual evolutionary improvements in blast furnace design and operation.

The demand for iron products, especially steel, will continue growing well into the future. Growth in steel many not translate into a related increase in demand for coking coals. Similar to continual improvements in reducing the heat rate in power plant boilers, reductions in the coke rate for blast furnaces will have a toll on coal demand. The rise of alternative technologies for iron ore smelting and steel production could reduce the need for coals even with rising demand for steel.

In the seventeenth and eighteenth centuries it seemed reasonable to think that it should be possible to convert coals to coke in the same way as converting wood to charcoal. This process for coal involved creating a mound of coal, piled so that there were passages through the mound which could be filled with wood. Igniting the wood generated heat that carbonized the coal; more heat came from burning the volatiles driven out of the coal. As the heat spread through the whole mound, the coal gradually converted to coke. Eventually, the hot coke was smothered with dirt. The mound process actually produced usable coke. Yields were low because some of the coal had to be burned to provide the heat to carbonize the rest. Concepts such as process control, quality control of the product, and environmental protection hadn't even been thought of in those days. There was no way of controlling the course of the process. There was no way to assure a uniform extent of coking throughout the mound so that the coke quality could be consistent. Temperature and the time to completion were at the mercy of the prevailing weather. Unburned tar was sticky, smelly, and contained compounds now recognized as suspect or proven carcinogens.

Enclosing the mound of carbonizing coal inside some kind of structure allowed coking to be done at consistent times and temperatures, controlled until all of the mass of coal had coked to about the same extent and tars or other vapors could be dealt with. The mass of coal was enclosed within a dome-like structures made of a heat-resistant material, looking like beehives, and thus giving us the name *beehive coke oven*. The beehive process made about 675 kilograms of coke per tonne of coal.

Beehive coke ovens served from the mid-nineteenth century to the beginning of the twentieth century. The fact that they burn the volatiles as a source of heat led to their demise, as it became increasingly apparent that the volatiles offered a cornucopia of useful, valuable chemical raw materials. The process for making coke evolved into one in which the volatiles could be captured and used rather than burned inside the oven. Consider what was learned almost two centuries ago: naphtha, derived from coal tar, will dissolve natural rubber, providing a way to cement two layers of cloth fabric into a rubberized cloth that also happened to be waterproof—the raincoat. An attempt to produce synthetic quinine from a compound found in coal tar yielded instead a material of beautiful purple shade, mauve, which quickly became popular as a synthetic dye. Phenol from coal tar, when used to treat a patient's skin, a surgeon's hands and instruments, and the air in the operating room, resulted in an extraordinary decrease in deaths from the infection of incisions or compound fractures. Solvent, synthetic dye, antiseptic—all from coal tar. Why burn this fruitful raw material just to heat the inside of a beehive oven?

A coke oven to capture these chemical materials from coals would have to be heated by some method other than burning the volatiles escaping from the coal and be able collect and condense them. In the late nineteenth century several inventors, notably Heinrich Koppers, developed approaches to collecting the byproducts of carbonization. Variations of their ideas evolved into the *byproduct recovery coke oven*. The byproduct ovens brought dramatic changes to the coke industry. By 1930, more than 12,000 byproduct ovens were operated in the United States alone, turning out about 95% of the coke produced in the United States.

A typical byproduct coke oven is a chamber 3–6 meters high, 10–15 meters long but only about 50 centimeters wide. Heat for the coking process comes from burning gas in flues between adjacent chambers. Heating occurs from the outside of the oven inward. The very narrow width of the ovens is because of the poor heat conduction of coals. Wider ovens would be difficult to heat through to the center. Numerous individual ovens—anywhere from 15 to a 100—are built side-by-side in a formation called a *battery*. A typical charge of coal ranges from 10 to 30 tonnes, depending on the size of the oven.

Heat from the hot walls slowly penetrates the mass of coal. In the oven, coal particles lying against the heated walls will be the first to soften, decompose, and resolidify to coke. The gradual transfer of heat from the outside-in establishes three zones inside the oven: a layer of coke closest to the walls, a layer of plastic coal undergoing coking, and an innermost layer of coal that has not yet reached coking temperature so that, for the moment, it remains unreacted. With time, the plastic layers move inward from the walls toward the center of the oven, where they eventually coalesce. Resolidification to coke causes a crack to form in the solid mass in the middle of the oven. Because of this, the largest coke pieces from a byproduct oven are just about half the width of the oven. The process ends when the temperature at the center of the oven becomes about equal to that at the walls and the entire charge of coal has been converted to coke. This takes some 15–20 hours. When coking has finished, a hydraulic pusher shoves the coke out of the oven. A spray of water cools the coke to keep it from catching fire. The quenched coke is moved to a wharf from which it can be taken for use.

All of the byproducts are gases or vapors piped out of the oven. Since no air is admitted, no byproducts are lost by burning. Gases leaving the oven pass through a spray of water that causes a rapid temperature drop, which causes condensation of coal tar. Though the tar can seem rather noxious, it once provided the raw materials for thousands of dyes, medicines, explosives, solvents, flavorings, and perfumes.

The byproduct recovery oven yields more coke per tonne of coal than the beehive oven, up to 75%, whereas the beehive oven converts about 66%. This difference might seem minor, but it gives the byproduct recovery oven an appreciable advantage when that small percentage difference is applied against the enormous tonnage of coal converted to coke each year.

A modern byproduct coke oven yields about 750 kilograms of coke, 300 cubic meters of gas, 30–40 liters of tar, 12–16 liters of light oil (mainly benzene, toluene, and xylenes), and 12 kilograms of other chemicals per tonne of coal. The exact amounts of these products depend on the compositions of the coals charged to the oven. The amount of "water" production—actually a solution of ammonia and small organic molecules in water—depends on the moisture content of the charge. Coke oven gas has high concentrations of hydrogen, carbon monoxide, and methane, which give it a high heat of combustion. Its primary use is to be burned in the flues between ovens in a battery. If there is more than enough gas for heating the ovens, it will be used elsewhere in the plant, such as for preheating the air blast to the blast furnaces or in the furnaces used for heat-treating finished steel. In the days when many cities had large distribution networks for synthetic fuel gases, coke oven gas was sold for off-site for home heating or illumination.

The demise of the beehive oven, certainly capable of making good coke, came from the recognition that byproducts could be the source of valuable chemical products. Today the coke industry is undergoing another transition. The role of chemicals from byproduct coke ovens, once the foundation of the organic chemical industry, has been taken over almost completely by chemicals from petroleum products. A byproduct recovery coke oven battery, with all its attendant operations for capturing and separating byproducts, can be the source of many emissions, some with unpleasant odors and others that are suspect carcinogens. In most industrialized nations, it's doubtful whether it would now be possible to obtain the necessary environmental permits to build a new byproduct recovery battery.

The modern approach to coke-making without producing and emitting volatiles is the *heat-recovery coke oven*, sometimes called *non-recovery oven*. Its name derives from the fact that burning the volatiles inside the oven recovers the heat released from their combustion. A typical oven might measure about 4 meters wide, 5 meters deep, and 3–4 meters high, quite a contrast to the byproduct recovery oven. An individual oven can accommodate 5–10 tonnes of coal per charge.

Ports in the oven door admit air into the oven. The limited air supply allows only partial combustion of the volatiles in the space above the coal charge. The gases, which contain combustible material, pass into flues beneath the floor of the oven, also called the *sole*. The remaining gases burn in the sole. The charge heats from two directions—from the top down, by partially burning volatiles inside the oven, and from the bottom up, as the previously unburned volatiles burn in the sole. Regulating the amount of air admitted to the oven and the flues equalizes the temperature between top and bottom of the coal charge.

It takes 2–3 days for complete coking of one charge of coal. The coke yield is slightly lower than from a byproduct recovery oven (Quanci 2011). When the last of the combustible volatiles has been burned in the sole, the gas in the flues could be as hot as 1,400°C (Sarna 2019). This heat can be recovered to make steam for use inside the coke plant or adjacent steel mill or used to drive a turbine to make electricity. Capturing and utilizing the heat in the exhaust gases gives it its name: heat-recovery coke oven.

The burning that occurs in the oven and in the sole destroys most of the potential emissions while at the same time supplying the heat needed to make the coke. Destroying potential emissions eliminates the expense of downstream equipment to treat tars, gases, and wastewater. Even the water used to quench the hot coke after the oven has been pushed goes to a holding pond to be recycled rather than being discharged to the environment.

A *blast furnace* is a vertical reactor in which four ingredients—iron ore, flux, coke, and air—react to produce three products: hot metal, slag, and gas. Air supplies the "blast" for the blast furnace. Of the four furnace ingredients, the one used in greatest quantity by weight is air. It's easy to overlook because coke, flux and ore are relatively dense solids, with clearly palpable masses, whereas

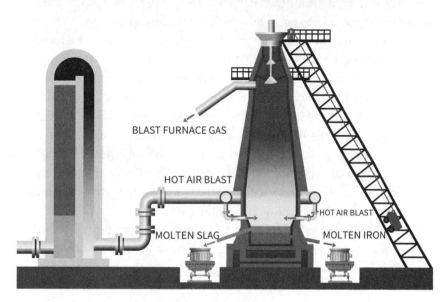

MECHANICAL EQUIPMENT OF METALLURGICAL PLANTS

BLAST FURNACE GAS

HOT AIR BLAST

HOT AIR BLAST

MOLTEN SLAG

MOLTEN IRON

BLAST FURNACE

Figure 10.3 The layout of a blast furnace operation. The cylindrical structure on the left is a stove used for heating the air blast being supplied to the furnace.
Courtesy of Newgena/123RF.COM.

air feels like nothing. Before it enters the furnace, the air blast is preheated by burning the gas leaving the top of the blast furnace, supplemented with coke oven gas if needed. A diagram is shown in Figure 10.3.

The air blast flows upward while the fuel, ore, and flux move downward. Limestone as a flux reacts with impurities in the ore and keeps them out of the metal by forming a relatively low-melting slag that is liquid at the temperatures inside the furnace. The ingredients needed to produce 1 tonne of iron are about 1.8 tonnes of iron ore, 600 kilograms of coke, and 250 kilograms of limestone.

Smelting of iron ore is by far the dominant application of metallurgical coke. A modern blast furnace can be more than 30 meters tall; contain some 3,200 tonnes of fuel, ore, and flux at a time; and can produce more than 9,000 tonnes of hot metal per day. Though the basic concept and design have not changed very much in 200 years, blast furnaces have evolved to a point at which a modern furnace will produce more hot metal in 2 days than some furnaces of a century ago would produce in a year.

As coke descends through the furnace, it experiences abrasion from furnace walls and from the other materials in the charge. Coke pieces must retain most of their size to keep passageways through the charge open so that the air blast and combustion gases can pass through and then exit the furnace. As coke approaches the bottom of the furnace, it must support the many tonnes of burden above it. High strength is even more important than porosity. Coke must resist crushing because maintaining a good flow of gases through the burden requires that the size of the coke pieces must be consistent. High carbon (generally 85–90%) or fixed carbon content gives a high calorific value. As calorific value increases, the amount of coke needed to achieve a desired heat output decreases. Day-to-day consistency—minimal variation—in coke properties allows furnace operators to reproduce the temperatures in the furnace and maintain the quality of the iron.

In the blast furnace, coke burns to carbon dioxide. But carbon dioxide undergoes a second reaction when it comes into contact with hot carbon (i.e., coke). This is the *Boudouard reaction*, another of the key reactions of coal technology:

$$CO_2 + C \leftrightharpoons 2\,CO$$

The double arrow in this equation indicates that the reaction can proceed in either direction: that is, it is an **equilibrium reaction**. Octave Boudouard, a French chemist, discovered this reaction in the early years of the twentieth century while studying chemical processes in blast furnaces.

Carbon monoxide from the Boudouard reaction is the critical reactant in the blast furnace. It is converted back to carbon dioxide by reacting with iron oxides in the ore. Because of the importance of the Boudouard reaction, the reactivity toward carbon dioxide is an important characteristic of cokes. As the coke is consumed, removal of some of the coke by reaction weakens the remaining solid. The change of coke strength as a result of reaction is an important piece of information for characterizing cokes and for blast-furnace operations.

Because of different designs and different operating conditions in blast furnaces, a specific set of values for specifications of ideal blast furnace coke does not exist. The recipe for a blast furnace charge can vary from one furnace to another because of differences in the quality and composition of the iron ore. Porosity, strength, size, carbon content, and consistency of quality are the most important properties. Porosity allows the coke to be permeated readily by the air blast, providing the rapid combustion desirable for establishing a high temperature in the bottom of the blast furnace for melting the iron and slag.

Again we see the issue of consistency, in this case in coke properties. Users of coal or coal products need to be able to design and operate plants, regardless of their purpose, with the confidence that their raw material will be the same—within set limits—for long periods of time. Providing this confidence starts at the preparation plant.

The blast furnace produces molten iron, referred to as *hot metal*. The name is apt: hot metal could be above 1,500°C. In an integrated steel mill that takes iron ore as input and sends finished steel products to market, hot metal would not be allowed to solidify but sent downstream immediately to steel-producing operations. Hot metal is converted to steel by reducing the level of impurities, especially carbon, and sometimes adding alloying agents such as nickel or chromium. Before the development of large integrated steel mills, hot metal from the blast furnace was allowed to solidify to a product called *pig iron*. The name derives from pouring the molten metal into a system of molds that looked like a sow nursing a litter of piglets.

Because more than 90% of coke is used to make hot metal, and more than 90% of hot metal is used to make steel, the coke industry and the steel industry are inextricably linked. If steel demand decreases or if steelworkers strike, mines producing coking coal suffer because of a lack of demand. If miners strike or mines close due to natural disasters, loss of supplies of coking coal could force the steel industry to look to coal imports or curtail production. Changes for the good or bad in the steel industry will cause significant effects on that part of the bituminous coal industry that mines and sells coking coal. A major change comes from increasing reliance on remelting of existing steel scrap for the production of new steel. This approach, called *secondary steelmaking*, relies on electric arc furnaces (Jones, Bowman and Lefrank 2012). It eliminates iron-ore smelting, blast furnaces, and almost all need for coke.

The carbon content of scrap steel used in secondary steelmaking will be lower than that of the hot metal from a blast furnace. Because downstream finishing operations that produce the final, marketable steel products remove some carbon from the metal, a source of carbon is added to the molten scrap, usually while it is still in the ladle. This ensures that the carbon content of the molten product will be higher than the desired level in the finished steel. Current rates of carbon addition are less than 15 kilograms per tonne of molten metal produced, much smaller than the coke rate of 0.356 tonnes per tonne of hot metal in a blast furnace.

The 10% of the coke not used in blast furnaces finds various applications. *Foundry coke* is the fuel used in iron foundries. In foundries, furnaces called *cupolas* are used to melt iron for casting. A cupola somewhat resembles a small version of a blast furnace, but no smelting is done in a cupola. Iron already exists in the metallic state—it just needs to be melted. The cupola charge consists of a mixture of coke, pig iron, scrap metal, and sometimes alloying agents.

Coke pieces sized about 12–20 millimeters are known as *coke breeze*. Only a small amount of coke breeze forms, roughly 50 kilograms of coke breeze per tonne of coal coked. The difference between blast furnace coke and coke breeze is in particle size. It has the same resistance to handling, high value, and low sulfur content as blast furnace coke. These properties make it a useful fuel. Coke breeze has been used to sinter iron ore into pellets for feeding to blast furnaces and as a boiler fuel for generating steam.

References

Jones, J. A. T., B. Bowman, and P. A. Lefrank. 2012. "Electric Furnace Steelmaking." In: *The Making, Shaping and Treating of Steel—Steelmaking and Refining Volume*, edited by Richard J Fruehan, Warrendale, PA: Association for Iron and Steel Technology.

Mackowsky, M. T. 1982. "The Application of Coal Petrography in Technical Processes." In: *Stach's Textbook of Coal Petrology*, edited by E. Stach, M. T. Mackowsky, M. Teichmüller, G. H. Taylor, D. Chandra, and R. Teichmüller, 413–476. Berlin: GebrüderBorntraeger.

Newman, L. L., W. A. Leech, M. H. Mawhinney, C. R. Velzy, C. O. Velzy, A. J. Tigges, H. Karlsson, and W. E. Lewis. 1967. "Fuels and Furnaces." In: *Standard Handbook for Mechanical Engineers*, edited by Theodore Baumeister and Lionel S. Marks, 702–717, New York: McGraw-Hill.

Quanci, John F. 2011. "Recent Trends in Heat-Recovery Cokemaking Processes." http://www.abmbrasil.com.br/cim/download/RecentTrendsinHeatRecoveryCokemaking_JohnQuanci.pdf.

Sarna, Satyendra K. 2019. "Carbonization of Coal in Heat Recovery Coke Oven Battery." Ispat Guru. https://www.ispatguru.com/carbonization-of-coal-in-heat-recovery-coke-oven-battery/.

Wang, Shaoqing, Yuegang Tang, Harold H. Schobert, Di Jiang, Yibo Sun, Yanan Guo, Yufei Su, and Shuopeng Yang. 2015. "Application and Thermal Properties of Hydrogen-Rich Bark Coal." *Fuel* 162: 121–127.

Wozek, Jeff, 2013. "The Sun Coke Energy Perspective." www.thecoalinstitute.org/ckfinder/userfiles/files/JWozekSunCokeNCCISpring2013.pdf.

11
Gasification

Coals are good fuels. A high-volatile A bituminous coal might have a heating value of about 32 megajoules per kilogram (MJ/kg) on an as-received basis, substantially better than an excellent fuel wood, such as oak, at 13 MJ/kg (dry basis), but not so good as petroleum fuels such as gasoline (44 MJ/kg). Coal occurs abundantly in many industrialized and developing nations, including China, India, Russia, the United States, South Africa, and Australia. This and the next two chapters discuss processes for conversion of coals into synthetic gaseous and liquid fuels, collectively called *synfuels*. Synfuels production adds more processing steps to the utilization of coal, which invariably means additional investment costs for equipment and additional ongoing costs for running the plant. Converting coal to a synthetic fuel will result in more expensive energy in terms of price per megajoule than simply burning the coal in the first place.

The industrialized world has an incalculable investment in existing devices that operate using liquid or gaseous fuels. They cannot be easily nor economically replaced. The world does not have enough manufacturing capacity to replace this vast array of devices in a reasonable period of time. Synthetic fuels from coal are an option to petroleum or gas if shortages of such fuels occur. Whether coal-based synfuels are the *best* option is a different issue, depending on what resources are available in a particular country or region and on what factors drive the need to consider alternatives to petroleum and gas.

Coal contains impurities undesirable in combustion processes, and that cause problems if emitted to the environment. They can be reduced or even eliminated when coal is being converted to liquid or gas. Synfuels burn more cleanly than coal, being of near-zero sulfur content and having no ash yield. Carbon dioxide (CO_2) capture can be done much more easily from the concentrated CO_2 streams encountered in synfuels processes than from the dilute flue gas of relatively low CO_2 concentrations produced during coal combustion. Coals are solids, difficult and laborious to handle. Gases or liquids can be handled easily in pipes, tanks, pumps, and valves. Synthetic fuels offer much greater ease and convenience of handling compared to solid coals.

Conversion of coals to synthetic gaseous fuels includes the processes that comprise the field of *coal gasification*. An important aspect of gasification lies in the versatility of what can be done with the primary product. Gaseous products can be burned directly as fuels. Coal-based *integrated gasification combined cycle* (IGCC) plants for electricity generation were discussed in Chapter 8. Products can be converted to methane (substitute natural gas); to methanol, which can be used directly as a fuel or converted further to gasoline; or to a variety of other liquid hydrocarbon fuels or chemical products.

It is easy to make at least some gas from coals simply by applying heat. As the molecular framework breaks apart, some of the fragments that come off are small enough to be gases. The gas can be used for various applications. This very simple technology produces an acceptable fuel. But the yield of gas is only a small percentage of the amount of coal being used, several hundred cubic meters per tonne of coal. The gas will be accompanied by some amount of oil and tar and a large amount of char. An enterprise established for the purpose of selling gas will find that it has a very large amount of waste material on its hands unless some applications can be found for the char and tar.

The ideal would be to convert the maximum possible amount of a coal into gas, leaving little or no char, oil, or tar. Making *water gas* offers a way to do this.[1] It takes advantage of the carbon–steam reaction:

$$C + H_2O \rightarrow CO + H_2.$$

This is the third of the principal reactions in coal technology. Water gas was a commercially viable product for decades in the nineteenth and into the early twentieth centuries. Its importance faded as electricity and petroleum and natural gas became increasingly available.

The challenge of making water gas is that the carbon–steam reaction is endothermic. Keeping the carbon–steam reaction running steadily requires that heat be applied to the reactor continuously. In an operation that already uses large quantities of coal, the most straightforward way to supply the necessary heat is to burn some of the coal. Burning takes advantage of the carbon–oxygen reaction

$$C + O_2 \rightarrow CO_2.$$

---

[1] The early gas industry developed a remarkable variety of names for its products: water gas, producer gas, town gas, illuminating gas, coal gas, and many more. These terms are not always used consistently in the literature. Fortunately, that doesn't matter much, because all of these products are largely obsolete.

This reaction is exothermic, making it possible to use the heat supplied by the carbon–oxygen reaction to drive the carbon–steam reaction and keep it going. It would be easy to heat the reactor by creating a coal fire underneath or around it. Only a fraction of the heat generated is likely to be effective in heating the reactor and its contents. Much of the heat will just heat up the surroundings.

A major step forward was the recognition that both the carbon–steam and the carbon–oxygen reactions could be run simultaneously inside the reaction vessel, the basis of the coal *gasifier*. By monitoring the temperature inside the gasifier and the rate of gas production, an operator can adjust the ratio of steam to air so that only enough coal is consumed in the carbon–oxygen reaction to ensure that the carbon–steam reaction is proceeding without interruption and at a desired rate. The gas product will be a mixture of carbon monoxide and hydrogen, both from the carbon–steam reaction; CO_2 from the carbon–oxygen reaction; and nitrogen, from the air. It can be a fairly good fuel, about 24 megajoules per cubic meter (MJ/m^3). With provision to feed coal into the gasifier and withdraw ash on a continual basis, here is a strategy that will produce gas for weeks or months on end. Two further steps will get us to the essence of modern gasifiers.

First, nitrogen, coming in with the air, serves no useful purpose. It adds nothing to the heat of combustion of the gas. Nitrogen has no role in chemical reactions of the gas. Nitrogen increases the total volume of gas, which means that the gasifier and all the ancillary piping and other equipment have to be larger and inevitably more expensive than if there were no need to accommodate the nitrogen. Compressing air to feed to the gasifier, or compressing the product gases for downstream processing, wastes considerable energy in compressing useless nitrogen that will just be returned to atmospheric pressure later.

We would be better off getting rid of the nitrogen by applying commercially available air-separation technology. An air-separation unit upstream of the gasifier would provide pure oxygen for feeding to the gasifier. Gasifiers that operate in this way are called *oxygen-blown* gasifiers. Similarly, gasifiers relying on air are said to be *air-blown*. A tradeoff exists between the costs of an air-separation unit and the benefits of working with nitrogen-free gas. There is a very clear consensus on this tradeoff: oxygen-blown gasifiers are today's state of the art.

The second issue concerns operating at elevated pressures. Operating at elevated pressure offers several advantages. For gasifiers of the same size, higher pressure provides a greater rate of coal processed, producing more marketable product per day. If the amount of gas needed is fixed, the necessary production can be met by using smaller, possibly less expensive, gasifiers. Gas to be fed into a pipeline for distribution must be compressed to the pressure needed—some 35–95 bars—in the pipeline. If the gas has already been produced at high pressure, it should be possible to reduce the expense and ongoing energy requirement for compressors needed to raise the gas pressure to the pressure of the pipeline. Use of high pressures favors formation of some methane. If the intended eventual product is methane, more methane coming directly from the gasifier reduces the burden on downstream processes. The advantages of pressurized operation must be balanced against the greater cost of a pressurized gasifier. The pressurized unit must be made much stronger to contain the pressure. Pressurized operation also has problems of how to get coal into, and ash out of, a vessel that is operating continuously at high pressure. Nonetheless, as with oxygen- versus air-blowing, the commercial market has spoken: all modern gasifiers are pressurized units.

Two approaches address the question of getting coal into an elevated-pressure gasifier without interrupting its operation. One is much like the system of air locks used on space vehicles. These so-called *lock hoppers* typically attach to the top of the gasifier and have two seals: one to isolate the lock hopper from the gasifier and one to isolate it from the ambient air. In operation, the lock hopper will be sealed from the atmosphere and open to the gasifier. The hopper will be at the same pressure as the gasifier. Coal feeds from the hopper into the gasifier. When that batch of coal is close to being depleted, the seal between the gasifier and the hopper is closed so that the pressure in the hopper can be let down to normal atmospheric pressure. The seal between hopper and air is opened, and a new batch of coal is put into the hopper. The hopper is sealed again and repressurized back to the pressure of the gasifier, allowing the seal between hopper and gasifier to be opened. Coal again begins to feed into the gasifier. This cycle of depressurizing, filling, and repressurizing the lock hopper works smoothly and reliably.

The second approach uses a slurry of finely ground coal in water. Although the coal particles are still solid, the slurry behaves as a liquid. This allows the slurry to be pumped to the operating pressure of the gasifier, to be injected directly into the gasifier. The mechanical complexity of the cycling of lock hoppers is eliminated in favor of a mixing tank to prepare the slurry and a pump.

All coal gasifiers are continuous-flow reactors. Gasification processes can be classified on the basis of the method used to bring the coal into contact with the gasifying medium (i.e., the steam–oxygen mixture). Fixed-bed processes use a vertical bed of coal supported on a grate, with the steam–oxygen mixture injected into the bottom. The gaseous products and the solid coal move in opposite directions. Gases rise through the bed while the coal descends as it is consumed at the bottom. The term "fixed-bed" is slightly misleading. What is "fixed" in a fixed-bed gasifier is the height of the coal bed. A specific piece of coal steadily descends through the gasifier until being consumed at the bottom. Fluidized-bed gasification resembles fluid-bed combustion (Chapter 8). The key difference comes from gasification using a mixture of air or oxygen and steam. Rather than relying on fluidization, finely pulverized coal is suspended in the medium, and then the suspension can be blown through the gasifier. An alternative involves slurrying fine coal in water and pumping the slurry into the gasifier. These operations are known as *entrained-flow gasifiers*.

Regardless of gasifier design, coal, steam, and oxygen are brought together to react. The crucial reactions are ones we have met before: the carbon–steam reaction and the carbon–oxygen reaction:

$$C + H_2O \rightarrow CO + H_2$$

$$C + O_2 \rightarrow CO_2.$$

With CO_2 from the carbon–oxygen reaction coming into contact with hot carbon (i.e., coal), there is also the possibility of the Boudouard reaction, which we saw in connection with coke reactions in blast furnaces:

$$CO_2 + C \rightleftharpoons 2\,CO.$$

In this high-pressure, high-temperature mixture of hot carbon, steam, oxygen, carbon monoxide, and CO_2, two additional reactions can occur, adding to the list of principal reactions in coal technology. One is the carbon–hydrogen reaction:

$$C + 2\,H_2 \rightleftharpoons CH_4.$$

The other is a reaction among the components of the gas phase, called the *water-gas shift* reaction:

$$H_2O + CO \rightleftharpoons H_2 + CO_2.$$

The water–gas shift is remarkably useful, as we will see in the next chapter.

These five reactions occur simultaneously inside a gasifier. The composition of the gaseous product leaving the gasifier depends on the balance among them. Some are exothermic; others are endothermic. The balance depends on the steam–oxygen ratio, temperature, and pressure. And, of course, the balance also depends on the specific coal put into the gasifier. High pressures will tend to suppress the Boudouard reaction but enhance the carbon–hydrogen reaction. These two effects are direct consequences of a principle that governs the behavior of systems at equilibrium, one developed by the French chemist Henri Louis Le Chatelier. *Le Chatelier's Principle* states that if a chemical system in equilibrium undergoes a change in concentration, temperature, or total pressure, the equilibrium will shift in such a way as to minimize that change. In the Boudouard reaction, there are more moles of gas on the right-hand side than on the left. An increase in pressure favors having less gas in the system, increasing CO_2 at the expense of CO. In contrast, the carbon–hydrogen reaction has more moles of gas on the left-hand side. In this case, increased pressure favors producing more methane.

If the temperature is kept below the initial deformation temperature, ash will remain a solid and can be removed as such. Removing solid ash from the bottom of a gasifier can be done via an ash lock hopper, which works just like the coal lock hopper. Gasifiers operating in this way are known as *dry-bottom*, or sometimes *dry-ash*, gasifiers. The alternative involves running the gasifier above the fluid temperature of the ash. Ash melts to form a slag, which can be removed from the reaction zone in its molten state. Such units are known as *slagging gasifiers*, or *slaggers*. Usually, the molten slag is quenched in a water bath, and then the solidified slag is removed via a lock hopper.

Once the design choice has been made of dry-bottom versus slagging operation, there is no flexibility. In a dry-bottom unit, operators keep the temperature below the initial deformation temperature by increasing the ratio of steam to oxygen, thus enhancing the endothermic carbon-steam reaction. Conversely, in a slagger, operators maintain adequately high temperatures by reducing the steam–oxygen ratio, with more oxygen favoring the exothermic carbon–oxygen reaction. In either case, if running slag appears in a dry-bottom unit or if slag solidifies inside a slagger there is a big problem.

The fixed-bed coal gasifier has enjoyed the most commercial success to date, though the design is about ninety years old.[2] All fixed-bed gasifiers in

---

[2] Many operating gasifiers worldwide are not actually *coal* gasifiers—they operate on such feedstocks as petroleum distillation residues, petroleum coke, or asphalt.

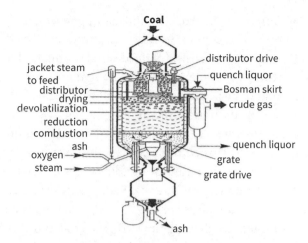

Figure 11.1 A diagram of a modern fixed-bed, dry bottom gasifier.
Courtesy of Professor John Bunt, North-West University.

commercial operation remove the ash as a solid and are known as *fixed-bed dry-bottom* (FBDB) gasifiers. FBDB gasifiers typically operate at 30–40 bars. A diagram of a modern FBDB is given in Figure 11.1.

Coal entering an FBDB gasifier is heated by hot gases ascending from the grate region. As the temperature of the coal rises, moisture is driven off. Dried coal descends further. Its temperature continues to increase and the coal begins to decompose, causing a wide variety of compounds to be vaporized and leave the gasifier with the gas stream. The remaining char descends still further, into a region of active gasification where the carbon–steam and carbon–oxygen reactions occur. The bottom-most zone of the gasifier is the combustion zone, where the principal reaction is the carbon–oxygen reaction supplying heat for the endothermic carbon–steam reaction. Some hydrogen reacts with hot char to produce methane.

Oxygen reacts rapidly with the hot char to produce CO_2. This reaction provides the heat needed for the carbon–steam reaction and the Boudouard reaction occurring higher in the bed. The char is consumed entirely. Hot ash drops through the grate into the ash lock, allowing it to be removed. The residence time of a given coal particle is approximately an hour, depending on the reactivity of the coal, temperature and pressure in the gasifier, and the steam–oxygen ratio. The cleaned gas composition depends on the specific coal being used and on operating conditions. It would likely be around 50% hydrogen, 35% carbon monoxide, and 15% methane (Schobert 1990, 276).

Moisture and organic vapors condense downstream of the gasifier, forming an aqueous solution of water-soluble products of thermal decomposition of the coal, such as ammonia, hydrogen sulfide, and phenols; and highly complex mixtures of organic compounds, called tars and oils. Oils float on water and tars sink. The gas stream consists of the desired products along with some CO_2, ammonia, hydrogen sulfide, and water vapor. It may contain particles of partially reacted coal or char. Treatment steps downstream of the gasifier are necessary to remove the undesirable components.

FBDB units can't operate well on mildly caking coals and not at on those that are strongly caking. This eliminates many bituminous coals as potential feedstocks. FBDB gasifiers cannot handle coal smaller than about 0.6 millimeter in size (Berkowitz 1979, 257). Such small particles could settle into the interstices between larger pieces of coal, reducing the permeability of the coal bed to the gases flowing upward. The coal and its char must be strong enough not to be crushed by the weight of the coal above it in the bed, like the issue of coke strength in a blast furnace. This limitation could be resolved by co-locating a gasification facility and an electricity-generating station using pulverized-coal–fired boilers. Coal too fine to be gasified can be sent to the power plant, and coal too coarse to be burned goes to the gasification plant.

Moisture content of the coal is not usually a problem. Lignites, which have high as-mined moisture contents, can be run successfully. The ash value is often not a significant limitation either. An undeniable advantage of the FBDB units is their long history of successful operation at commercial scale. Gasifiers that are descendants of original designs dating from the 1930s are in use in synthetic fuel plants in South Africa and the United States.

The most successful fixed-bed gasifier is the Sasol-Lurgi gasifier. The gas produced is mostly a mixture of hydrogen and carbon monoxide, which is used as synthesis gas or is converted to methane. The basic components are a coal lock, a water-cooled pressurized reaction vessel, and an ash lock. Although the basic design is now about 90 years old, it is still used in commercial-scale synthetic fuel plants, including the Sasol plants in South Africa and the Dakota Gasification plant in the United States. Various size units are built, but a typical gasifier without the coal and ash locks is 4 meters in diameter and about 8 meters high. Including the coal and ash locks doubles the overall height. A gasifier of this size can convert as much as 45 tonnes of coal to about 45,000 cubic meters of gas per hour.

Entrained-flow operation involves coal and the steam–oxygen mixture flowing together through the gasifier. The coal must be ground to sizes smaller than 0.1 millimeter to entrain in the gas stream and react completely. Both upward and downward flow designs have been developed. Entrained-flow gasifiers operate as slaggers, well above the ash fusibility temperatures of most coals. Very short residence times, on the order of a few seconds, mean that the coal does not have time to experience thermal breakdown. Byproduct tars do not form. Very short residence times also mean that coal particles do not have time to pass through the plastic state associated with caking behavior. Use of caking coals is not a problem in an entrained-flow gasifier.

Entrained-flow gasifiers can, in principle, use any coal as feedstock. Because of the high reaction temperatures, coals of any rank and virtually any degree of reactivity can be gasified. Even highly caking coals or anthracites of low reactivity can be accommodated. (This does not mean that one specific entrained-flow gasifier can be switched from running on one kind of coal to a second, very different coal.) Lack of tar formation means that a plant using entrained-flow gasifiers does not need provisions for handling byproduct tars. Slag is quenched in a water bath before being removed from the gasifier. Quenching produces a glassy material, often in the form of a grit.

The General Electric Energy gasifier consists of two main sections, one for the gasification reactions and the other for quenching the slag. Preheated and pressurized slurry and a separate stream of preheated oxygen are blown into the gasifier. Because of the very short residence time, there is no production of byproduct tars. Slag is solidified in the quench section and removed via an ash lock. Temperatures are controlled to be above the ash fusibility temperature and high enough to keep the viscosity of the slag low enough that it will drain easily.

Entrained-flow gasifiers are used by Eastman Chemicals to produce synthesis gas for chemical manufacture. About 40 General Electric gasifiers running on coal are in commercial use worldwide; more than 70 more additional units use either oil or natural gas. The General Electric gasifier is also used in combined-cycle plants for electricity generation. The gas has other applications, such as ammonia production for fertilizers and making methanol for the chemical industry. A few of these gasifiers, mainly in China, make town gas[3] as a commercial fuel gas, seemingly returning to the bygone days of the beginnings of coal gasification.

The gasification reactions do not require that the coal actually has to be out of the ground. Gasifying coal in its seam would completely eliminate mining, preparation, and transportation, along with their attendant problems. Underground coal gasification (UCG) provides a fourth option to gasifier reactor design.

---

[3] Primarily a mixture of carbon monoxide and hydrogen, usually also containing some methane or other light gaseous hydrocarbons.

UCG requires at least two wells to be bored into the seam: an injection well for supplying air and possibly steam and a production well for removing the gas produced. These wells must be linked, either by taking advantage of the natural porosity and permeability of the coal or through a connecting channel bored between the two wells. Injecting air starts the combustion reaction. As combustion proceeds, the still-unreacted coal lying ahead of the combustion front will be heated enough to allow the reaction of carbon in the coal with CO_2 produced by the combustion, producing carbon monoxide. The carbon–steam reaction also occurs with coals of high moisture content or if steam is injected with air.

Coal is consumed laterally away from the injection well toward the production well. As coal is consumed, a cavity forms that eventually exposes the roof rock to the gasification area. The exposure of increasingly more roof rock results in decreasing quality of the gas. The operation is ended when gas quality has declined to a point where it is no longer acceptable.

UCG is potentially less expensive than above-ground gasification because it eliminates mining and transportation (Fallows 2010). There is no need for coal crushing and handling equipment, gasifiers themselves, or downstream clean-up equipment. UCG could, in principle, be applied to conversion of coal in seams that either are too deep or that pitch at too great an angle to be mined at all. UCG also eliminates most of the temporary environmental degradation associated with coal mining. The gas produced from UCG could be cleaned and used in the various applications discussed later.

UCG generally relies on air-blown operation, so that the product gas is highly diluted with nitrogen. Its heat of combustion is low, about 6 MJ/m^3, comparable to producer gas and well below the value for natural gas. Using oxygen instead of air raises the combustion temperature and improves the gas by increasing the yield of methane and decreasing CO_2 (Pearce 2016). If there is an extensive system of fractures in the coal seam and in the overburden, it might be easier for gas to migrate through the natural fractures than to move laterally through the coal seam. Then methane and CO_2—both of which are greenhouse gases—would be emitted directly into the atmosphere. The potential contamination of groundwater must be considered.

Above-ground gasifiers can be of standardized designs, with only slight modifications to accommodate the characteristics of the specific coal being used. UCG installations are more likely to be one-of-a-kind designs, customized to fit the local geological conditions. Groundwater could become contaminated by organic compounds formed during coal pyrolysis and by various inorganic compounds liberated as the minerals in the coal convert to ash. These contaminants create problems for environmental quality, human health, or both. Groundwater could also percolate into the gasification zone, reducing the temperature and possibly stopping the process. Water-related issues are of particular concern when the UCG system is emplaced below the natural water table. Subsidence of the ground above could sever the link between production and injection wells and allow gas to escape to the surface.

The cavity created as the coal is gasified could possibly be used for sequestration of CO_2 (Pearce 2016). With appropriate advance planning, a pipeline system that delivers the gas to users could be configured to accommodate return of CO_2 for sequestration. The ability to provide on-site CO_2 sequestration might prove to be another selling point for UCG.

Currently the most successful underground coal gasification operation is in Uzbekistan (Fei 2016). The city of Angren, in eastern Uzbekistan, receives electricity from a plant running on gas from a UCG project near Tashkent (Pearce 2016). Other UCG projects seem to be in various stages—announced, on the drawing boards, or in some scale of testing—in at least a dozen countries around the world. China is thought to have the potential of producing three times more substitute natural gas from UCG than conventional natural gas (Zou et al. 2019). Other coal-producing countries, such as South Africa and Australia, have also shown interest.

Depending on gasifier design and operating conditions, the gas will contain various amounts of CO_2, water vapor, ammonia, gaseous compounds of sulfur, condensable tars and oils, and fine particles of unreacted coal or of ash. Most or all of these will have to be removed before the gas is put to its final use. The gas treatment and clean-up section of a coal gasification facility fits between gasification upstream and whatever is to be done with the gas downstream. Because there are four types of gasifiers (counting UCG) and numerous potential applications of the gas, there is no unique choice of gas clean-up operations. Here we will see some examples of common approaches to gas treatment.

One of the first things that can be done in treatment of the raw gas is to remove fine particles of ash or partially reacted coal. This can be accomplished using a cyclone separator.

In fixed-bed gasification, cooling the gas or quenching it with a water spray produces a condensate of organic tars and oils and an aqueous liquor of ammonia and water-soluble organic compounds. Oil and tar are separated on the basis of their specific gravities: oil floats, tar sinks. Then, dissolved organic compounds are extracted from the water by a solvent. These operations produce marketable byproducts. Phenols can be sold for chemical use, as can some components of the tars and oils. Alternatively, tars and oils can be burned on site as a fuel to help produce steam. Since the tars and oils are going to be produced anyway, they are, in a sense, free fuel.

The next step in water treatment removes ammonia. Ammonia has important industrial applications, especially in fertilizer production. Other uses include production of synthetic fibers and the ever-popular explosives. Recovery and sale of byproduct ammonia produces revenue that helps the overall economics. As a last step, water will be sent to the plant's cooling towers, similar to those discussed in Chapter 8 and serving the same purpose, thus reducing thermal pollution.

After removal of the condensable water, oil, and tar, the gas still contains CO_2, various sulfur compounds, and some small organic compounds that remain as vapors. Taken together, hydrogen sulfide and CO_2 are known as *acid gases*. The acid gases can be absorbed into a liquid, either by simple physical dissolution or by some chemical reaction.

Dissolution is the most important technique for acid gas removal. An example involves contacting the gas with cold methanol, known as the Rectisol process. It takes advantage of the facts that any gas will be more soluble in a liquid at high pressure and less soluble at higher temperatures. Hydrogen sulfide, CO_2, and some related compounds all readily dissolve into cold methanol. Their solubilities in methanol increase with as the temperature decreases. In contrast, the desirable products—carbon monoxide, hydrogen, and methane—are relatively insoluble in methanol, and their solubilities do not change much with temperature. Taking advantage of these differences, extremely cold methanol, at about −40°C, does a good job of dissolving acid gas from a gas stream. The Rectisol absorber operates at high pressure and low temperature. The resulting solution of acid gases in cold methanol is then sent to a regenerator, where the pressure is brought down and methanol is reclaimed.

The Claus process converts hydrogen sulfide to elemental sulfur. It relies on two reactions. A portion of the hydrogen sulfide—as might be recovered from the Rectisol process, for example—is burned, producing water vapor and sulfur dioxide as products. In the second step, the sulfur dioxide is reacted with the rest of the hydrogen sulfide. This reaction produces elemental sulfur and more water vapor. Sulfur can be a very valuable byproduct of a coal gasification facility. Its most important application is in the manufacture of sulfuric acid, which has a huge array of uses: in the electrolyte in automobile batteries; in making fertilizers, fungicides, and pesticides; and pharmaceuticals.

In the past, CO_2 has been allowed to escape to the atmosphere. We are now seeing increasing demands worldwide for CO_2 capture, storage and sequestration to address concerns about global climate change. Gasification can provide a stream of gas—as might come from the Rectisol process—that is highly concentrated in CO_2. In any separation process, the higher the concentration of the stuff we want to remove, the easier it will be to do the job.

The main components of the gasification product, hydrogen and carbon monoxide, burn easily with very good heats of combustion, each about 12 MJ/m^3. Decades ago, many countries had significant infrastructure for making and distributing fuel gases from coal. Until syntheses of liquid fuels from carbon monoxide and H_2 were developed in the early part of the twentieth century, the major purpose of converting coal to gas was to make gas for heating and illumination. That market has been taken over almost completely by natural gas. In localized applications, small-scale gasifiers can be used to generate gas that would be useful on site for generating process heat or raising steam in factories or industrial parks.

Integrated gasification combined-cycle plants offer a relatively new application for coal gasification. Entrained-flow gasifiers are used. An IGCC plant has a much higher net efficiency (Miller 2017, 299) than do pulverized-coal–fired power plants. This efficiency gain can be viewed as a way to generate more electricity from a given quantity of coal or to consume less coal—and produce less CO_2—to meet a given demand for electricity. Gas treatment operations upstream of the combustion turbine remove many potential pollutants. CO_2 separation can be integrated into an IGCC system, offering the potential of capturing and sequestering CO_2 (LePree 2013). In the future, it may be that the most important ultimate product of a coal gasifier is electricity rather than gas. A diagram of such an operation is given in Figure 11.2.

In times when supplies of natural gas are constrained or in areas lacking adequate supplies or distribution infrastructure for natural gas, using coal to produce a substitute natural gas (SNG) could be an attractive energy option. Most natural gas supplies, as delivered to the customer, contain 95% or more of methane. Production of SNG involves *methanation*, the synthesis of methane. Methane synthesis requires three molecules of hydrogen for every molecule of carbon monoxide (i.e., a hydrogen to carbon monoxide ratio of 3). But the carbon–steam reaction, the critical chemical step in gasification, doesn't do this for us. It produces a hydrogen-to-carbon monoxide ratio of 1. Fortunately, the mismatch in ratios is relatively easy to rectify via the water–gas shift reaction.

"Shifting" the cleaned synthesis gas from the gasifier adjusts the amounts of hydrogen and carbon monoxide to the values needed for methanation. In this case, the water–gas shift is run deliberately as a separate process step rather

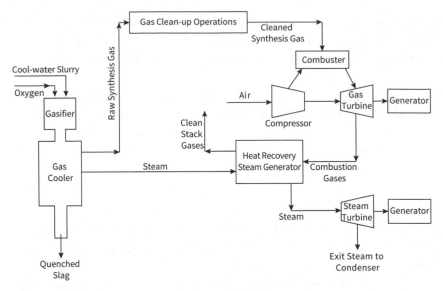

Figure 11.2 The basic layout of an integrated gasification combined cycle power plant. The cleaned gas is fired in a turbine that operates the generator on the upper right. The hot exhaust gases from that turbine are used to produce steam for a second turbine that operates the generator on the lower right.

Artwork by Lindsay Findley, from the author's sketch.

than simply having it occur as one of several reactions inside the gasifier. Another benefit is that hydrogen and CO_2 are relatively easy to separate from each other. Removing CO_2 allows the H_2 to increase the hydrogen-to-carbon monoxide ratio to the desired value of 3. Taking the CO_2 from the water–gas shift unit provides a highly concentrated stream of CO_2 that is much more amenable for CO_2 capture operations.

The largest plant for SNG production, in Beulah, North Dakota, uses lignite, with fixed-bed gasifiers to make the synthesis gas. About 20,000 tonnes of lignite come into the plant per day. It is separated by size so that larger pieces are retained for the gasifiers and smaller pieces are sent next door to a power plant. About 820 tonnes of ash are produced each day, eventually returned to the mine for disposal in the pit. The 400,000 cubic meters of SNG are cooled, dried, and compressed to 100 bars for feeding into gas distribution pipelines.

Fixed-bed gasifiers produce a variety of substances in addition to the desired gas. These byproducts can be converted into additional revenue. Originally, the tars, oils, and phenols were burned as the heat source for the steam boiler in the plant. Now phenol is sold as a chemical product, and the tars and oils are treated to produce a marketable fuel oil. Some of the plant's gas is now used in the boiler. Dakota Gasification's sales have been divided almost evenly between SNG, for which the plant was built in the first place, and various chemical products. They include phenol, of which a major use is production to plastics; ammonia and ammonium sulfate, for fertilizers; and naphtha, used in refineries. Sulfur-containing compounds produce ammonium sulfate, an excellent fertilizer, sold under the name of Dak Sul 45 (Dakota Gasification 2021). A granulated urea fertilizer plant on site has been running since 2018 (Maize 2018). The project draws on anhydrous ammonia currently made from synthetic gas when gas prices are low. Urea synthesis also relies on methane, readily available in an SNG plant (Maize 2018). CO_2 from the Rectisol process is captured and shipped via pipeline about 330 kilometers to the region of Weyburn, Saskatchewan. There, the gas is pumped into the ground to achieve enhanced recovery of oil. The process also sequesters the CO_2. About 4.5 million cubic meters of CO_2 are handled every day.

It is possible to shift the gas composition to remove all of the carbon monoxide. Then, when the CO_2 is removed, the product is pure hydrogen. Hydrogen is a very attractive fuel, especially for producing electricity in fuel cells, because the only product of consuming hydrogen is water. Currently, hydrogen production is dominated by hydrogen made from natural gas. It remains to be seen whether coal gasification will have a role in this market. The approach of using gasification combined with the water–gas shift is sometimes referred to as "making hydrogen from coal." This term is misleading. Most of the hydrogen produced by coal gasification is not actually *from* coal. Coals are a rather poor source of hydrogen. It is the carbon in coals that is important. Carbon serves as the reducing agent for splitting water molecules to hydrogen and oxygen. Oxygen promptly reacts with hot carbon to make carbon monoxide, which reacts with steam, ultimately producing CO_2. CO_2 can be separated from hydrogen, captured and sequestered, or be put to uses that do not involve emitting it to the environment.

Along with IGCC plants, the other major application of coal gasification is to produce a mixture of carbon monoxide and hydrogen called synthesis gas (or syngas). It is an exceptionally versatile reagent. Synthesis gas can be used to produce methanol and higher alcohols as well as hydrocarbons ranging from methane up to high-molecular-weight waxes. Production of commodity chemicals is also an important application (LaPree 2013). Since most of the hydrocarbon products and the methanol can be used as liquid fuels, they will be discussed in Chapter 12.

Methane is produced during the coalification process. It is formed both in diagenesis and later in catagenesis. Though the gases are chemically identical, the former is known as biogenic gas and the latter as thermogenic gas. The amount of biogenic gas produced can be about 34 cubic meters per tonne (m^3/t) of coalifying material (Murray 1991). From high-volatile bituminous rank to anthracite, the volume of thermogenic methane could be in excess of 250 m^3/t (Murray 1991). Much of this methane comes off first in the transition from low-volatile bituminous to semianthracite and then even more in the coalification from semianthracite into the higher ranks of anthracite (Francis 1961, 441).[4]

High concentrations of methane are produced and trapped in coal seams when coalification proceeds to a point at which the local geological temperatures and pressures allow gas to be expelled as a separate phase (Murray 1991). Because coals are porous, with extensive internal surface areas, high-rank coals can adsorb methane as well as the sandstone reservoirs in which natural gas is often found—sometimes even doing a better job than sandstone. Methane can also occur in cleats or other fractures in the seam and as gas dissolved in water in the coal. Methane adsorbed onto the coal accounts for most of the gas. Because methane is also a greenhouse gas, reducing its emissions from coal seams into the environment can qualify for carbon credits as part of the Clean Development Mechanism in the Kyoto Protocol (Pareek 2008).

Coal bed methane (CBM) has potential for augmenting supplies of natural gas—or indeed for providing gas supply to countries that are coal-rich but gas-poor. Global reserves of CBM are estimated to be equivalent to about 10% of conventional natural gas resources (Maize 2018). Large-scale CBM production has been implemented in several countries, including the United States, Canada, and Australia (Mastalerz and Drobniak 2014). Interest is increasing in numerous other countries, such as India and China (World Coal Institute 2017). CBM wells could produce for about 20–40 years (Murray 1991), roughly comparable to a shale gas well produced by fracking (Chapter 14). The surface of most coals has a greater affinity for CO_2 than for methane. Pumping CO_2 into a coal seam being used as a CBM source helps with gas production by displacing adsorbed methane from the coal. At the same time, it helps sequester CO_2.

---

[4] Estimated at about 80 cubic meters per tonne (m^3/t) for low-volume bituminous transitioning to semianthracite (88–92% carbon) and about 160 m^3/t in the anthracite range (92–95% carbon). These data have been converted from work originally published in English units.

References

Berkowitz, Norbert. 1979. *Introduction to Coal Technology*. New York: Academic Press.

Dakota Gasification Company. 2021. "Ammonium Sulfate (DakSul 45)." https://www.dakota gas.com/products/fertilizers/ammonium-sulfate-dak-sul-45

Fallows, James. 2010. "Dirty Coal, Clean Future." *The Atlantic* 306 (8): 64–78.

Fei, Mao. 2016. "Underground Coal Gasification (UCG): A New Trend of Supply-Side Economics of Fossil Fuels." *Natural Gas Industry B* 3: 312–322.

Francis, Wilfrid. 1961. *Coal: Its Formation and Composition*. London: Edward Arnold Publishers Ltd.

LePree, J. 2013. "Innovations in Gasification." *Chemical Engineering* 120(12): 24–30.

Maize, Kennedy. 2018. "Advanced Coal Technologies Improve Emissions and Efficiency." *Power* 162(11): 42–43.

Mastalerz, Maria, and Agnieska Drobniak. 2014. "Coalbed Methane; Reserves, Production and Future Outlook." In: *Future Energy*, edited by Trevor Letcher, 145–158. Amsterdam: Elsevier.

Miller, Bruce. G. 2017. *Clean Coal Engineering Technology*. Amsterdam: Butterworth-Heinemann.

Murray, D. Keith. 1991. "Coalbed Methane: Natural Gas Resources from Coal Seams." In: *Geology in Coal Resource Utilization*, edited by Douglas C. Peters, 97–104. Fairfax, VA: TechBooks.

Pareek, H. S. 2008. *Coal in India*. Bangalore: Geological Society of India.

Pearce, Fred. 2016. "Hello, Cool World." *New Scientist*, 229(3061): 30–33.

Schobert, Harold H. 1990. *The Chemistry of Hydrocarbon Fuels*. London: Butterworths.

World Coal Institute. 2007. *Coal Meeting the Climate Challenge*. World Coal Institute: Richmond, UK.

Zou, Caineng, Yanpeng Chen, Lingfeng Kong, Fenjin Sun, Shanshan Chen, and Zhen Dong. 2019. "Underground Coal Gasification and Its Strategic Significance to the Development of Natural Gas Industry in China." *Petroleum Exploration and Development* 46: 205–215.

12
Synthesis

Mixtures of carbon monoxide (CO) and hydrogen (H_2) are commonly known as synthesis gas, for the straightforward reason that the gas can be used, on an industrial scale, to synthesize many valuable fuel and chemical products. In this chapter we will see some examples.

After the raw gas from a coal gasifier has been cleaned, it may not be ready to be sent directly to synthesis processes. We know from the carbon–steam reaction that we should expect equal amounts of CO and hydrogen in the product:

$$C + H_2O \rightarrow CO + H_2.$$

The syntheses of substitute natural gas (i.e., methane) requires three molecules of hydrogen for each molecule of CO, an H_2:CO ratio of three. Methanol, a useful industrial chemical, solvent, and fuel, can be synthesized using an H_2:CO ratio of two.[1] The necessary H_2:CO ratios for these two processes are different from the value for cleaned synthesis gas, and they are also different from each other. It would be remarkably inefficient if we had to tie methanol synthesis exclusively to one specific gasifier design and coal supply and use a completely different gasifier and coal feedstock to make substitute natural gas. Fortunately, a solution comes from using the water–gas shift reaction.

The water–gas shift reaction is represented by the equation:

$$CO + H_2O \leftrightarrows CO_2 + H_2.$$

The double arrows signifying that the water–gas shift is an equilibrium process turns out to be good news indeed. If the temperature of the reaction is kept higher than 100°C, so that steam does not condense, it is easy to shift the reaction in one direction or the other by employing Le Chatelier's Principle. If

---

[1] The relevant chemical equations are $CO + 3H_2 \rightarrow CH_4 + H_2O$ for methane synthesis, and $CO + 2H_2 \rightarrow CH_3OH$ for methanol.

we consider a reactor that was at equilibrium and then injected some carbon dioxide (CO_2), the equilibrium would shift to the left, producing more CO and steam, consuming some of the CO_2. Alternatively, if we inject steam, the equilibrium will shift in the other direction, consuming some of the added steam. Le Chatelier's Principle gives us great flexibility in directing reactions to favor the formation of products that we specifically want.

In the examples of methane and methanol production, it is necessary to increase the proportion of hydrogen in the synthesis gas. By injecting steam into the gas, the equilibrium will shift to use up some of the added steam, shifting to the right-hand side to increase the concentrations of CO_2 and hydrogen. Not only does this change increase the amount of desired hydrogen, it also provides an opportunity to capture CO_2 from a gas stream that will be rich in CO_2, an easier job than capturing it from a dilute stream of combustion products.

As long as water is kept in the gas phase, the number of moles of gas is the same on both sides of the water–gas shift equation. Because of this, the position of equilibrium is not affected by pressure. This is helpful because it means that the pressure in the shift reactor can be whatever value is best in the context of the overall process. The water–gas shift reactor can be operated either at the pressure of the incoming cleaned synthesis gas or at the pressure needed downstream in whatever synthesis reaction is being conducted. There is no need to compress the gas or reduce its pressure before it goes to the shift reactor. Then there is also no need for the additional capital equipment and maintenance costs for compressors or pressure let-down equipment.

Changes in temperature or composition will affect the equilibrium of the water–gas shift. Referring to the equation as it is written above, removing water vapor by condensing it to liquid or absorbing it with a dehydrating agent will shift the reaction to the left (i.e., to try to make more water vapor). Trapping CO_2 will cause a shift to the right, to favor making more CO_2. The reaction that produces $CO_2 + H_2$ is mildly exothermic; reducing temperature favors a shift to the right, while increasing temperature will move the equilibrium to the left.

The importance of Le Chatelier's Principle as it applies to the water–gas shift reaction is this: we can start with a synthesis gas of whatever hydrogen-to-carbon ratio comes out of the gasifier and shift the composition of the gas to whatever ratio is needed for the intended application. In principle, we can make any gas composition from pure hydrogen to pure CO or any ratio in between. It doesn't matter that there might be variations in the H_2:CO ratio of the raw gas produced from coals of different compositions. In fact, it doesn't matter if we use coal at all. Natural gas, biomass, petroleum fractions, or waste materials containing carbon compounds can all be subjected to the carbon–steam reaction—though not all in the same piece of equipment—and produce synthesis gas. Once the gas has been shifted, the source of the CO becomes irrelevant.

In its broadest sense, the term *liquefaction* refers to conversion of a gas or a solid into a liquid. Processes for making synthetic liquid fuels from coal fall into two camps. In *indirect liquefaction*, coal is first gasified to produce synthesis gas; then, as a separate process step, synthesis gas is converted to liquid hydrocarbons. The indirectness comes because the liquid products are formed from synthesis gas rather than from coal itself. Two distinct steps are involved: gasification followed by synthesis. The world's largest complex for producing synthetic liquid fuels from coal, the Sasol Synfuels plant in South Africa, is based on indirect liquefaction. In *direct liquefaction*, coal is reacted to produce a liquid fuel directly from the coal without the intermediate gasification step. Direct liquefaction is discussed in Chapter 13.

In indirect liquefaction, coal contributes the carbon that is in the CO and a small amount of the hydrogen to synthesis gas. Indirect liquefaction of coal starts with gasification. The raw gas is cleaned and shifted to get its hydrogen-to-CO ratio to the value needed for what happens next.

All coal liquefaction processes, indirect or direct, intend to produce petroleum-like liquids as replacements for gasoline, jet fuel, diesel fuel, and fuel oil. Liquid fuels made from petroleum have about 1.8 hydrogen atoms for every carbon atom in their molecules. Most coals have about 0.8 hydrogen atoms on average for each carbon atom. The essence of coal liquefaction is this: How do we get from 0.8 to 1.8 hydrogen-to-carbon ratio and do it as efficiently and cheaply as possible?

The most extensively studied and used indirect liquefaction method is the Fischer-Tropsch (F-T) process, developed in Germany in the 1920s by Franz Fischer and his colleague Hans Tropsch. The F-T process reacts hydrogen with CO in the presence of a catalyst. The generic synthesis reaction is

$$nCO + (2n+1)H_2 \rightarrow C_nH_{2n+2} + nH_2O.$$

In this equation, n can have, in principle, any value. If n is 1, the product is methane. For values of n between 5 and about 10, the product is gasoline. The hydrogen and CO are in the cleaned synthesis gas obtained from coal gasification. The products are determined by the H_2:CO ratio after the gas has been shifted, as well as by the temperature, pressure, and type of catalyst selected for the synthesis reaction. Depending on the conditions selected, the products can include hydrocarbons that could be useful liquid fuels, as well as a wide variety of chemical products. F-T synthesis dominates the possible applications of synthesis gas made from coal and represents the major route to synthetic liquid fuels from coal. It so thoroughly leads the field of coal-to-liquids technology that some people seem to believe that "coal-to-liquids" *is* F-T synthesis.

Fischer at first sought to produce light hydrocarbons as a possible replacement for petroleum-derived gasoline. Hydrocarbons with straight chains of carbon atoms, the usual structures produced in the F-T process, have lower octane ratings than needed for today's internal combustion engines. Several approaches can fix this, including changing the synthesis conditions to produce higher-octane branched chains (Song and Sayari 1994) or treating the F-T products in petroleum refinery operations that convert straight chains to branched chains (Schobert 1990, 237–248).

Many variations of the basic process concept have been tested. Operating at high pressures produces hydrocarbon waxes. They can be converted into ethylene, the world's most important industrial organic chemical and the starting point for making a variety of polymers. When the ratio of hydrogen to CO is kept low and an unsaturated hydrocarbon is added with the synthesis gas, products containing oxygen can form.[2] *Oxygenates* have uses in their own right, as solvents, for example, or they can serve as intermediate materials in the production of many commercial products such as detergents.

In the F-T synthesis, the reaction between CO and hydrogen takes place on the surface of a catalyst. Making hydrocarbon products means that the carbon-to-oxygen bonds in the CO have to be broken to remove the oxygen atom and that the hydrogen-to-hydrogen bond in molecular H_2 must also be broken to supply hydrogen atoms to the carbon atoms. This is a problem because both of these bonds are very strong. Direct reactions between CO molecules and H_2 molecules would be unlikely. Catalysts are used to increase the rate of a chemical reaction, usually by providing an alternative chemical pathway for the desired reaction. Events that occur on the catalyst surface determine the products that form. Potential products range from molecules containing only one carbon atom, methane or methanol, all the way up to the wax-like molecules.[3]

---

[2] This approach to using synthesis gas is called the *Oxo synthesis*. The initial products of the reaction are aldehydes, which have some commercial uses. The aldehydes are then converted to alcohols. Alcohols containing about 12–18 carbon atoms are very useful in the synthesis of detergents. Smaller alcohols, with about eight carbon atoms, are used in the production of plasticizers.

[3] This does not mean that all possible products can be made with the same catalyst and conditions of temperature and pressure.

F-T synthesis starts with molecules of CO and hydrogen becoming adsorbed onto the surface of the catalyst (Schobert 2013, 381–389). On the catalyst surface, the H_2 molecule comes apart to produce single hydrogen atoms. The carbon atom attached to the catalyst surface reacts with hydrogen atoms to form a methylene ($-CH_2-$) group. Further reaction with one more hydrogen atom produces a methyl group ($-CH_3$). Then, one of two things can happen. If the methyl group reacts with a fourth hydrogen atom, the product will be methane, which will leave the surface of the catalyst and flow out of the reactor. This sequence of steps is illustrated in Figure 12.1. We met this methane synthesis in the previous chapter, for making substitute natural gas. To make synthetic, petroleum-like liquid fuels instead of gas, chains of carbon atoms have to form on the catalyst.

Making chains of carbon atoms on the catalyst surface requires that something else happen. Rather than forming methane, the methyl group can react with methylene, a reaction called *methylene insertion*, to form a chain of two carbon atoms, $-CH_2-CH_3$, called an *ethyl group*, on the catalyst. At this

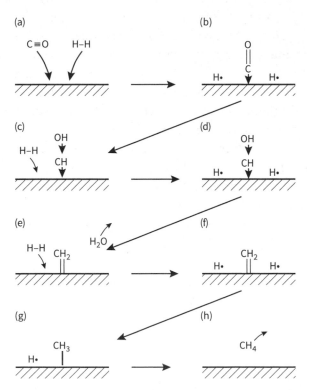

Figure 12.1 The sequence of steps by which carbon monoxide and hydrogen (upper left) react on a catalyst surface eventually to produce methane (lower right). Artwork by Lindsay Findley, from the author's sketch.

point, the ethyl group can experience two possible fates: it can react with a hydrogen atom to form a stable ethane molecule, CH_3—CH_3 or it can experience another methylene insertion to form a chain containing three carbon atoms, —CH_2—CH_2—CH_3, called a *propyl group*, on the catalyst surface. At each step of the way, the addition of one last hydrogen atom will form stable molecules and end the growth of the carbon chain. These stable molecules leave the catalyst surface and exit the reactor. Alternatively, yet another methylene insertion adds one more carbon atom to a growing chain. Repetition of methylene insertions builds increasingly longer chains of carbon atoms on the catalyst surface. Adding the final hydrogen atom to carbon-atom chains that have five or more carbon atoms leads to stable molecules that will be liquid at ordinary temperatures.

The reaction conditions, the hydrogen-to-CO ratio, and the specific catalyst determine how many repetitive methylene insertion reactions can occur. Fuel engineers can change any or all of these variables to modify the kinds of products that form. In every case, the objective is to achieve a balance between growth of the chain of carbon atoms by methylene insertion and formation of stable products by hydrogen-atom addition. The balance we are trying to establish is not unique to the F-T synthesis. This concern—growth versus termination—is just the same as that in a process for making polymers (e.g., polyethylene).

This balance has been worked out in the field of polymer science. It can be described in terms of the Anderson-Schulz-Flory distribution, named in part for the American chemist and Nobel Laureate Paul Flory, one of the leading polymer scientists of the twentieth century. The distribution is based on two assumptions. First, for a given set of reaction conditions the probability that a chain of carbon atoms will grow by methylene insertion, as opposed to its forming a stable compound by adding a hydrogen atom, is the same regardless of the length of the chain. Second, the amount of a compound having a given number—let's say six—of carbon atoms will be a fraction of the amount of the compound with one fewer carbon atom (i.e., five, in this example). The actual numerical value of the fraction of C_6 to C_5 compounds depends on the outcome of competition between methylene insertion and hydrogen atom addition. That outcome depends on the relative rates of these two processes. But those rates are established by the specific conditions of the reaction. No matter what reaction conditions are selected, the F-T reaction never results in a single, pure compound as product. Rather, it inevitably results in products having a distribution of molecular sizes.

The distribution of products represents a potential disadvantage of the F-T process. The maximum yield of synthetic gasoline, in which the dominant molecules have between 5 and 10 carbon atoms, will be about 45% (Rehnlund 2007). Changing the operating conditions, the catalyst, or both will change the product distribution, but, regardless of what we do, there will always be a balance between growth and termination. Consequently, separation and refining operations are needed downstream of the F-T synthesis reactor. Finding F-T catalysts that will produce narrower ranges of products represents a challenge for fuel researchers. F-T products might be distilled into fractions of various boiling ranges, corresponding to such commercial products as gasoline, diesel fuel, and heavier fuel oils, to make them fit for use in the engines or combustors.

The F-T process was a vital component of the German economy during World War II. Nine F-T plants produced about 800 million liters of fuel per year at their peak of production in 1943. They provided gasoline, kerosene (jet fuel), and diesel fuel. These plants gave Germany a route to liquid fuels to augment or replace the related fuels as made from petroleum. The fuels were particularly vital for armored vehicles and aircraft.

After the war, American and British scientific teams investigating German wartime technology became enthusiastic about the remarkable potential for using coal to produce synthetic fuels and chemicals via gasification combined with F-T synthesis. Subsequently the United States Bureau of Mines tested a very inexpensive catalyst, mill scale, a mixture of iron oxides and carbonates that forms on hot steel in rolling mills. The Bureau of Mines approach produced a synthetic gasoline with octane ratings[4] higher than 70, along with other compounds potentially useful in the chemical industry (Stranges 1997). Despite the promise of this effort, there is always the reality check provided by economics. Processes such as the F-T that were run in wartime conditions when economics was a secondary—probably minimal—consideration could not compete with inexpensive, abundant petroleum in the postwar world. West European countries could get motor fuels and chemicals more cheaply from Middle Eastern oil. In the United States, postwar development of oil fields in the southwest provided fuel at lower cost than liquids could be made from coal.

During the synfuels boom of 40 years ago, there was a thought that plants needed to be built, not necessarily because petroleum fuels were in short supply just then, but because a day might come when the plant infrastructure,

---

[4] Three scales are used to express octane rating: the research octane number (RON), the motor octane number (MON), and the anti-knock index (AKI), also known as the pump octane number. The RON is measured in a specially instrumented test engine under carefully controlled conditions. The MON is measured in the same way, except that the fuel is preheated, the ignition timing is varied, and the motor speed is higher. It is about 8–10 points lower than the RON. The AKI is the average of RON and MON.

know-how, and experienced personnel would be needed. Even if a plant site and coal supply had been selected and all of the necessary construction and environmental permits were in hand, it would still take several years to erect the plant and get all of the operations smoothly integrated. Doing it in the midst of a severe national emergency would be all the harder.

After a decade and a half of economic depression and war, by 1950, much of the world was beginning to see steadily increasing demand for consumer goods, growing industrialization, and more opportunities for travel. All of these factors translated into growing need for liquid fuels. South Africa is richly endowed with coal but has very little indigenous petroleum. The early 1950s were also a time of unrest in the oil-rich countries of the Middle East. In 1950, the South African government decided to proceed with construction of a synthetic liquids plant. F-T technology was selected because it had been used successfully in Germany and because the alternative, direct liquefaction, had not been proved on the scale envisioned by the South Africans. The plant was ordered by the South African Coal Oil and Gas Corporation, commonly known as Sasol. The plant was commissioned in 1954, at Sasolburg, outside Johannesburg. A successful coal-to-liquids operation requires access to abundant coal supplies and abundant amounts of water. It's helpful if a plant can be near a center of population and industry. There were coal supplies south of Vereeniging, with the Vaal River nearby. Johannesburg was not far. So, as it's said, "Sasolburg located itself" (Norman and Whitfield 2006, 109).

The oil price shock of 1973 prompted a decision to proceed with a second plant, at the time called Sasol-2. This plant was constructed in Secunda, a new town that was built to support the plant,[5] about 200 kilometers from Sasolburg. While Sasol-2 was under construction, a decision was made to proceed with Sasol-3, which would duplicate the second plant. Currently, the facility in Sasolburg is running on natural gas, partly because resistance to expanding the coal mining operations in the Sasolburg area. The Secunda facility is now known as Sasol Synfuels, with no distinction between Sasol-2 and Sasol-3.

The heart of the Sasol plant in Secunda is a battery of Sasol-Lurgi fixed-bed dry-bottom (FBDB) gasifiers. FBDB gasifiers were selected for the Sasol plant because they are known to operate successfully on a large scale and therefore provide less technological risk, even though they also produce byproducts such as methane and tars.

[5] Its name derives from the Latin for *second*.

The Secunda plant consists of two halves, each with 40 gasifiers, and it consumes about 10,000 tonnes of coal each day. The plant supplies synthetic liquid fuels equivalent to about one-fourth of the South African market for gasoline and diesel fuel. The F-T synthesis relies on iron-based catalysts. The primary products are hydrocarbons containing 1 to about 20 carbon atoms per molecule. Liquid fuel production is about 25,000 cubic meters per day. In addition to liquid fuels, FT products are converted into polyethylene, polypropylene, acrylates, and the alcohols for synthesis of detergents. The total number of chemical products either made and sold by Sasol, or made by others from Sasol-supplied raw materials, is close to 200.

All of these fuels and chemicals contribute significantly to the South African economy. Sasol seem intents on wringing every possible marketable product from the coal fed into the plant. Reasonably so—marketable products translate into revenue. The Sasol experience provided useful insights for the design of the Dakota Gasification plant.

Sasol is a technical and economic success. The technical success of Sasol is unquestioned. This is a tribute to the contributions of generations of scientists and engineers who have steadily improved the plant operations. From the time that the Secunda plant came on line in the 1980s, there have been continual improvements both in gasification for synthesis gas production and in the F-T processes. If a commercial plant were to be built using the latest gasification and F-T know-how, combined with CO_2 capture and sequestration, it would represent a very large step forward in indirect liquefaction of coal.

Methanol synthesis is an alternative to the F-T processes for liquid fuels. Methanol is already a valuable component of the worldwide chemical industry, made in amounts of more than 100 million tonnes per year, mostly from natural gas via synthesis gas. The dominant uses of methanol lie in its conversion to a wide variety of chemicals, including formaldehyde and acetic acid. It is also a very useful solvent. In fuel technology, methanol has potential applications in several ways: as a possible liquid fuel, as an additive to gasoline, or as an intermediate material for further conversion to gasoline.

The first industrial production of methanol from synthesis gas occurred in 1923, by Badische Anilin und Soda Fabrik, now known as BASF. The original process used very high pressures, about 300 bars. At first the synthesis gas came from coal gasification. The preferred route now uses natural gas as the raw material, producing virtually all the methanol used in the world. An enormous infrastructure already exists for methanol production from synthesis gas. Since both coal gasification and methanol synthesis are quite mature technologies, there seem to be no technical barriers to developing a methanol industry based on coal.

High pressures favor methanol production because the reaction proceeds with a substantial reduction in number of moles of gas, from three moles (CO plus hydrogen) to one (vapor-phase methanol). Methanol synthesis is highly exothermic, which adds the engineering challenge of designing reactors that deal with the excess heat. There has been much interest in developing catalysts that will reduce the necessary reaction temperature and pressure. Doing so could result in reduced costs for reactors that would need to cope with lower pressures and result in savings of energy costs because of the lower temperatures.

Several arguments have been advanced in favor of methanol as an alternative liquid fuel. First, it is a liquid. This may sound trivial, but it is actually an important point. If a fuel handling system has materials that are compatible with methanol, and if care is taken to prevent accidental adulteration of methanol by water, then methanol could be distributed via the same transportation, storage, and handling infrastructure as currently exists for gasoline. It would be far easier to modify the existing fuel infrastructure to accommodate methanol than to create an entirely new infrastructure for a fuel in a different physical form. Second, because methanol is produced from synthesis gas, in principle it could be made from any carbon-containing resource, not just the natural gas that is used currently, but also coal, biomass, oil sands, or municipal waste. Third, many countries already have a large infrastructure for producing, handling, and storing methanol. In an energy crisis, methanol could be diverted from its existing chemical markets to supply the transportation sector, though granted with significant dislocations to the chemical industry. But such a diversion might seem preferable to waiting years to build new synthetic fuel plants. Fourth, existence of proven technology for converting methanol to gasoline (MTG, discussed below) can completely eliminate any concerns about using methanol itself as a fuel. The MTG product is not a liquid alternative to gasoline or a replacement for gasoline; it *is* gasoline. Finally, methanol synthesis catalysts can also convert CO_2 via the reaction.

$$CO_2 + 3\,H_2 \rightarrow CH_3OH + H_2O$$

Such a reaction makes it possible to convert CO_2, captured from other processes, into a useful fuel or chemical products.

Methanol has its ardent supporters (Olah et al. 2006). It also has vigorous detractors. The carbon–steam, water–gas shift, and methanol synthesis reactions used together provide an excellent route for producing methanol from coal. Stopping at this point provides a liquid fuel that already has a worldwide manufacturing base and distribution infrastructure. Its large-scale introduction into the fuel market would be controversial, especially based on environmental and safety issues. But we don't need to stop with methanol—we can turn it directly into actual gasoline.

The MTG process uses a special catalyst called ZSM-5. The initials stand for Zeolite-Sucony-Mobil, the latter two being the companies involved in its development. ZSM-5 is one of a vast family of natural and synthetic minerals called zeolites. The zeolites have crystal frameworks of silicon, aluminum, and oxygen atoms and incorporating other elements. In ZSM-5 the other element is sodium.

The high porosity of zeolites means that the reactions occur mainly on the internal surface. The zeolite structure consists of pores of very specific sizes. The size of the pores limits access, such that only molecules of a limited range of sizes can enter the zeolite structure and undergo the intended chemical reaction on the internal surface. Most zeolite catalysts are synthetic and are made for a specific purpose, controlling both the pore size and the reactivity of the pore surfaces.

The pores in ZSM-5 are designed so that only molecules with 10 or fewer carbon atoms can move through the zeolite structure. A molecule having more than 10 carbon atoms that forms inside ZSM-5 has to break apart before it can exit. Because of this size restriction, about 80% of the product consists of hydrocarbon molecules containing 5–10 carbon atoms. Hydrocarbons in this size range are major components of gasoline.

The overall MTG process is highly exothermic. Control of the large heat release is achieved by separating the process into two steps. First, methanol is converted into dimethyl ether (DME). It is possible to stop at this point, taking DME as the final product. DME works well as a fuel for compression-ignition (Diesel) engines. It is normally a gas, so requires different handling from conventional liquid Diesel fuel. However, it uses much the same handling infrastructure as liquefied petroleum gas (LPG), which is already successfully used as a vehicle fuel. DME-fueled vehicles have been demonstrated successfully in several places around the world (Semelsberger et al. 2006). Such vehicles have met the tightest emission standards in Japan, Europe, and the United States. In the large truck market, Volvo has been one of the leading proponents of DME fuel. A molecule of dimethyl ether contains no direct carbon-to-carbon bonds. When burned in an engine, DME produces very little carbonaceous soot, which can be a problem with conventional Diesel engines.

In the full MTG process, the methanol-DME-water mixture formed in the first stage flows to a second-stage reactor, which contains the ZSM-5 catalyst. There, DME is converted to hydrocarbons having 10 or fewer carbon atoms.

Unlike the F-T synthesis, the MTG process makes a liquid that is almost entirely gasoline. The dominant 5- to 10-carbon-atom products of the MTG process have especially high octane ratings. So, the MTG process gives a liquid product that is not just gasoline, it is high-octane, premium gasoline, with octane ratings of about 92. MTG gasoline can be blended compatibly with conventional petroleum-derived gasoline in any proportion. It can be stored, shipped, and handled in today's gasoline distribution infrastructure with no changes needed. MTG gasoline contains no sulfur. Today, China leads the world in aspects of methanol use in fuels, with most of the methanol coming from coal via synthesis gas. Most of the fuel methanol is used for blending with gasoline, but China also has about a dozen MTG plants (Green Car Congress 2017).

The extraordinary versatility of synthesis gas production and water gas shift means that in the future coal could be used as a raw material for an MTG plant if natural gas became in short supply. In countries or regions lacking indigenous natural gas, an MTG plant could be designed to run on coal from the outset. The MTG process could take advantage of the existing worldwide infrastructure for manufacturing methanol. Large methanol plants can make up to 5,000 tonnes per day (Methanol Institute 2020).

The extraordinary power of synthesis gas chemistry is difficult to overstate. Once synthesis gas has been made, cleaned, and shifted, the actual nature of the raw material used for producing it is irrelevant. A CO molecule will react exactly the same regardless of the source of its carbon atom—whether from coal or from other options.

The power of synthesis gas chemistry derives from three key points: first, any carbon source can be converted to synthesis gas via the carbon–steam reaction. The list includes coals, petroleum fractions, natural gas, oil sands, biomass, oil shale, municipal waste, and agricultural or forestry wastes. Second, any H_2:CO ratio that is needed can be made from synthesis gas using the water–gas shift reaction, thanks to Le Chatelier's Principle. If desired, the gas can be shifted all the way to pure hydrogen. Third, any liquid or gaseous fuel and many chemical products can be made by appropriate choice of the synthesis reaction. These products include hydrogen, substitute natural gas,

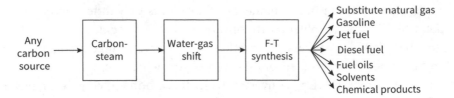

Figure 12.2 The power of gasification, water–gas shift, and synthesis reactions provides the potential to any carbonaceous material into a range of gaseous and liquid fuels plus chemical products.
Artwork by Lindsay Findley, from the author's sketch.

methanol, gasoline, aviation fuel, diesel fuel, fuel oil, and raw materials for the chemicals and plastics industries. All major fuel products and many of the chemical products now in daily use worldwide can be made from any available carbon source, using three reactions: carbon–steam, water–gas shift, and synthesis. Figure 12.2 captures this idea. Finally, CO_2 will be easier to capture from gasification or shift reactions than from the flue gases in conventional combustion processes.

This tremendous versatility cannot be overemphasized. The raw material and process for synthesis gas production can be selected depending on local availability of resources. More than one kind of raw material could be used, either in parallel process units or by co-feeding, such as feeding mixtures of coal and biomass to gasifiers. In addition, the choice of process for synthesis gas production can evolve with time. It might be desirable to build a plant based on natural gas, but if gas supplies dwindle or prices rise in the decades to come, the plant could be retrofitted with coal gasifiers and the necessary gas clean-up operations. As Sasol has shown, the coal to synthesis gas to fuels and chemicals technology works. A plant can be designed and operated with minimal technical risk.

So, the combination of carbon–steam, water–gas shift, and F-T syntheses provides an opportunity to use any cheap, abundant carbon source to make a range of gaseous and liquid fuels, chemical products, and polymers as needed to fit market opportunities. What's not to like? CO_2.

Suppose we want to make F-T gasoline. For simplicity, we presume that gasoline consists entirely of C_7H_{16} molecules—heptane and its isomers. From the synthesis equation, making this hydrocarbon requires 15 hydrogen molecules and 7 molecules of CO, equivalent to a hydrogen-to-CO ratio of 15:7 (i.e., 2.1). The carbon–steam reaction produces a H_2:CO ratio of about 1. A fraction of the cleaned gas has to be shifted to increase hydrogen at the expense of CO.

We get seven molecules of hydrogen from the carbon–steam reaction, which says that we require eight more molecules of "shift hydrogen."[6] We still have the necessary 15 molecules of hydrogen, but they are split apart conceptually to indicate that seven come from the gasifier in the cleaned synthesis gas, and we make eight more in the shift reaction. These two apparent kinds of hydrogen molecules are exactly identical chemically, and the synthesis reactor certainly cannot differentiate between them. The shift reaction produces a molecule of CO_2 along with every molecule of hydrogen. Making eight molecules of "shift hydrogen" means that we will also, whether we want it or not, make eight molecules of CO_2. In terms of moles rather than molecules, one mole of C_7H_{16} that we want for our gasoline weighs 100 grams per mole. CO_2 weighs only 44 grams per mole, but we make eight moles—a total of 352 grams. By weight, we produce three and a half times more CO_2 than the amount of our desired gasoline.

There are two more sources of CO_2. The raw gas from the gasifier contains some CO_2, from the carbon–oxygen reaction. In a complex handling thousands of tonnes of coal per day, it would make sense to use coal-fired boilers to make the steam needed for gasification and water–gas shift, producing still more CO_2. The Sasol facility in Secunda, South Africa is among the largest single point sources of CO_2 on our planet (Schrag 2009). In a plenary address at the 2009 International Conference on Coal Science and Technology, Dr. Sven Godorr estimated that, for every 3 tonnes of carbon (not coal) entering the plant, 2 tonnes go up the stack as CO_2 and 1 tonne winds up in the products. A 2007 study by the National Petroleum Council in the United States estimates emissions to be 700 kilograms of CO_2 for every barrel of F-T liquid produced (Bellman et al. 2007). Depending on the exact mix of products—gasoline, jet fuel, diesel fuel, and heating oil—a barrel of liquids could be some 120–140 kg.

A F-T plant is a CO_2 factory that produces liquid fuels and chemicals as byproducts—*if* the plant has been designed and built to the standards of the 1970s. An F-T plant equipped with provisions for CO_2 capture and reuse or sequestration could exploit all of the extraordinary versatility of the combination of carbon–steam/shift/synthesis processing with minimal CO_2 emissions. It would take advantage of the comparative ease of capturing CO_2 from a highly concentrated gas stream, rather than from one in which CO_2 is a minor component.

There is one more factor that seems often to be overlooked in considering synthetic fuels from coals. Gasification is based on the carbon–steam

---

[6] This comes from writing the synthesis reaction as $7 CO + 7$ "gasifier H_2" $+ 8$ "shift H_2" $\rightarrow C_7H_{16} + 7 H_2O$, instead of the way it would normally be written for $n = 7$: $7 CO + 15 H_2 \rightarrow C_7H_{16} + 7 H_2O$.

reaction. Increasing the H_2:CO ratio via the water–gas shift reaction requires adding steam. But steam implies water. For an F-T plant, it's estimated that the water requirement would be somewhere between 1 and 8 cubic meters per cubic meter of product (Ramage and Tilman 2009). Based on our profligate use of water in the United States (Reubold 2019), a hypothetical F-T plant producing 25,000 cubic meters per day of liquid fuels would consume water at a rate equivalent to a city of nearly a quarter-million people.[7]

Where does the water come from? If a plant were to be located on a river or other large natural water source, then water availability might be thought to be unlimited. We've already seen this with regard to power plants. In such places, likely the water-handling systems for a synthetic fuels plant would not differ much from those we've discussed in power plants. Many parts of the world lack abundant water supplies. In arid regions, plant design would need to accommodate the recycling of water for reuse and, wherever feasible, use air cooling rather than water cooling.

Sasol Synfuels is a testament to the fact that gasification coupled with F-T synthesis is technically feasible on a large, commercial scale. It supplies about 30% of the liquid fuel needs of a nation of some 60 million people. On a molecular scale, the complex chemical framework of coals is torn down to molecules containing only one carbon atom each—CO. In subsequent process steps, the single carbon atom molecules are put back together into chains containing 5 to about 20 carbon atoms to get the array of liquid fuels desired. In the next chapter, we will explore the concept of breaking down the macromolecular framework of coals to get liquid products directly.

References

Bellman, David K., James E. Burns, Frank A. Clemente, Michael L. Eastman, James R Katzer, Gregory J. Kawalkin, George G. Muntean, Hubert Schenk, Joseph P. Strakey, and Connie S. Trecazzi. 2007. "Coal to Liquids and Gas." In: *Hard Truths: Facing the Hard Truths About Energy*. Topic Paper 18. Washington, DC: National Petroleum Council.

Green Car Congress (GCC). 2017. "EIA: China's Use of Methanol in Liquid Fuels Has Grown Rapidly since 2000: >500K bpd by 2016." Green Car Congress. https://www.greencarcongress.com/2017/02/20170223-methanol.html.

Methanol Institute. 2020. "Methanol Production." https://www.methanol.org/methanol-production/.

Norman, Nick, and Gavin Whitfield. 2006. *Geological Journeys*. Cape Town: Struik.

Olah, George A., Alain Goeppert, and G. K. Surya Prakash. 2006. *Beyond Oil and Gas: The Methanol Economy*. Weinhein: Wiley-VCH.

---

[7] This is based on assuming a water consumption of 4 cubic meters per cubic meter of product, a daily US water consumption of 1.1 cubic meters per day per household, and an average of 2.6 persons per household.

Ramage, Michael P., and G. David Tilman, eds. 2009. *Liquid Transportation Fuels from Coal and Biomass*. Washington, DC: National Academies Press.

Rehnlund, Björn. 2007. *Synthetic Gasoline and Diesel Oil Produced by Fischer-Tropsch Technology. A Possibility for the Future?* IEA/AMF Research Report. Göteborg, Sweden: AtraxEnergi AB.

Reubold, Todd. 2019. "America Uses 322 Billion Gallons of Water Each Day. Here's Where It Goes." Ensia. https://ensia.com/articles/water-use/.

Schobert, Harold H. 1990. *The Chemistry of Hydrocarbon Fuels*. London: Butterworths.

Schobert, Harold H. 2013. *Chemistry of Fossil Fuels and Biofuels*. Cambridge: Cambridge University Press.

Schrag, Dan. 2009. "Coal as a Low-Carbon Fuel?" *Nature Geoscience* 2: 818–820.

Semelsberger, Troy A., Rodney L. Borup, and Howard L. Greene. 2006. "Dimethyl Ether (DME) as an Alternative Fuel." *Journal of Power Sources* 156: 497–511.

Song, Xuemin, and Abdelhamid Sayari. 1994. "A. Direct Synthesis of Isoalkanes through Fischer-Tropsch Reaction on Hybrid Catalysts." *Applied Catalysis A: General* 110: 121–136.

Stranges, Anthony. 1997. "The U.S. Bureau of Mines's Synthetic Fuel Programme, 1920–1950s: German Connections and American Advances." *Annals of Science* 54: 29–68.

13

Liquefaction

Indirect coal liquefaction gets its name because the liquid products actually come from synthesis gas. The role of the coal is only to make the synthesis gas. Other carbonaceous raw materials could be used instead of coals. In contrast, the term ***direct liquefaction*** means that the liquids are made from the coal itself without intermediate synthesis gas. With either process, the reason for the efforts in liquefying coal has always been to produce liquids that could augment or replace petroleum liquids.

The molecules in petroleum liquids have about 1.8 hydrogen atoms for every carbon atom. Humic coals have less than one hydrogen atom per carbon atom. The challenge is to increase the H:C ratio of approximately 0.8 in coals to the approximately 1.8 of petroleum liquids. We have two choices: either increase the numerator by adding hydrogen or decrease the denominator by "rejecting" carbon.

Adding hydrogen to coal is called *coal hydrogenation*, or ***hydroliquefaction***. Just as indirect liquefaction is sometimes taken to mean the Fischer-Tropsch (F-T) synthesis and nothing else, the term "direct liquefaction" is often understood to refer to hydroliquefaction, without considering other possibilities. Coal carbonization produces tars and oils that can be treated to make liquid fuels. Carbon-rich char is left behind. Alternatively, hydrogen-rich products can be dissolved, or extracted, from coal by various solvents. The insoluble portion consists of carbon-rich material.

The simplest way to make liquids from coals is to heat coals in the absence of air. These processes also produce a char or coke residue and gases. Often the liquid yield will be in the range of 20%. Without outside markets or uses inside the plant for the solid and gases, liquid production will be very wasteful and likely uneconomic. In many parts of the world, such an operation would also have trouble meeting environmental regulations. Such ***carbonization*** processes must rely on an internal transfer of hydrogen from some portions of the molecular framework of the coal to the fragments that eventually become the liquid products. Without an externally added source of hydrogen, the amount of liquid that can be produced is necessarily limited by the amount of hydrogen originally present in the coal being used.

In the nineteenth century, producers of *coal oil* were keen on the use of cannel or boghead coals as raw material. The desirability of cannel coals was known by the early 1850s (Lucier 2008, 41). Sapropelic coals have higher hydrogen contents than humic coals, so there is more hydrogen available for the internal hydrogen transfer processes. Cannel coal from Kentucky has an atomic hydrogen-to-carbon ratio of about 1.08 (Haslam and Russell 1926, 49),[1] very comparable to the average value of 1.12 for low-temperature tars produced from carbonization of bituminous coals (Rhodes 1945, 1293).[2]

Carbonization represents a simple, technically mature process and straightforward equipment design. There would be very little technical risk in establishing a carbonization plant. But gasification and direct liquefaction provide high yields of a single kind of product—a gas or a liquid. Strong market demand for such products outweighs the added investment and operating costs of a gasification or liquefaction facility.

Tars from low-temperature carbonization can be refined successfully into gasoline or diesel fuel. About 10–15 years ago, stories appeared on the internet proclaiming that we could obtain "free oil" from coal via carbonization. These stories usually referred to a process developed in the United States nearly a century ago by Lewis Karrick. Conspiracy theorists argue that all that is needed is to dig up coal, heat it, and collect the liquids. All that stands between us and unlimited supplies of free gasoline is Big Oil, or else Big Government, or else, for the true believer, the nefarious collusion of Big Oil with Big Government. What really stand in the way are facts.

Much of Karrick's process development work was done using coals having unusually high amounts of resinite, which would yield an aliphatic oil roughly resembling petroleum products. The Karrick process indeed produces a liquid. Under favorable circumstances, the yield is about 150–175 liters per tonne of coal processed. Without markets for the char and gas, the waste would be enormous. An accountant might be able to figure out how this liquid could be "free," provided that high-value markets existed for the large amounts of byproduct char and gas and provided that these byproducts could be sold at a high enough price to compensate for giving away the liquid. The questions remain as to who gets to pay for mining the coal in the first place and for refining the tar into specification-grade, environmentally compliant, marketable products. Lewis Karrick was a fine and very competent fuel engineer. There is no evidence that he was a nut.

---

[1] Calculated from the cited data for 10 tars produced by carbonization at 600°C.
[2] Run at 700°C or less.

Solvent extraction provides a route to liquid products from coals. Many common organic solvents, such as chloroform or acetone, dissolve less than 10% of the coal. Most of the extracted material seems to be waxes or resins that survived coalification. A second set of solvents can extract up to about 40% apparently without breaking chemical bonds in the macromolecular framework of the coal. Examples are pyridine and quinoline.

Above 300°C, several compounds—such as phenanthrene—can be very effective solvents (Orchin 1951). This is the temperature range in which most coals start to experience active thermal decomposition. At such temperatures the solvent could be "dissolving" the coal, dissolving molecular fragments produced from the molecular framework as it breaks apart, or could even be actively participating in chemically breaking down the framework.

Solvent extraction has been used for a century and a half to study the composition of coals. Commercially, the interest has been the prospect of making useful liquid products. A selected coal can be treated with a solvent to obtain an extract of the coal components. The solvent can be distilled off to obtain the extract and recover the solvent for recycling, or the solvent–extract mixture could be processed further without separation. Subsequent refining steps will likely be needed to produce acceptable specification-grade fuels.

If the combustion turbines in integrated gasification combined cycle (IGCC) plants could be fired directly with coal, the gasification and gas clean-up operations could be eliminated, saving the substantial sums needed for the investment and running costs of the gasifiers and gas clean-up equipment. The problem is mineral matter. Ash particles of hard minerals could erode turbine blades, sand-blasting the blades as the turbine operates. Particles with low fusibility temperatures could stick to turbine blades, causing corrosion of the blade underneath the accumulated ash. Different amounts of ash accumulating on different turbine blades cause the turbine to become unbalanced. If so, or if a blade weakened by corrosion or erosion breaks off, a catastrophic failure of the turbine will result. Such an accident would cost many millions of dollars, even without being compounded by possible human injury or death. We could use coal in a combustion turbine if only we had ash-free coal.

A form of solvent extraction, the HyperCoal process, was developed to produce an ash-free solid for use as a fuel for turbines. Removing the solvent from the filtered liquid leaves a product having virtually the same ultimate analysis as the parent coal but with very low mineral matter content. Thirty to seventy percent of a particular coal can be converted to HyperCoal. The process gives a solid with an ash yield of 0.02% or less, about as close to ash-free as one could hope to get.

Solvent extraction of bituminous coals has been developed for the production of clean diesel and aviation fuels. This work has been done in the United

States at Penn State University and in South Africa at North-West University. The Penn State process focused mainly on using light cycle oil as the starting solvent,[3] while the North-West process used residue oil, obtained from the byproduct tars from Sasol-Lurgi gasifiers. In the best case, both variations of this process give extract yields of about 70%. The clean liquid is reacted with hydrogen in two steps: first to remove most of the sulfur and part of the nitrogen and, second, to convert the aromatic compounds to saturated, aliphatic compounds.

The Penn State version has been successfully tested in two pilot plant campaigns. The dominant product, amounting to about 80%, is clean jet, diesel, and light fuel oils. Small amounts of gasoline and heavy fuel oil account for the rest. The products make superb jet fuel (Balster et al. 2008), clean diesel fuel that can be used as-is or blended with conventional diesel fuel made from petroleum (Beaver et al. 2009), and a liquid that can be fed straight to solid-oxide fuel cells for electricity generation (Zhou et al. 2004). These liquids are comparable to the petroleum products collectively known as *middle-distillate fuels*.[4] The addition of several operations—all of which are demonstratively feasible—could make this a nearly emission-free coal-to-liquids process, shown in Figure 13.1 (Schobert 2015).

Both direct liquefaction and the development of fluidity in coking coals rely on shuttling hydrogen inside the coal, particularly in the early stages. The development of a fluid phase in coke-making involves the pyrolytic breaking apart of the coal framework and forming stable products by internal transfer of hydrogen. Repetitive reduction of molecular size and stabilization of products eventually provides molecules that are stable in the liquid phase— the tars and oils. Likely the very earliest stages of direct liquefaction follow the same mechanism. The difference between these technologies comes at the point in coke-making at which it becomes increasingly more difficult to stabilize products by internal hydrogen transfer because we are running out of available hydrogen. In direct liquefaction, we try never to run out of hydrogen by supplying it ourselves, rather than relying entirely on hydrogen originally in the coal. In coking, we observe a temperature of maximum fluidity, followed

---

[3] Light cycle oil (LCO) comes from petroleum. Many oil refineries use a fluidized catalytic cracking process to increase the amount of light products, such as gasoline, at the expense of heavier ones, such as fuel oil. Not all of the material fed to the cracking unit converts to gasoline. Some heavier products remain, LCO being one of them. LCO can be recycled through the cracker or can be blended with other process streams to become part of the products sold by the refinery.

[4] They get that name because their boiling ranges are in the middle of those of very volatile products, such as gasoline, and very high-boiling products, such as heavy fuel oils. This boiling behavior means that they also are removed from the middle of the distillation column in a refinery.

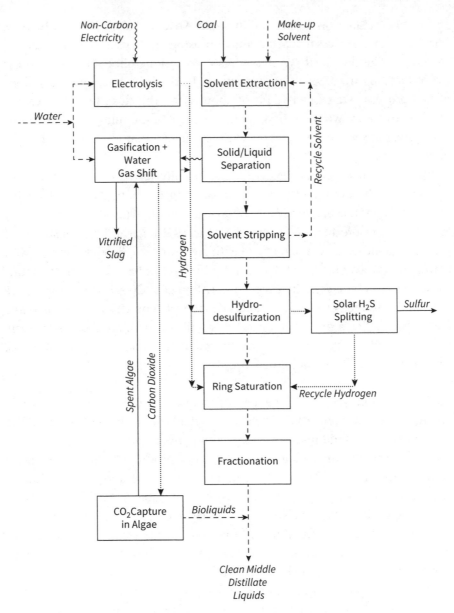

Figure 13.1 A flow diagram of a conceptual emission-free process for making clean liquid fuels from coal. All steps have been proved at some scale but never integrated into this layout. Coal, water, non-carbon electricity go into the plant; clean liquid fuels, vitrified slag, and sulfur come out.

Artwork by Lindsay Findley, from the author's sketch.

eventually by resolidification. In direct liquefaction, a continuous supply of hydrogen assures, in turn, a continuous production of liquid. The difference is that we might observe 70% of the coal being converted to solid (coke) in a coke oven, compared with 90% of the coal being converted to liquids and gases in a liquefaction reactor.

Put crudely, in direct liquefaction, we somehow have to pound hydrogen into the system to increase the hydrogen-to-carbon ratio. At the same time, it's necessary to break apart the macromolecular framework of coals to a molecular weight range characteristic of molecules that would exist as liquids at ordinary temperatures. The breaking apart is easily handled by heat. In most coals, bond-breaking begins in earnest at temperatures around 350°C. Heating a coal sample to a temperature somewhat above the onset of bond-breaking, say 400°C, should suffice to begin to dismantle the molecular framework. Then we must consider why we need hydrogen and how best to supply it.

As we've seen with coking, bond-breaking produces two radicals. In liquefaction, the most crucial radical reaction stabilizes the radical by "capping" it with a hydrogen atom. Using R· to represent a generic radical produced by bond-breaking in the coal structure, the process is R· + H· → RH. Since R· represents a fragment broken off the macromolecular coal framework, necessarily it has to be of lower molecular weight than the parent coal. Capping the radical with hydrogen increases the hydrogen-to-carbon ratio. We are reducing molecular weight and increasing hydrogen at the same time. We are headed in the right direction—smaller molecules and higher H:C ratio. Repetitive bond-breaking and hydrogen-capping reactions will lead to molecules small enough to be liquids at normal temperatures.

A second reaction is recombination of two radicals derived from the coal. Because they do not have to be identical, we denote one as R′. The reaction is R· + ·R′ → R–R′. Often the new bond formed in the R–R′ product is stronger than the bonds that broke apart to produce R· and R′·. R–R′ has to have a higher molecular weight than either R· or R′· itself. No hydrogen has been added. Radical recombination reactions turn the coal into a new solid that is less reactive than was the original coal. We are going backward: higher molecular weight and no hydrogen addition. Such reactions are often called *retrogressive reactions*.

Maximizing the yield of useful liquid products requires maximizing hydrogen capping and minimizing (or totally eliminating if possible) retrogressive recombination. This can be achieved by providing H· to the coal-derived radicals immediately as they form. Doing so intercepts and shuts down retrogressive reactions. There are three possible sources of H·. Hydrogen already present in the coal can be transferred internally. This works, but we know from fluidity measurements that the supply of hydrogen is limited. A second source can be gaseous hydrogen, H_2. Before discussing the third possible source, let's consider some ramifications of using H_2.

Overlooking the not-insignificant question of where H_2 actually comes from in the first place, using H_2 introduces two issues. The H–H bond in the H_2 is strong. H_2 will not easily break apart without a catalyst. H_2 is a gas, which we want to react with coal radicals in solution. To get H·, the process that should happen is $H_2(g) \leftrightarrows 2\ H\cdot(cat)$, where g represents the gas phase and *cat* indicates a species on the surface of a catalyst. This equilibrium is subject to Le Chatelier's Principle, which in this case requires high pressures to push the reaction to the right-hand side, favoring formation of H·. In addition, the solubility of a gas in a liquid is proportional to its pressure. We need high pressures for good solubility of H_2 in a liquid medium. These reasons make it necessary to use gaseous hydrogen at high pressures.

Direct liquefaction is usually done at 400°C or higher, to facilitate breaking bonds in the coal. The solubilities of all gases in liquids decrease with increasing temperature. Combined with the gas solubility and equilibrium considerations just mentioned, this means that we must operate at very high pressures to get hydrogen into the liquid. A gasifier might run at about 30 bar. A direct liquefaction reactor using gas-phase hydrogen might operate at nearly 200 bar. A high-temperature, high-pressure operation has economic consequences. Reactors and ancillary equipment need to be made of materials able to withstand such conditions and will have higher maintenance costs. High temperatures lead to high energy costs.

If the chosen catalyst is expensive, it may be economically important to recover and recycle it. This adds more operations to the process and therefore more cost. Two alternatives are available. The first takes advantage of Mother Nature's Free Catalyst—pyrite. At the temperatures of direct liquefaction, pyrite decomposes partially to pyrrhotite, thought to be the active form of the catalyst. Its chemical formula is $Fe_{1-x}S$, where x is usually 0–0.2.[5] Many better catalysts are known, but pyrite or pyrrhotite has the undeniable advantage of being free, and there is little incentive to recover them. A second alternative is to use a very cheap catalyst that does not come with the coal but is still

---

[5] Pyrrhotite is a classic example of a non-stoichiometric compound. Such compounds are almost always inorganic solids. As a class, they are characterized by the fact that the proportions of elements in the compound cannot be represented by ratios of small natural numbers, as is otherwise the case with almost all known compounds. Non-stoichiometric compounds can be visualized has having a small fraction of atoms missing, or a few extra atoms added to, their solid structures. Non-stoichiometric compounds are very important in two classes of materials: catalysts and superconductors.

so cheap that we needn't recover it. These are called *throw-away catalysts* or *once-through catalysts*. The only materials both cheap enough and catalytically active enough to use are compounds of iron: red mud, a byproduct of the aluminum industry, or mill scale, which we've seen as an F-T catalyst.[6]

A different route to H· is available. We've seen that hydrogen can transfer internally from hydrogen-rich structures or from hydrogen-rich macerals, already in the coal. However, we can deliberately add a supply of hydrogen-rich structures that will provide H· in the same way as internal hydrogen transfer. These compounds are known as *hydrogen donors*, or as donor solvents. When discussing development of fluidity in caking coals, we used tetralin as a simple model of hydrogen-rich structures that transfer hydrogen internally. Tetralin—or compounds that act similarly to tetralin—can be supplied externally to do the same job. If we supply enough external hydrogen, there is no danger of running out. The reaction system will not revert to coke, nor will it stop with some 20–30% yield of liquid product.

Externally added tetralin functions in the same way as the tetralin-like structures discussed in Chapter 10. This reaction converts tetralin to naphthalene. Tetralin can be regenerated, either right in the liquefaction reactor or in a separate hydrogenation step. Using pure tetralin, or any pure compound, would be too expensive for an industrial process. In a practical operation, much of the solvent will be derived from the process itself and be continuously recycled. The hydrogen-donating ability of the solvent can be restored in a separate hydrogenation operation, downstream of the liquefaction reactor. This allows regeneration of the solvent to be done in the same kinds of reactors as those used in hydrogenations in oil refineries. Using a hydrogen-donor solvent and regenerating with hydrogen does not eliminate the need for hydrogen in the overall coal-to-liquid process; it only changes the point in the process at which hydrogen is added. This concept was the basis of the Exxon Donor Solvent process for direct liquefaction (Schobert 1987, 243–246), successfully tested in a pilot plant that processed nominally 225 tonnes of coal per day.

As we saw with gasification, introducing solid coal into a pressurized reactor is challenging. One solution involves using a slurry of coal in some liquid vehicle, such as water, and pumping it to the pressure of the reactor. The

[6] Aluminum is produced from its oxide, which occurs in nature as bauxite. Bauxites contain appreciable quantities of oxides of other metals, most notably iron. To obtain pure aluminum oxide for smelting to aluminum, native bauxite is treated with sodium hydroxide solution. The only oxide to dissolve is aluminum oxide; the others remain behind as a mud-like mass. The iron compounds give the mud a distinctive red color, hence the name.

same approach is used in direct liquefaction, though relying on hydrocarbon-based liquids as the slurry vehicle.

There is no doubt that direct liquefaction works, from a technical standpoint, and indeed works well. The problem is whether direct coal liquefaction can work from an economic standpoint, meaning that the cost of the coal-derived liquids must be competitive with the cost of liquid fuels from petroleum. We have yet to solve this problem.

The first significant work on direct liquefaction was done at the time of World War I by the German scientist Friedrich Bergius. He worked in the laboratory of Fritz Haber, one of whose achievements was development of a way to synthesize ammonia for fertilizer production starting with nitrogen and hydrogen. Haber's laboratory at the University of Karlsruhe also had a strong influence on coal chemistry. It was a place with the infrastructure and facilities for experimental work on high-temperature, high-pressure chemistry. Those who worked there became adept at doing such experiments. Working at temperatures close to 500°C and pressures of about 200 bar can be a tricky business.

In Haber's laboratory, Bergius gained experience in working with high-pressure equipment and working with hydrogen at high pressures. He had also investigated processes for enhancing the yield of gasoline from petroleum by thermally cracking large molecules to smaller ones typical of gasoline components. Such reactions often are accompanied by production of hydrogen-rich hydrocarbon gases, which represent a waste—from the perspective of gasoline production—of hydrogen internally available in the system. These reactions also produced carbon-rich coke (Schobert 2013, 286). Bergius discovered that using externally added, high-pressure hydrogen would suppress coke formation and suppress the production of undesired unsaturated compounds (Stranges 1997).

Bergius began a research program on coals by conducting basic scientific research on the physical properties and chemical constitution of coals, spending at least 2 years doing this. He also conducted experiments on artificial coalification of sawdust, studying the conditions by which it can be converted to coal-like materials. Bergius's researches on artificial and natural coals led him to the conclusion, somewhat erroneous, that coals have molecular structures similar to terpenes (Stranges 1997) but possibly with aromatic rings attached. The terpenes are a large family of natural hydrocarbons that have molecular formulas beginning with $C_{10}H_{16}$ and increasing by five carbon and eight hydrogen atoms, to $C_{15}H_{24}$, $C_{20}H_{32}$ . . . at least to $C_{40}H_{64}$ (Howard 1947, 20). Natural rubber is a polyterpene.

Based on his concept of terpene-like structures, Bergius reasoned that adding hydrogen to coals would produce materials not far different from heavy petroleum oils, with molecular weights of ~300 (Stranges 1997). Further reaction should produce liquids of fewer than 15 carbon atoms, similar to gasoline, naphtha, or light kerosene. He was absolutely right. Reacting pulverized coals with hydrogen at very high pressures, above 400°C, converted 80% of the coal to products that were soluble in benzene, along with some gases. That's not too bad, even by today's standards. George Box might have been pleased: the terpene model was wrong, but useful.

Bergius first made synthetic gasoline from coal hydrogenation in 1909 (Borkin 1978, 26). By 1910, he was operating at 450°C and 300 bar (Stranges 1997). The German military and economic conditions of 1916 provided the impetus to scale up his laboratory work to industrial production. What was needed to move forward with the scaling-up was a skilled process engineer, particularly someone experienced in scaling up high-pressure processes—someone like Carl Bosch, for example, who had been responsible for scaling up and commercializing the Haber process for ammonia synthesis.

The practical engineering difficulties of scaling up Bergius's laboratory work were such that it made no contribution at all to the German war effort (at that time). Before a set of reactions can transition from the laboratory to a pilot plant and then to a successful industrial process much additional developmental work is needed and often additional scientific discoveries are needed as well. As the scale of operations gets larger, so, too, does the price tag. It took Bergius about 10 years to get to a point of having a coal hydrogenation process that would operate on a continuous basis. The cost of going further was so great that Bergius reached a financial agreement with Badische Anilin- und Soda-Fabrik (BASF) to industrialize his process. Fortunately, BASF had someone on staff who would be just the person for the job—Bosch himself.

The first plant began operating in 1927, at Leuna, in southeastern Germany. It experienced numerous problems with the technology and with equipment operation (Borkin 1978, 53). Costs were far over budget. At the time, the cost of the synthetic gasoline was about seven times higher than the world price of gasoline from petroleum (Borkin 1978, 26). Bosch argued that it had taken 15 years to expand the Haber synthetic ammonia process to the commercial levels of the early 1930s and that it would likely take even longer to achieve fully successful industrial-scale synthetic gasoline (Borkin 1978, 49–50).

Bergius's laboratory work was a one-stage process, breaking apart the coal macromolecules and adding hydrogen at the same time, in the same reactor. Processes aimed at producing commodity materials, such as gasoline, have a much greater chance of being economically viable if run as continuous-flow reactions, rather than one batch at a time. Industrial-scale success came from converting the laboratory batch process into a successful flow-reactor system and from separating the overall conversion into two stages. For the Bergius-Bosch process, the first stage adds hydrogen to coal to convert it only into a heavy oil with a consistency almost like pitch. The second stage cracks the heavy oil into smaller hydrocarbon molecules that would be comparable to familiar gasoline and kerosene.

By 1931, the plant at Leuna was turning out 300,000 tonnes of synthetic petroleum per year (Stranges 1997). Largely through Carl Bosch's efforts, there were 12 plants running by 1944. At their peak in 1944, the plants produced at least 3 million tonnes. Two-thirds of this output, some 2 million tonnes, was used to produce 100-octane gasoline by adding tetraethyl lead to the liquid product.[7] In World War II these plants provided most of the liquid fuel needs of the German military (Stranges 1997). The cost was high, estimated to be $0.24 per US gallon (Stranges 1997) or $0.06 per liter in 1944. In today's dollars, this is equivalent to $0.92 per liter, at least 1.25 times the price of regular-grade gasoline on the US market in the spring of 2021.

Both bituminous and low-rank coals were used as feedstocks. Bergius could not hydrogenate anthracite or bituminous coals having carbon contents exceeding 85%, on a moisture-and-ash-free (maf) basis. Bituminous coals required very severe reaction conditions, up to 485°C and 700 bar. Severe indeed. The first stage achieved conversion of 60% of the maf coal to a product called middle oil. Second-stage hydrogenation in the vapor phase, in catalytic reactors, converted about 50% of the middle oil to 70–75 octane gasoline. Addition of tetraethyl lead boosted the octane rating to 100.

Pulverized coal was made into a slurry in an oil derived from the process. Red mud was commonly used as a once-through, throw-away catalyst. The Bergius-Bosch process converted 1 tonne of coal to about 170–190 liters of gasoline, 210 liters of diesel fuel, and 145 liters of fuel oil. The gasoline contained about 20–35% of aromatic compounds. Another 40% of the liquid product, heavy oil, was recycled to be the liquid vehicle for making the slurry of coal. Further trips through the reactor with each recycle pass helped to convert some of the components of heavy oil into lighter products.

---

[7] For decades, tetraethyl lead was added to gasolines as an anti-knock agent (i.e., to prevent the explosion of the unburned gasoline/air mixture in the cylinder of an internal combustion engine). Engine knock wastes fuel, reduces acceleration and fuel economy, and can eventually lead to serious engine maintenance problems. Fuels with this additive were sold as "leaded gasoline" or "ethyl gasoline." By the 1970s, it was recognized that chronic exposure to low levels of lead from automobile exhausts results in a variety of health problems. Most countries phased out the use of tetraethyl lead. By about 2015, only three Middle Eastern countries were still using leaded gasoline.

The heavy oil also absorbed and distributed heat produced in the exothermic hydrogenation reactions, preventing overheating that would lead to coke formation.

The high pressures of hydrogen help to produce higher conversions, putting more H_2 into solution and favoring formation of more H· on the catalyst. The 700 bar pressure used for liquefaction of bituminous coals is phenomenally high. Hydrogen, the smallest stable molecule, has an insidious ability to work its way into and through pores, fissures, pinholes, and leaks. Hydrogen is flammable in air. Hydrogen–air mixtures containing 4–74% hydrogen can ignite; mixtures of 18–60% in air can explode. With the extraordinary ability of hydrogen to find and exploit even the tiniest leaks, it seems remarkable that these plants didn't simply blow themselves up without any assistance from the Allied bombing campaign.

Friedrich Bergius is particularly noteworthy for being the only coal scientist ever to win the Nobel Prize. Bergius won the Prize for chemistry in 1931, sharing the award with Carl Bosch. The American philosopher Alfred North Whitehead once remarked that "the safest general characterization of the European philosophical tradition is that it consists of a series of footnotes to Plato" (Whitehead 1979). Well, to paraphrase, just about all of direct liquefaction consists of a series of footnotes to Bergius. Most of the research and development work in this area in the past 80 years has focused on modifying the original work of Bergius to reduce the temperature, pressure, and reaction time.

Operation of Bergius-Bosch plants ended in 1944, or shortly thereafter. Worldwide interest in research and development on the production of synthetic liquids directly from coals became somnolent by the 1960s and was only revived by the OPEC oil embargo of 1973.

Renewed interest in direct liquefaction resulted in development of various concepts collectively lumped under the term "second-generation processes." These were modifications aimed at operating at lower temperatures and pressures, improving process efficiency, and improving economics. None of the second-generation processes ever progressed beyond the pilot-plant scale, except for a commercial-scale plant in China, discussed below. The worldwide decline in petroleum prices from mid-1980s, hitting bottom in mid-1998, destroyed the likelihood of building commercial direct-liquefaction plants. It also ended most of the supporting research and development work. By the turn of the century, the large pilot plants that had operated at times from the mid-1970s to mid-1990s had been closed, and virtually all had been dismantled.

Synthetic fuels from coal seem able to compete economically only in times of petroleum scarcity or high prices. The British destroyed Romanian oil wells in 1916, hampering German access to petroleum, and along came Bergius. It took a long time, a lot of effort, and another world war to establish a significant liquefaction industry in Germany. Cheap oil after the World War II ended most interest in coal conversion for about 25 years. Then came the OPEC oil embargo and the Iranian revolution. Coal liquefaction was suddenly back in business. A period of declining oil prices followed by the shale revolution has taken liquefaction out of play again.

Virtually all second-generation processes have several features in common: pulverized coal is introduced as a slurry, reacted typically at 400–500°C and at least 140 bar in multiple stages. The intent is to achieve at least 90% conversion of the coal (maf basis) to liquids. In many processes the liquid medium must be recovered and its hydrogen-donor ability replenished and recycled. Unconverted or partially converted coal and mineral matter must be separated from the liquid product. If a catalyst is used, it has to be recovered, reactivated, and recycled. The liquid, a synthetic crude oil, is fractionated. Heavy distillation fractions are recycled through the process as the slurry vehicle. Lighter fractions can be upgraded to marketable products similar to ones obtained from petroleum refining.

We have so far glossed over the question of where the hydrogen comes from. The most straightforward approach would be to make hydrogen on site, using coal gasification, gas clean-up, and shifting. On-site gasification adds substantial capital cost for air separation, steam generation, gasifiers, gas clean-up, and shift reactors. It could account for more than half of the total cost of a liquefaction plant—just to make hydrogen. There will be accompanying production of carbon dioxide (CO_2) in the shift reactor, which could be captured for sequestration. Otherwise, it represents a significant emission from the plant. Steadily increasing concerns about CO_2 emissions from make it worth considering alternatives for hydrogen production. The most likely seems to be production of hydrogen by electrolysis of water, especially using a "non-carbon" electricity source, such as solar, wind, hydropower, or nuclear power.

Process development studies reduced reaction severities and improved conversions. Conversions were in the 90–95% range (maf basis). Bosch and Bergius were reaffirmed: the best approach uses two-stage processes. A catalyst could be present in one or both stages. The intent was to maximize the yield of distillable liquids while minimizing the hydrocarbon gases and heavy liquids produced and minimizing hydrogen consumption.

The last major coal liquefaction pilot plant to remain in operation in the United States was located in Wilsonville, Alabama.[8] At the end of the pilot-plant campaign, Wilsonville operated as a catalytic two-stage liquefaction (CTSL) process. The first stage dissolved fragments of the macromolecular structure of the coal into a recycled, rehydrogenated solvent. The rate of rehydrogenation was matched with the rate of coal conversion, so that an ample supply of hydrogen donors was always available. In the second stage, liquids produced in the first stage were cracked into distillable liquids via reaction with hydrogen over a catalyst. The goal was to have second-stage products that could be upgraded to marketable liquid fuels by standard petroleum refining technology. CSTL achieved yields of distillates of up to 880 liters per tonne of coal. Using high-volatile bituminous coal from Illinois, a yield of nearly 80% distillate liquids and a total coal conversion of 97% (maf basis) were the best results achieved. For comparison, the best results from the Bergius plants also achieved 92–97% conversion, but at substantially higher pressures and higher temperatures[9] (Probstein and Hicks 1982, 292). Residence times in the Bergius plants were also much longer. The direct liquefaction program of the 1970s and '80s succeeded in obtaining significant reductions in reaction severities. The pilot plant in Wilsonville was shut down in the mid-1990s.

By the mid-1980s, large pilot plants processing about 200 tonnes of coal per day had been operated successfully in several countries. It should be possible to commercialize these processes with little technical risk. We could start tomorrow. However, the economic risk would be very great. These processes could be considered for commercialization only when the world price of oil and its availability on world markets would make liquefaction viable. As world oil prices dropped in the 1990s, government and industrial funding for direct liquefaction dropped significantly and focused more on laboratory and small pilot-plant tests, rather than running big pilot plants on the scale of Wilsonville. The subsequent impact of the shale revolution has made it all the more likely that conventional direct liquefaction is not going to be revived soon. What is needed is a breakout idea that will substantially impact process economics, whether this be new chemical reaction pathways, new ways of providing energy to break bonds, new solvents, or new something.

Three crucial ingredients led to the technical success of direct liquefaction in the mid-70s to mid-90s: time, money, and people. The worldwide

---

[8] This plant was formally known as the Advanced Coal Liquefaction R&D Facility, but is almost universally known to the coal liquefaction community simply as the Wilsonville pilot plant. It was operated by Catalytic, Inc.

[9] At the time this chapter was in final editing, it is about $108.

program took about two decades, from the first OPEC embargo to the shutdown of Wilsonville. The investment in the United States alone was nearly $6 billion dollars in 2021 dollars (Burke et al. 2001). Counting all the research and pilot-scale testing in numerous other countries, the collective investment would surely double that figure, likely even higher. These accomplishments reflect the contributions of an international army of scientists and engineers in industry, government laboratories, and academia aided by untold numbers of technicians, postdoctoral scholars, graduate students, and "lab rats." Investors, policymakers, and politicians in every country must realize that research and development efforts on this scale are not like light switches—you can't just flip it on and expect something to happen, right now. Success takes a sustained, serious commitment for the long haul. We need to be mindful of Carl Bosch's admonition about the time that was required to industrialize the synthetic ammonia process, as well as the time involved to get Bergius process plants running on a commercial scale.[10]

In 1993, I had the opportunity to visit Yuzo Sanada, one of many great Japanese coal scientists. At that time, both public and private investment in coal conversion was already in decline as a result of decreasing oil prices. When Dr. Sanada was asked why he was continuing with liquefaction research, he responded, "Because the last player to stay in the game wins all the stakes."

The world's only commercial-scale direct liquefaction plant, the last player in the game, is the so-called Shenhua plant, built by the Shenhua Group Corporation. It is located in Ejin Horo Banner, Ordos City, in China's Inner Mongolia region. The Shenhua process represents evolutionary development of earlier work beginning with the H-Coal process (Nowacki 1979, 123). In this process, pulverized coal is made into a slurry with a recycled oil. The slurry, mixed with hydrogen, is heated and passed through a reactor that contains a bed of catalyst particles. Rapid flow of the three-phase mixture— solid coal, liquid oil, gaseous hydrogen—keeps the bed in a churning motion that looks similar to boiling. The bubbling or boiling is referred to as ebullition; thus, this is an *ebullated-bed reactor*. The coal is ground to a much finer size than the catalyst particles so that unreacted particles of coal or mineral matter should be swept up and out of the top of the reactor without

---

[10] Though it might seem that the COVID-19 vaccine was developed and being used for large-scale vaccinations within only about a year of the recognition of the pandemic, that achievement—as exemplary and praiseworthy as it is—would have been impossible without the cadre of immunologists, virologists, epidemiologists, and other professionals and without the infrastructure for testing and producing vaccines already in place. We would be in quite a mess otherwise.

taking catalyst particles with them. The heavy, non-distillable portion of the product is gasified to make the necessary hydrogen. H-Coal was eventually scaled up to a 550-tonne per day pilot plant in Catlettsburg, Kentucky.

In the Shenhua plant, bituminous coal is slurried with recycle solvent and catalyst. The slurry is fed to a liquefaction reactor (the largest one ever built, with a 6,000 tonne per day capacity), followed by solid–liquid separation. The primary liquids are reacted further with hydrogen to produce diesel fuel and naphtha. Construction started in 2004, with an initial investment of about $2 billion dollars (Woodward 2014), or $2.8 billion in today's dollars. The plant became fully operational late in 2010. By 2013, the plant was producing nearly a million tonnes of liquids per year and running 315 days (Shu 2014). An analysis completed in 2015 suggested that the energy return on investment—the ratio of the energy in all of the plant's marketable products to the energy invested in making those products—is "low, or even negative" (Kong et al. 2015). That is, such plants do not add significantly to the total amount of energy available to society. Also, life-cycle CO_2 emissions for the production of liquids from coal are "much higher" than from conventional fuels (Kong et al. 2015), so that future plants must incorporate methods for dealing with CO_2.

In two-stage process concepts, the second stage operates with catalytic reactors to upgrade the first-stage liquid into marketable products. An oil refinery employs numerous catalytic reactors for basically the same purpose but using crude oil rather than a first-stage coal liquid. We could consider the refinery operations as the "second stage" of liquefaction. This means that the first-stage coal liquefaction products would feed directly into an existing oil refinery. The liquefaction plant would be retrofitted to existing refinery infrastructure. Certainly there are opportunities and possibilities for integrating coal liquid streams with petroleum refinery operations (Burgess-Clifford et al. 2008).

Direct liquefaction could be used for chemical production as the primary product goal. Direct liquefaction preserves vestiges of the molecular structure of the coal. It is a potentially good source of aromatic chemicals (Schobert and Song 2002). Innovative thinking might lead to a processing plant with a dual-product strategy, making liquid fuels and chemicals as two separate streams—possibly incorporating an IGCC plant as a third option.

Like F-T plants, direct liquefaction plants will require significant amounts of water. The needs had been estimated to be about 2 cubic meters of water for each cubic meter of liquid fuel product (Ramage and Tilman 2009, 214). The Shenhua experience suggests that the water requirement is about 6 tonnes of water for every tonne of product (Shu, 2014).

A 2007 study listed 12 coal-to-liquids plants being considered for construction (Bellman et al. 2007, 13). Nearly 15 years later, none has been built. Why not? Regardless of the specific liquefaction technology, the plants will be very complex and therefore very expensive. Building a commercial plant would require an investment in multiple billions of dollars. It will not be easy to come up with this kind of money, especially with several major funding sources announcing that they will no longer finance coal projects.

A second factor is the swings in world prices for petroleum in the past 30 years. If all the money needed were in the bank, and if all the necessary permits and other regulatory hurdles had been cleared, it would take at least 5 years to build the plant—to put concrete on the ground and steel on the concrete. Years could be consumed beforehand just getting the permits. It's thought that synthetic liquids from coal make sense only when petroleum is at or above $60 per barrel (O'Sullivan 2018, 150).[11] But then it has to stay at that level through many years of finding a plant site, getting the necessary permits, constructing the plant, and getting product onto the market. Who can tell what the oil price will be 10 years from now?[12]

The substantial increase in petroleum production as a result of hydraulic fracturing has, at least temporarily, allayed fears of actually "running out of oil" and fears of artificial shortages and price hikes caused by geopolitical events. The shale revolution has become a major game-changer and will be discussed in the next chapter.

After the 2001 terror attack on the United States, followed by invasions of Iraq and Afghanistan, I used to hear comments like, "If our supply of Mideast oil is cut off, we'll build coal-to-liquids plants just like the Germans did in World War II." Well, not quite. Of course the plants would not be just like the German plants—we have actually learned some things about liquefaction in the past 80 years. But, much more importantly, those who offered these comments were probably not aware of the long construction times needed for large facilities running complicated processes. The decision to build the gasification plant in North Dakota was made in 1969. The plant came online in 1985. Much of that time was spent in courtrooms or hearing rooms, not in pouring concrete or connecting pipes. I have been told that Sasol's Secunda facility took 6 years to build, under a government that certainly had no tolerance for lawsuits, demonstrations, appeals, and counterappeals. Shenhua was commissioned 4 years after it had been started and was fully operational in

---

[11] At the moment this is being written, it is about $62—just barely high enough.
[12] Furthermore, anyone who can accurately forecast the price of petroleum on the world market a decade into the future has his or her financial future secured—no need to be fooling around with coal liquefaction.

6 years. There's no question that liquefaction technology works. Any country, given a good supply of coal and the will to do so, could build and run liquefaction plants. The intriguing question is what we would do in the 6 years or so while the plants are being built.

> During the Second World War, while we trained soldiers in a few weeks and military officers in a few months, the major technological developments were made by scientists and engineers who were recruited, educated, and professionally seasoned before the war. The requisite establishment could not have been created de novo in anything less than at least a generation. When the blizzard arrives, it's a little late to place an order for a municipal snowplow. (Vogel 1992, 268)

References

Balster, Lori M., E. Corporan, Matt J. DeWitt, J. Timothy Edwards, J. S. Ervin, J. L. Graham, S. Y. Lee, Sibtosh Pal, Donald K. Phelps, Leslie R. Rudnick, Robert J. Santoro, Harold H. Schobert, L. M. Shafer, Richard C. Striebich, Z. J. West, Geoffrey R. Wilson, R. Woodward, and Steven Zabarnick. 2008. "Development of an Advanced, Thermally Stable, Coal-Based Jet Fuel." *Fuel Processing Technology* 89: 364–378.

Beaver, Bruce, André Boehman, Caroline B. Clifford, Robert J. Santoro, and Harold H. Schobert. 2009. "Final Report for Contract FA9550-07-01-0451." U.S. Air Force Office of Scientific Research.

Bellman, David K., James E. Burns, Frank A. Clemente, Michael L. Eastman, James R Katzer, Gregory J. Kawalkin, George G. Muntean, Hubert Schenk, Joseph P. Strakey, and Connie S. Trecazzi. 2007. "Coal to Liquids and Gas." In: *Hard Truths: Facing the Hard Truths about Energy*, 13. Topic Paper 18. Washington, DC: National Petroleum Council.

Borkin, Joseph. 1978. *The Crime and Punishment of I. G. Farben*. New York: Barnes and Noble.

Burgess-Clifford, Caroline E., André Boehman, Chunshan Song, Bruce Miller, and Gareth Mitchell. 2008. "Refinery Integration of By-Products from Coal-Derived Jet Fuels." U.S. Department of Energy Report, No. DE-FC26-03NT41828.

Burke, Frank P., Susan D. Brandes, Duane C. McCoy, Richard A Winschel, David Gray, and Glen Tomlinson. 2001. "DOE Direct Liquefaction Process Development Campaign of the Late Twentieth Century: Topical Report." U.S. Department of Energy Report. No. DOE/PC 93054-94.

Haslam, Robert T., and Robert P. Russell. 1926. *Fuels and Their Combustion*, 49. New York: McGraw-Hill.

Howard, Frank A. 1947. *Buna Rubber*. New York: D. Van Nostrand and Company.

Kong, Zhaoyang, Xiuchen Dong, Bo Xu, Rui Li, Qiang Yin, and Cuifang Song. 2015. "EROI Analysis for Direct Coal Liquefaction without and with CCS: The Case of the Shenhua DCL Project in China." *Energies* 8: 786–807.

Lucier, Paul. 2008. *Scientists and Swindlers*. Baltimore: Johns Hopkins University Press.

Nowacki, Perry. 1979. *Coal Liquefaction Processes*. Park Ridge, NJ: Noyes Data Corporation.

Orchin, Milton, C. Golumbic, J. E. Anderson, and Henry H. Storch. 1951. "Studies of the Extraction and Coking of Coal and Their Significance in Relation to Its Structure." U.S. Bureau of Mines Bulletin No. 505.

O'Sullivan, Meghan L. 2018. *Windfall*. New York: Simon and Schuster.

Probstein, Ronald F., and R. Edwin Hicks. 1982. *Synthetic Fuels*. New York: McGraw-Hill.

Ramage, Michael P., and G. David Tilman, eds. 2009. *Liquid Transportation Fuels from Coal and Biomass*. Washington, DC: The National Academies Press.

Rhodes, E. O. 1945. "The Chemical Nature of Coal Tar." In: *Chemistry of Coal Utilization*, edited by Homer H. Lowry, 1293. New York: John Wiley & Sons.

Schobert, Harold H. 1987. *Coal: The Energy Source of Past and Future*. Washington, DC: American Chemical Society.

Schobert, Harold H. 2013. *Chemistry of Fossil Fuels and Biofuels*. Cambridge: Cambridge University Press.

Schobert, Harold. H. 2015. "Toward the Zero-Emission Coal-to-Liquids Plant." *Technology* 3: 147–153.

Schobert, Harold H., and Chunshan Song. 2002. "Chemicals and Materials from Coal in the 21st Century." *Fuel* 81: 15–32.

Shu, Ge. 2014. "Recent Progress of Shenhua Direct Coal Liquefaction Project." China Shenhua Coal to Liquid and Chemical Co. Ltd. https://iea.blob.core.windows.net.

Stranges, Anthony. 1997. "The U.S. Bureau of Mines's Synthetic Fuel Programme, 1920–1950s: German Connections and American Advances." *Annals of Science* 54: 29–68.

Vogel, Steven. 1992. *Vital Circuits*. New York: Oxford University Press.

Whitehead, Alfred North. 1979. *Process and Reality: An Essay in Cosmology*. New York: Simon and Schuster.

Woodward, Katie. 2014. "Results from China's Coal to Oil Project." World Coal. https://www.worldcoal.com/coal/29012014/results_from_chinese_coal_liquefaction_project_462

Zhou, Z. F., Christina Gallo, Michael B. Pague, Harold H. Schobert, and Serguei N. Lvov. 2004. "Direct Oxidation of Jet Fuels and Pennsylvania Crude Oil in a Solid Oxide Fuel Cell." *Journal of Power Sources* 133: 181–187.

14

Competition

Coal is suffering from competition with large quantities of inexpensive natural gas and from the increasingly attractive costs of wind and solar energy. Coal is under assault for its contributions to human-made climate change because, of all the useful fuels, coals produce the most carbon dioxide (CO_2) emissions per unit of useful energy.

The past decade has seen a significant decline in the use of coal in many countries, mines closing, jobs lost, bankruptcy filings, and some leading financial institutions announcing that they will no longer fund coal-related projects. There is indeed a war on coal. It's just not the one the coal industry has been fighting. There is a vigorous assault on coal, a war on two fronts. On one, there is steadily increasing availability and decreasing cost of energy forms perceived to be cleaner and more convenient than coal. Natural gas and oil have been made cheaper and more available by new methods of extraction. Renewable energy forms, wind and solar, are becoming more economical. These are discussed in this chapter. The other front, the subject of the next chapter, is the growing appreciation that the world's climate is changing, that there is a human contribution to climate change, and that the long-term effects could be calamitous.

Coal mining jobs had been disappearing for decades in response to market forces—the decreasing costs of natural gas, solar energy, and wind energy (Anton 2017). These forces combine with increasing mining productivity and steady reductions in the heat rate of power plant boilers and the coke rate of blast furnaces. Nearly half of the decline in US coal production is due to natural gas—increasing availability coupled with decreasing prices—and another one-fifth of the decline is due to various forms of renewable energy (Houser et al. 2017). Governmental regulations established prior to 2016 could be linked only to a 4% decline in coal production (Houser et al. 2017).

The availability of new supplies of natural gas and oil may continue to impact coal use for several decades. The principal cause is the development of new extraction technology that allows gas and oil to be recovered from geological formations, mainly shale, that were previously thought too expensive or too difficult to exploit. Bringing these new supplies to market has had such an impact that it is sometimes referred to as the *shale revolution*. The revolution has not only affected the oil and gas industry; it is clearly affecting the coal industry as well. For coal in the twenty-first century, the shale revolution is a major game-changer.

Around the turn of the century, there seemed to be a consensus that world supplies of oil and natural gas were steadily diminishing and that shortages of both were looming. There seems to be a consensus of sorts that, in the long—whatever long means—term, the energy economy will have to be based primarily on a mixture of renewable energy sources: solar and wind energy, biofuels, and possibly hydrogen.[1] But, as we were moving into the first decade of the present century, it seemed likely that shortages of oil and gas would have an impact well before these alternatives could be fully deployed. What do we do when caught in the middle, with oil and gas declining but renewables not yet picking up the slack? Coal remains abundant. Producing substitute natural gas and synthetic liquid fuels from coal is unquestionably technically feasible. A few decades ago, gasification and liquefaction appeared to offer a "bridge" between the declining availability of oil and gas, on the one hand, and the steadily but slowly increasing availability of alternative energy sources, on the other hand. Now that bridge is natural gas.

In the past 10–15 years, this picture has changed very dramatically because of the emergence of enormous new quantities of oil and gas. They became available thanks to the maturing and commercialization of methods for extracting oil or gas from shales that could not have been exploited by conventional technology. This rapid, remarkable change in the energy scene is the shale revolution.

Until this century, most oil and natural gas produced worldwide has come from accumulations in reservoirs of porous rocks, such as sandstone or limestone. The reservoirs lie sealed under layers of impermeable rocks that prevent the hydrocarbon fluids from percolating all the way to the surface and escaping. Once a deposit has been located—an activity requiring considerable skill and knowledge—the oil or gas can be extracted via wells drilled into the reservoir. Thanks to the permeability of the rocks in the reservoir, the fluids can migrate, or can be encouraged to migrate, to the well bore, from which we can bring them to the surface.

A good reservoir rock needs two characteristics: porosity and permeability. *Porosity* indicates the number of void spaces in the rock, as with coals. *Permeability* indicates how readily the rock allows the oil or gas to travel

---

[1] That's because we will at some point have used up all of the economically available sources of coal, oil, and gas.

through it. For a rock to have good permeability, its pores must interconnect, forming passageways that allow the fluid to travel. The pores must also be large enough to permit the movement of oil or gas molecules. A rock could have a high porosity, but its pores could be so small as to be permeable only to the small molecular components of natural gas (e.g., methane) but not allow passage of the much larger molecules in petroleum. Some rocks could be porous but have virtually no permeability at all because of the extremely small size of pores—shale, for example.

Oil or gas are relatively easy and inexpensive to produce from rocks having good porosity and permeability. Such deposits have come to be known as *conventional oil* or *conventional gas*. This term suggests that there must be other kinds of oil and gas that are somehow unconventional. The adjective "unconventional" indicates deposits of oil or gas that do not occur in the porous, permeable rocks from which they can be extracted by via vertical wells drilled into the reservoir. The world contains vast amounts of unconventional oil and gas. The Orinoco oil belt in Venezuela possibly contains a trillion barrels of unconventional oil. This quantity of oil alone would supply the entire world with oil, at the 2019 consumption rate, for nearly 30 years.

Unconventional oil includes several types of materials. Particular interest nowadays attaches to "tight oil." or *shale oil*.[2] Shale oil might eventually prove to be the most important of the unconventional liquid hydrocarbon sources, at least in North America. The term "unconventional gas" refers to gas in formations of low permeability, including not only rocks such as shales, but also to coals, as in coalbed methane. Enormous quantities of shale gas occur in all of the North American countries and many other countries as well, such as China, Argentina, Australia, and South Africa.

Shale oil and shale gas are held in formations much different from the conventional porous, permeable reservoir rocks. Methods for recovering and producing these materials will have to be unconventional as well. The extraction method that has led to the shale revolution relies on a process called *hydraulic fracturing*, commonly known as *fracking*.

Every shale is different. Oil or gas might exist in shales not amenable to recovery by present technology. Different layers of shale might contain different quantities of recoverable oil or gas. The shale to be exploited must be at relatively shallow depth to keep drilling costs low but deep enough to have

---

[2] Shale oil is petroleum that is tightly held in shale rock of low permeability. The seemingly similar term "oil shale" refers to shales or other rocks that hold significant amounts of type I or type II kerogen. Extracting shale oil requires fracking of the shale to free the oil. Obtaining oil from oil shale requires heating the shale to convert its kerogen to hydrocarbon liquids.

an ample reservoir pressure at the bottom of the well bore.[3] Fracking requires large quantities of water, so gas or oil recovery could be difficult in arid regions. China has the largest shale reserves in the world, but those shales are much more fragmented, more resistant to fracturing, and are found in such places as Sichuan Province, which is earthquake-prone, or in the desert regions of Xinjiang Province (Brooks 2013).

The very low permeability of shale makes it difficult for enough gas or oil to flow to a conventional, vertical well in quantities that make it commercially worthwhile. The surface area of the well, going through the shale, is limited; only a small percentage of the shale is exposed to the well. The combined effect of small rates of flow of oil or gas to a small surface area presented by the wellbore results in production too low to be economically viable. The solution comes from substantially increasing the amount of well casing that contacts the shale. This can be achieved by drilling the well horizontally through the shale formation. This is shown in Figure 14.1.

Hydraulic fracturing uses water pumped into a well at high pressures to break some of the surrounding rock. The technique was originally developed for typical vertical wells passing through reservoir rocks to increase their porosity and permeability (Levorsen 1958, 131). The process consists of forcing a liquid containing suspended grains of sand into the pores of rocks using extremely high pressures. The resulting fractures provide connecting passageways between the oil- or gas-bearing rock and the wellbore. The technology that we know today evolved as a result of bringing together three techniques: the use of three-dimensional seismic data, the ability to do horizontal drilling, and hydraulic fracturing. Applying these techniques to shale rather than to conventional reservoir rocks has opened up the great quantities of oil and natural gas that constitute the shale revolution, with their dramatic effect on the energy economy.

By 2015, the United States and Canada accounted for almost all of the production of oil from shales, but it is expected that several other countries will contribute significantly by about 2040 (O'Sullivan 2018, 47). The large shale oil fields in North Dakota, Oklahoma, and Texas have sometimes been called "Cowboyistan" (Krauss 2015), a facetious reference to the names of such

---

[3] Two factors contribute to reservoir pressure: *hydrostatic pressure*, caused by fluid in the pores, and *lithostatic pressure*, the stress applied by the weight of the overburden. Hydrostatic pressure determines how much gas can be held in the shale. Lithostatic pressure is one of the factors that control permeability.

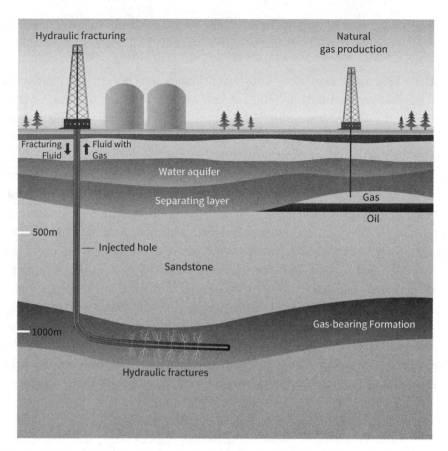

Figure 14.1 A schematic of hydraulic fracking of a horizontally drilled well. Such operations are the source of the present abundance of natural gas.
Courtesy of medicalstocks/Shutterstock.com.

oil-producing nations as Kazakhstan.[4] By comparison to conventional wells, shale oil production involves drilling hundreds, even thousands, of wells into shale, wells that are quick to drill and comparatively inexpensive. The average cost to drill and complete a well in American shale oil basins in 2016 was about $5 million dollars (O'Sullivan 2018, 32). The huge amounts of oil and natural gas made available by fracking restored the global competitiveness of the American petrochemical industry. The United States became the world's leading producer of natural gas and petroleum, reducing reliance on

---

[4] The term derives from the fact that many of the oil-producing nations around the Caspian Sea have names ending in –stan and are sometimes collectively called "The Stans." The odd nickname of "Cowboyistan" at least is better than the one applied to the Kashagan oil field in the Caspian Sea, which took more than a hundred billion dollars and 10 years longer than expected to develop: it's known as the "Cash All Gone" field.

imports. Manufacturing costs have been reduced as a result of lower costs for energy. Possibly the United States could dominate oil markets for decades (Krauss 2015).

For a long time, it has been known that shales can contain gas, but the oil and gas industry focused on shallower reserves that are easier to extract. Many people in the oil and gas industry thought that trying to exploit shale gas was a waste of time and money. Thanks to the oil embargo and oil price shocks of the 1970s, the US government began funding research into horizontal drilling and fracking. It took more than 20 years to perfect fracking, a time scale that should call to mind experiences with the Bergius-Bosch direct liquefaction process. The US government funded more than $100 million dollars' worth of research and development effort and probably allowed several billion dollars in tax breaks to participating companies. The continual improvements in the technology that resulted in its extensive applications by the turn of the century are largely due to the persistence and dedication of George Mitchell, who combined the method of hydraulic fracturing with improvements in the technology of horizontal drilling. Mitchell had spent 17 years and many tens of millions of dollars finally to achieve success. He was close to 80 years old at the time. Fracking provides another example of the payoff that comes from a willingness to fund research and stick with it for the long term—even for decades.

Cheap natural gas has done much more damage to coal's place in the energy market than have environmental regulations (Webber 2015). The low cost of natural gas in the United States has made it attractive for manufacturers to switch to the use of gas instead of coal. Utilities have been switching from coal to gas for electricity production. In 2016, natural gas surpassed coal as the fuel source responsible for generating the largest fraction of electricity in the United States. Most large power plants under construction in the United States will rely on gas or renewables. What changes the game are cheaper alternatives to coal, not government edicts or investment decisions (Helm 2017, 200). In 2005, natural gas in the United States cost $15 per gigajoule, but had dropped to about $3 by 2015 (O'Sullivan 2018, 151). The Great River Energy Coal Creekpower station near Underwood, North Dakota, underscores these developments. It will be closed earlier than originally planned due in part to historically high levels of natural gas production in that state. Great River went so far as to offer to give away the plant, but there were no takers (Hughlett 2020a).

New power plants being built in the region of the Marcellus and Utica shales are designed for gas, not coal. The 940-megawatt (MW) Lordstown Energy Center in Ohio uses two gas-fired combined-cycle units, operating with an efficiency of about 61% (Larson 2019). Low gas prices, combined with the fact that zero-sulfur, zero-ash gas eliminates the need for sulfur and particulate emission control equipment, make gas more attractive than coal. Similarly, shale gas exported from the United States to Europe and South Korea as liquefied natural gas (LNG) can have a similar effect, shifting electricity generation and industrial operations away from coal toward gas.

China has the largest reserves of shale gas (Economist 2014). Production is low, and coal still provides about 70% of China's energy needs. Chinese shale deposits are not as amenable to current fracking technology as those of the United States (Brooks 2013). In China, coalbed methane might be easier and less expensive to develop than shale gas. If China were to make a serious effort to reduce CO_2 emissions, natural gas could potentially displace much of its coal consumption. Concerns about air quality in China might bring a time when the residential and industrial energy sectors transition from coal toward gas. Much of China's methanol production derives from coal via the technologies discussed in Chapter 12. In most countries, including the United States, methanol is made from natural gas by reacting it with steam (very much like gasification, but in this business called *steam reforming*), shifting, and synthesis. The low price of gas in the United States has made it cost-competitive to produce methanol to be shipped to China, even with the burden of the transportation cost, rather than making methanol from coal in China. In the Chinese market, much of the methanol is converted to ethylene and propylene, raw materials for polyethylene and polypropylene. Methanol is also used for blending in Chinese gasoline.

India is second among the nations of Southeast Asia in terms of shale gas reserves (ElSakka et al. 2018). India is also richly endowed with coal. The impact of huge resources of shale gas—some 50 trillion cubic meters total—on Chinese and Indian coal consumption remains to be seen. Both countries may continue to rely heavily on coal for decades to come.

There can be no question that the shale revolution has quickly and dramatically changed the North American energy economy. The shale revolution will have worldwide effects, though mainly in developed countries (Paylor 2017). It has also shifted the competitiveness of the chemical industry. The lower CO_2 emissions per unit of energy also help to make shale gas an attractive commodity. These changes have a negative impact on the coal industry right now, maybe for decades to come and possibly forever.

As has been the case for many a political revolution, the shale revolution also has its dark side. Methane leaking into the atmosphere from shale gas wells or pipelines contributes to global warming. In some areas, excess gas is simply burned—"flared"—at the well. Flaring takes care of the methane all right, but combustion produces CO_2, the most notorious of the greenhouse gases.[5] However, shale gas has a heat of combustion about twice as high as most coals, yet forms only half the CO_2 per unit of energy produced. Proponents argue that, on balance, use of shale gas or conventional natural gas is less harmful to the climate than is use of coal to produce the same amount of energy. Air quality data from numerous public agencies has shown that the air quality around shale operations meets safety standards (Condo and Klaber 2017).

Fracking uses enormous volumes of water. The amount required depends on the geological nature of the area being fracked, but an order-of-magnitude figure would be about 45 million liters of water per well (Erickson 2019). By US standards, this is as much water as nearly 400 people would use in a year. The total water consumption for fracking in the United States alone would supply 3.8 million people (Hambling 2015). Water usage is a particular concern in arid or semi-arid regions, where water availability is already crucial for homes, businesses, and agriculture.

Much of the public opposition to fracking seems to be based on the fear that groundwater supplies will be contaminated by the fracking operation releasing chemicals into local water supplies. Fracked shale is usually separated from aquifers by hundreds of meters of low-porosity, low-permeability rock. However, the US Environmental Protection Agency (EPA) has found evidence that hydraulic fracturing activities can impact drinking water resources under some circumstances (EPA 2016).

Some people living where fracking is taking place have encountered methane in their water supply. The most notorious example appears in the documentary film *Gasland*, in a scene in which a man is shown igniting water from his kitchen tap (Fox 2010). In 2010, an analysis conducted by the EPA found methane in groundwater matching the gas produced from two nearby fracking sites in Parker County, Texas. This was shown by carefully matching the proportions of ^{12}C and ^{13}C isotopes in the various methane samples. Drinking water wells within 1 kilometer of a shale-gas drilling site in the Marcellus Shale contained 17 times more methane than samples from water wells farther away (Aldhous 2012). The same water samples did not contain detectable amounts of the chemicals used in the fracking fluid or of mineral salts in the shale. The methane in this case did not come from fracking.

There is also the issue of what to do with the thousands of cubic meters of water coming out of the well. Most water contamination issues associated

---

[5] While CO_2 indeed seems to be the best-known and most notorious of the greenhouse gases, in terms of the ability to absorb infrared radiation, methane is 20 times more powerful than CO_2.

with fracking come from what happens to all this water aboveground, not what occurs in the fracking underground (Ritter 2014). The water has to be safely treated and disposed of somehow. It can be hauled to a wastewater treatment facility. It may be possible to hold the water in tanks or open ponds to recycle and use again. Reuse seems to be increasingly favored (Redden 2020). Or, it can be injected into extremely deep wells with the thought that then we would be rid of it forever. This may not be the best of ideas.

Fracking has been linked to minor earthquakes in the United States, the United Kingdom, and Canada (Bao and Eaton 2016). An estimated 5–10% of fracked wells might be associated with earthquakes (Maddow 2019, 291). Prior to the significant increase in fracking activity around 2008, Oklahoma typically saw two magnitude 3.0 earthquakes[6] in a year (Maddow 2019, 291). In the first half of 2014, Oklahoma experienced 16 earthquakes of this magnitude (Maddow 2019, 291). Including these were 284 earthquakes above magnitude 3.0 in that year in Oklahoma (Maddow 2019, 291). There was a magnitude 5.7 quake in Wilzetta County, referred to as a "boomer" (Maddow 2019, 159). Most are not caused by fracking, but rather relate to the reinjection of fracking fluids into deep formations. Pumping millions of cubic meters of fracking wastewater deep underground at high pressures causes faults—fractures that allow one side of the break to be displaced parallel to the other side—that may have existed for millennia to become tectonically active again (Ingersoll 2015). One side of the fault may start to move relative to the other. The great volume of high-pressure water changes the pattern of underground pressures and stresses. When one side of a fault suddenly slips past the other, an earthquake can occur (Wald 2017). Generating earthquakes by human activities such as this is a process called *induced seismicity*.

Concerns about fracking have led at least to public opposition and, in some places, to moratoria or outright bans. These concerns include the issues of possible contamination of resources, induced seismicity, construction of new pipelines, and accidents or spills with transportation of gas, oil, or wastewater. On the other hand, in countries such as Lithuania, Ukraine, and Poland, which currently depend on gas imports from Russia but have domestic shale reserves, fracking could look rather attractive compared to being vulnerable to Russian geopolitical goals.

---

[6] The severity of earthquakes is expressed using a scale developed by the American physicist and seismologist Charles Richter. The Richter scale is logarithmic, meaning that an earthquake of, say, magnitude 5.0 will be 10 times worse than one of magnitude 4.0. As a rough rule, a magnitude 3.0 earthquake can be felt by people but rarely causes damage; at 4.0 there is shaking indoors and some objects might topple or fall over, but no structural damage; and at 5.7, old or poorly constructed buildings could experience significant damage, with light damage to other buildings.

The term *renewables* refers to various energy sources that occur in nature and that are considered to be inexhaustible. This contrasts renewables with energy sources that cannot be renewed, at least on a human time scale, which are thought of as being *finite resources*. Finite resources—mainly coal, oil, and gas—will eventually reach a point at which the energy expended to recover the resource is greater than the energy it will produce. Many kinds of energy sources are considered to be renewables: wind, solar, biomass, geothermal, and the energy in falling water or in tides. Renewables perpetuate themselves, or with help from us in the case of biomass. Among the renewables, wind and solar currently seem to be making the most inroads into territory once occupied by coal. On April 21, 2017, none of the electricity used in the United Kingdom was generated using coal (LePage 2017). The last time this had happened was back in 1882.

Using windmills to produce electricity dates from the late 1880s. The prolific Danish engineer Poul la Cour recognized the virtue of using some of the wind-generated electricity to produce hydrogen from water for energy storage. His work on using wind to generate electricity is his most enduring achievement. In response to the oil price shocks of the early 1970s, the Danish government took several steps to encourage a shift back to the use of wind. Those steps included a 30% subsidy on the purchase of wind machines and a requirement that electricity utilities purchase the electricity that these devices produced. This provided the impetus for the revitalization of the wind industry.

Most of the world's electricity comes from generators that are operated by turbines. Since wind is a fluid (air) in motion, it, too, can be used as the working fluid in a turbine. We use the term *wind turbine* to refer to machines that use the energy of wind to operate an electricity generator. Today, the term "windmill" is usually restricted to mean those devices that use the wind for other mechanical applications, such as pumping water.

The blades of wind turbines are commonly made of fiberglass-reinforced composite materials. Carbon fibers offer some advantages relative to fiberglass, particularly higher stiffness with lower density (Mishnaevsky et al. 2017). Carbon-fiber reinforced blades will be lighter and thinner yet stiffer than the traditional blades. At present, the carbon fiber blades are likely to be more expensive. However, we will see in Chapter 17 that there are some good opportunities for making carbon fibers from coals.

Wind turbines can operate with wind speeds from a little under 15 to about 100 kilometers per hour. Great improvements have been made in reliability since the 1980s, when some wind turbines were out of service more than half the time. Currently the "down time" for a wind turbine is in the range of 2%.

In 1919, the German physicist Albert Betz derived the mathematical relationship that establishes the maximum theoretical amount of the kinetic energy of wind that can be converted to mechanical work (Betz 1966, 205). This is known as *Betz's law*, or the *Betz limit*, and has the value $^{16}/_{27}$, about 59%. Wind turbines currently operate at efficiencies of about 45%.

Worldwide, the total wind flow is ample to supply several hundred times the total annual energy consumption. It is scarcely likely that every potential site for wind-to-electricity energy conversion will be exploited. Land-use questions may be the most contentious issues for wind energy development. Some wind sites may be too remote to be practical or cost effective for connecting to a distribution grid. Others may be in environmentally sensitive areas where development might be limited. Still others might be in areas where the costs of acquiring the land and developing the site might be prohibitive. Some of the best locations are over oceans, though near to shore. Open expanses of flat or gently rolling plains provide high potential for wind energy development. Ridge-and-valley topography is also attractive because the ridges act as channels or conduits for the wind flow.

Commercial generation of electricity from wind nowadays relies on the *wind farm* concept. A wind farm is a location containing large numbers—from dozens to hundreds—of wind turbines, each generating 100–500 kilowatts. Worldwide, the wind industry is steadily growing. It is the fastest growing form of renewable energy in the United States (Outcalt 2020). The current world leader in terms of percentage of energy demand generated by wind is Denmark, appropriate as the home country of Poul la Cour. Currently, Denmark's wind turbines supply close to 48% (4,000 MW) of Danish electricity demand. China and the United States lead the world in terms of installed generating capacity. China is producing about 210 gigawatts from wind power (Proctor 2017). The planned Gansu wind farm will have a capacity of 20 gigawatts. The largest single facility in the United States is the Alta wind energy center in California, producing about 1.5 gigawatts. The Glenrock wind farm in Wyoming is built on reclaimed land once used for coal mining (Outcalt 2020), suggesting that other mined land reclamation sites might also be suitable. Wind power became competitive with the cost of new coal-burning power plants in the United States around 2014 (Malm 2016, 159).

Britain has the largest single offshore wind stations, one at 370 MW and another about 300 MW, and leads the world in offshore generation. The first offshore facility in the United States, producing 30 MW, is the Block Island Offshore Wind Farm, off the coast of Rhode Island (Harvey 2017a). The European Union has been projecting costs of $0.11 per kilowatt-hour (kWh) for electricity generated in offshore wind installations (Harvey 2017a).

Generally, electricity from coal costs between $0.07 and $0.15 per kWh; from natural gas, between $0.04 and $0.07; and from onshore wind, between $0.03 and $0.05 (Outcalt 2020). In 2017, the cost of generation without subsidies—known as *levelized costs*[7]—showed onshore wind to be only slightly behind natural gas as the cheapest form of electricity generation in the United States, at about $50 per megawatt-hour (Hughlett 2020b). In Germany, wind is less expensive than natural gas, coal, or photovoltaics; in China, where it will probably take a long time to displace coal, onshore wind generation is in second place behind coal.

Wind energy is essentially pollution free. There is no air or water pollution, no acid precipitation or smog, no radioactive waste, fly ash, or scrubber sludge. There are no emissions of CO_2 or of any other greenhouse gases. A wind farm poses no significant threat to the safety of the public.

Individual wind turbines require a relatively short time to build. In fact, they can be built on an assembly line. In comparison, the construction time to build a coal-fired power plant is several years because coal-fired plants are generally one-off designs which must be erected on site, with components brought in by truck or rail. Furthermore, wind turbine installations are modular. Once a site for a wind farm has been selected, new turbines can be added quickly and relatively inexpensively as electricity demand increases. Expansion of coal-fired plants would likely be in increments of several hundred megawatts at a time, costing tens or hundreds of millions of dollars. Small, natural-gas–fired gas turbine plants can be built relatively quickly and can be especially desirable for exploiting shale gas, but even these plants do not offer flexibility comparable to expanding a wind farm a few turbines at a time.

Electricity production using wind turbines requires a great deal of space. Because of the turbulence created by the rotating blades, turbines have to be placed 100–300 meters apart. A reasonably sized wind farm will have at least a hundred wind turbines. Though the total expanse of a wind farm may be large, the turbines themselves use no more than about 5% of the total land. The remaining land can be put to other uses, such as farming, ranching, forestry, or recreation. Because the turbines occupy only a small fraction of the land area, wind energy development is compatible with farming. Farmers can plant crops right up to the base of the turbine towers. Sheep may safely graze. Cows also graze close to the turbine towers. Leasing land for wind turbine

---

[7] Levelized costs omit a number of factors that will vary from one place to another but which could have important impacts on the final cost of electricity to the consumer. Some of the factors omitted include the cost of conventionally generated electricity or of batteries as back-up energy sources, and the additional cost of managing a system in which the amount of electricity coming from wind or solar can fluctuate, sometimes greatly, over short time scales.

installations provides extra income and increased land values for the land-owners. Concerns arise when the total area set aside for the wind farm must be weighed against other competing uses of the same land, such as housing developments.

Wind also has an advantage relative to solar energy. Solar suffers the undeniable disadvantage that the sun doesn't shine at night. On the other hand, wind can blow day or night and on sunny or cloudy days. Winds are often at their strongest and most reliable during the dark and stormy nights of winter, precisely when most of us have our greatest demand for electricity. But, even in a suitably windy site, wind does not blow all the time, nor does it usually blow steadily. Coal and shale gas plants can operate 24 hours per day, every day. Wind farms, necessarily dependent on wind, might only produce an average output of one-fifth of the rated capacity.

Because of the fluctuating nature of wind and the possibility of long periods of calm, a wind energy system needs back-up generating capacity or at least an effective energy storage system. An energy storage system is needed in areas where there is a mismatch between the times of peak generation from the wind farm and times of peak demand by customers. The costs of back-up and energy storage systems have to be factored into the total cost of the wind installation, as they do with solar energy. Energy storage systems include rechargeable batteries and supercapacitors, which we will revisit in Chapter 17.

Noise is another issue. Nobody has yet designed a silent turbine. Regulations might be established to limit the noise production from a wind farm and determine how close turbines may be sited to occupied buildings.

Birds can be killed by flying into the spinning blades of a turbine. Bird kills are of special concern for wind farms that happen to lie on bird migration routes. Bird kills are certainly not unique to wind energy systems. Birds are also killed or injured by flying into transmission and distribution lines from other kinds of electricity-generating plants. But by far the most effective killer of birds is glass—windowpanes.

A final argument is "visual pollution." Though they produce no emissions to the air or water, many people simply object to the sight of tens or hundreds of wind turbines and their associated transmission lines on the landscape. Offshore wind farms could negate many concerns relating to land use and noise, yet they, too, are accused of being a form of visual pollution.

The *capacity factor* is the amount of electricity actually generated (e.g., over the course of a year) divided by the maximum generation at the peak capacity of the plant. For example, a plant with a maximum capacity of 200 million kWh per year that actually generated 100 million kWh, would have a capacity factor of 50%. Capacity factors for wind farms were about 34% in 2019. A solar photovoltaic site is about 24% (EIA 2020). Similar values for coal-fired plants in the United States are about 50%. *Availability* represents the percentage of time the unit is actually operating (i.e., not including scheduled maintenance shutdowns and unexpected outages). If a plant operates for 7,000 hours in a year, its availability would 80%, for 8,760 hours in a year. Availability is about 95% or better for both wind and solar, compared with 80–90% for coal plants.

Two strategies allow us to take advantage of solar energy. One converts the heat of solar radiation into steam, which then gets us into the familiar territory of turbines and generators to produce electricity. The other, much less hardware-intensive, approach is to convert the electromagnetic energy of solar radiation directly to electricity.

We all observe that solar energy provides heat, sometimes painfully so in the form of sunburns. Heat can certainly be used to raise steam for a steam turbine/generator set. Since the energy in sunlight is not converted directly into electricity, this technology is referred to as *indirect solar conversion*, and the facilities are sometimes called *solar thermal power stations*. All of the principles are the same as those for a coal-fired plant. A source of heat produces steam that drives a turbine coupled to a generator that produces electricity. Heat in, electricity out. The only significant difference is the source of heat: combustion of coal or capturing the heat in sunlight.

Fossil fuel, nuclear, and indirect solar plants all use large and technologically complicated devices basically to boil water. The components of different plants—boiler, turbine, generator, condenser, cooling tower—will differ somewhat in particulars, such as temperatures and pressures, but share the same basic principles. All of this hardware involved in electricity generation, much of it rather expensive, could be dispensed with by simply converting the energy of sunlight directly into electricity. This approach is called direct solar electricity generation, or *photovoltaics*.

Conversion of light directly to electricity by the photovoltaic effect was discovered by the French physicist Edmond Becquerel, in 1839. He discovered the photovoltaic effect at age 19, while working in his father's laboratory. An explanation of the physical basis of the photoelectric[8] effect—how the wavelengths of light determine their ability to knock electrons out of the atoms—was published by Albert Einstein, in 1904. Einstein's Nobel Prize in physics was based on this research, not on his theory of relativity.

---

[8] Though "photovoltaic" and "photoelectric" are sometimes used as synonyms, strictly speaking, in the photoelectric effect, an electron is knocked away from an atom into a surrounding vacuum, but, in the photovoltaic effect, the electron remains in the material.

Development of the first practical photovoltaic cell occurred in 1954, at Bell Laboratories in the United States. These devices are now often referred to as *solar cells*.

Photovoltaic cells rely on semiconductors. Unlike familiar electric conductors such as copper, in which electricity flows easily, semiconductors require a modest input of energy to stimulate conduction. In solar conversion that needed input is the energy in sunlight. Silicon, often tweaked by the addition of tiny amounts of other elements such as gallium or arsenic, has electronic properties such that the energy in sunlight can stimulate conduction. It has been known for about 80 years that anthracites also are semiconductors (van Krevelen 1993, 415). They do not appear to be appropriate for direct photovoltaic conversion of sunlight. But, the early observations apparently have not been followed up by investigations of how one might alter the semiconducting properties by deliberately changing the anthracite molecular framework, incorporating small amounts of other elements, or both.

Single photovoltaic cells made of crystalline silicon have a theoretical maximum efficiency of 33.7%. Steady improvements in photovoltaic technology have brought practical silicon-based cells into the mid-20% range of efficiencies. With current technology, it would take an area of 15 square kilometers covered with photovoltaic cells to replace a 2,000 MW power station. An array of photovoltaic cells with an area less than a square meter could provide some tens of watts of power, enough for several small electric lights or for operating portable radios. With about twice that area, sufficient electricity is produced to drive a small electric motor, as, for example, in a water pump.

In the United States, the cost of a residential photovoltaic system dropped by half from 2010 to 2018, while the capacity of rooftop installations increased by a factor of 10 (Kennedy 2021). The next step forward might require a similar drop in the cost of batteries for an energy storage system. The alternative to building-by-building installations of solar panels is to set up centralized photovoltaic generating stations. Many countries now have this technology in place. A small installation in the United Kingdom, near Flitwick, uses 31,240 solar panels for a 10 MW capacity (Harvey 2017b). In the United States, the Topaz Solar Station is a 550 MW plant relying on photovoltaic technology. In 2017, China had already reached its 2020 target of producing 110 gigawatts from solar energy (Proctor 2017). One of the largest projects, the Longyangxia Dam Solar Park, uses 4 million solar panels over 25 square kilometers to generate 850 MW (Proctor 2017). The rapid acceleration of solar energy in China is due to the coming together of three factors: first is the dreadful air quality in some of the large cities, a good portion of which comes from coal combustion; second is a recognition that solar power can be installed relatively quickly, especially in comparison with coal or nuclear facilities; and the third is that a growing renewable energy industry is also a significant source of new jobs (Proctor 2017). China leads the world in having the most extensive solar infrastructure, with 175 gigawatts of capacity (Witt 2020). Even bigger than the Longyangxia facility is the Kurnool Ultra Mega Solar Park in Andhra Pradesh, India, a 1 gigawatt plant (Proctor 2017). India's original goal had been to have 20 gigawatts of solar capacity by 2022, but, with declining costs of solar, the country achieved its goal in 2018.

Photovoltaic systems do not use water for cooling or for steam generation so are well suited to remote or arid regions. The Topaz station uses no water at all. Photovoltaics can operate on any scale, from small hand-held consumer devices to full-scale electricity plants. Neglecting the issues of manufacturing inexpensive photovoltaic cells, there is really only one fundamental problem with solar energy: the sun doesn't shine at night. An energy storage system or a back-up means of generating electricity (or both) is needed.

The cost of solar electricity depends on three principal factors: The first is efficiency—the fraction of energy in the sunlight striking the cells that is converted to useful electrical energy. The higher the efficiency, the fewer the cells required to produce a given amount of electricity, or the more electricity can be generated with a given number of cells. The second factor is cost of producing the cells. The third is the overall expenditure required to install, operate, and maintain a solar energy facility. The situation is complicated by the fact that the more efficient cells are also the more expensive. Thus, it may prove better, for overall cost considerations, to use a large number of cheap but low-efficiency cells.

Since 2010, the cost of generating electricity with solar energy has dropped by about 80% (Witt 2017). Clearly, this represents a tremendous improvement. But, to be competitive with electricity generated in conventional fossil fuel plants, it's the cost of electricity delivered to consumers that must be competitive. In terms of levelized costs, solar photovoltaics and onshore wind are among the least expensive sources of electricity (Rozenblat 2019). Not only is the cost of producing electricity dropping, but so, too, is the price of the cells. This price is generally expressed as *dollars per peak watt*, where a peak watt is defined as 1 watt of power generated when the solar radiation power is 1,000 watts per square meter. Forty years ago, the cost of a solar cell was $1,000 per peak watt. It hit $1 in 2009. By 2020, costs of solar electricity were some five to eight times lower than had originally been forecast (Haggerty 2020).

As is the case with wind energy, solar modules and arrays can be made in assembly line fashion. This provides significant advantages in cost and delivery times compared to large, one-off designs for coal plants. The Flitwick 10 MW station in the United Kingdom was installed in 12 weeks (Harvey 2017b).

Solar energy is available anywhere on Earth. It can be collected easily with portable devices, thus minimizing problems of transporting fuel or transmitting electricity from point of production to point of use. Solar energy is nonpolluting and clean, and it produces no air or water pollution and no hazardous wastes. No mining or drilling is needed, other than the mining of a suitable source of silicon in the first place. The energy source—the sun— cannot be owned and monopolized, as fossil fuels have become because of the unequal geographic distribution of known reserves. The most important advantage of solar energy is that it is inexhaustible.

Like any energy source, solar energy also has its drawbacks. Sunlight is a dilute energy source, in contrast to concentrated energy in a flame. It requires large areas of land for centralized electricity generation. In many parts of the world direct sunlight is not consistent and nowhere is it available at night. Clouds and seasonal changes affect the rate at which direct sunlight can be collected. Energy storage or back-up systems using other energy sources are needed.

A renewable energy system that combines wind and solar (as in Figure 14.2) offers the possibility of relying mainly on solar during daylight hours and

Figure 14.2 A solar park with wind turbines. Developments such as these are providing renewable energy at less and less cost.
Courtesy of Soonthorn Wongsaita/Shutterstock.com.

wind at night, the time when winds often pick up. There is always the chance of a night when there is little or no wind, leading to a condition described by the German neologism *dunkelflauten*—the "dark doldrums" (Fairley 2020). Avoiding the dark doldrums creates a need for an energy storage system or for a hybrid with another energy source not dependent on sunshine or wind. For example, a hybrid solar system could use a low-cost fossil fuel system added to a photovoltaic system to compensate for variations in sunlight. A combined solar–natural gas hybrid would be one of the most environmentally benign ways of using fossil fuels to generate electricity.

The concept of pumped storage uses some of the wind or solar electricity generated during the daytime to pump water to large ponds on the top of a mountain. At night, the water is then allowed to flow back down hill to generate electricity using water-driven turbines. The Longyangxia Dam Solar Park also includes a 1,280 MW hydroelectric plant, which helps balance the variable output of the solar installation (Proctor 2017). Alternatively, some of the electricity generated from renewables could be used to electrolyze water, producing hydrogen. When needed, the hydrogen would be used in a combustion turbine much like the turbines used in shale gas plants or integrated gasification combined cycle (IGCC) plants.

In the United States, coal-fired power plants currently provide about 20% of the electricity, compared with nearly half in 2010. Natural gas supplies 40%, and renewable energy sources add another 20% (Penn 2020). Worldwide, solar accounts for about 9% of installed capacity, which results from an investment of about a trillion dollars (Witt 2020). It may now be less expensive to build a new solar plant than to operate an old coal-fired plant (Witt 2020). In 2017, China was installing at least one wind turbine and enough solar panels to cover a soccer pitch—about 7,000 square meters—every hour. Often, a solar or wind farm can be built in 6 months or less, quite a change from the 5–10 years that might be required for a coal-fired plant (Klein 2017).

Sites with features favorable for wind energy generation can now compete with the costs of coal-fired power (Edenhofer 2015). Solar photovoltaic energy can match the costs of coal generation. However, there are additional costs because both wind and solar are intermittent sources of energy, therefore requiring some kind of back-up capacity. These additional costs could make coal a more attractive investment than renewable-energy sources in some countries. The price of coal-based electricity generation remains lower than that of renewable power when the costs of intermittency are taken into account (Edenhofer 2015).

Worldwide, all forms of renewable energy, including geothermal and biomass, along with wind and solar, currently account for about 8% of electricity generation (LePage 2017). Historically, it takes decades for any new energy source to gain a significant fraction of the market, and it never totally displaces the older sources. In 1954, Lewis Strauss, chairman of the US Atomic Energy Commission, stated that a time would come when electricity from nuclear energy would be "too cheap to meter."[9] Well, it hasn't happened yet. While many look forward to a time when all of our energy needs will be supplied entirely by renewables, others caution that the road might be long and rough.

Certainly the availability of enormous new amounts of gas and oil from the shale revolution, on the one hand, and the steadily dropping costs of wind and solar, on the other hand, represents formidable competition for coal as a fuel for electricity generation. Using the US upper Midwest as an example, a huge surge in natural gas production from North Dakota combined with a steadily increasing number of wind farms in the region resulted in closures of coal-fired plants earlier than planned (Hughlett 2020a). It seems very unlikely that the downward trend in renewable energy costs is likely to reverse (Hausfather and Peters 2020). These factors are by no means confined to heavily industrialized nations. By 2018, such countries as Chile, Egypt, Mexico, and Morocco were producing electricity from solar installations at less cost than even natural gas (Figueres et al. 2018).

As this chapter was in its final editing in April 2021, the United Mine Workers of America (UMW)—the labor union that represents coal miners—announced that it would support President Biden's green energy policies in exchange for Congressional funding for the training that will provide good-paying jobs and benefits to coal miners who transition into the renewable energy sector. This is excellent foresight by the UMW. The writing seems to be on the wall.

The shale revolution, combined with the decreasing cost, plus increasing reliability, of wind and solar energy combine to put tremendous economic pressure on the traditional uses of coals. That alone would represent a difficult situation. At the same time, there is steadily growing concern about the effects of climate change. Although it is not correct to attribute climate change solely to increasing concentrations of CO_2 in the atmosphere, CO_2 certainly is a major factor. While there are many sources of CO_2 emissions to the air—both

---

[9] What Strauss was really referring to, but was unable to reveal at the time, was energy from nuclear fusion, not the fission reactors in use in his day and yet today. The historical background is available in an article by Thomas Wellock, at https://public-blog.nrc-gateway.gov/2016/06/03/too-cheap-to-meter-a-history-of-the-phrase/

natural and human-generated—the carbon–oxygen reaction as applied to fossil fuels is a major one. Of the various fossil fuels, coals produce more CO_2 per unit of useful energy (e.g., per megajoule) than any of the others. This situation puts a second intense pressure on coal utilization.

References

Aldhous, Peter. 2012. "Drilling into the Unknown." *New Scientist* 213(2849): 8–10.

Anton, P. 2017. "Hopeful Coal Miners are Pawns in an Economic Scam." *Minneapolis Star-Tribune*. April 4.

Bao, Xuewei, and David W. Eaton. 2016. "Fault Activation by Hydraulic Fracturing in Western Canada." *Science* 354: 1406–1409.

Betz, Albert. 1966. *Introduction to the Theory of Flow Machines*. Oxford: Pergamon Press.

Brooks, Michael. 2013. "Frack to the Future." *New Scientist* 219(2929): 36–41.

Condo, Katherine, and Katherine Klaber. 2017. "Air Emission Data Show Safety of US Northeast Shale Work." *Oil & Gas Journal* 115(3): 29–31.

The Economist. 2014. "Shale Game: Natural Gas in China." August 30, 60.

Edenhofer, Ottmar. 2015. "King Coal and the Queen of Subsidies." *Science* 349: 1286–1287.

EIA. 2020. "Capacity Factors for Utility Scale Generators Primarily Using Non-Fossil Fuels." U.S. Energy Information Administration. https://www.eia.gov/electricity/monthly/epm_table_grapher.php?t=epmt_6_07_b.

ElSakka, Ahmed, Ghareb M. Hamada, Eswaran Padmanabhan, and Ahmed M. Salim. 2018. "South East Asia Contains Abundant, Untapped Shale Reservoirs." *Oil & Gas Journal* 116(3): 34–44.

EPA. 2016. "Hydraulic Fracturing for Oil and Gas: Impacts from the Hydraulic Fracturing Water Cycle on Drinking Water Resources in the United States. Executive Summary." U.S. Environmental Protection Agency Report No. EPA-600-R-16-236ES.

Erickson, Britt E. 2019. "Wastewater from Fracking: Disposal Challenge or Resource?" *Chemical and Engineering News* 97(45): 22–25.

Fairley, Peter. 2020. "The H2 Solution." *Scientific American* 322(2): 36–43.

Figueres, Christiana, Corrine Le Quéré, Anand Mahindra, Oliver Bäte, Gail Whiteman, Glen Peters, and Dabo Guan. 2018. "Emissions Are Still Rising: Ramp up the Cuts." *Nature* 564: 27–30.

Fox, Josh. 2010. *Gasland*. Brooklyn: International WOW Company.

Haggerty, Jean. 2020. "Sunny Places Could See Average Solar Prices of $0.01 or $0.02 per Kilowatt-Hour Within 15 Years." PV Magazine. https://pv-magazine-usa.com/2020/05/19/sunny-places-could-see-average-solar-prices-of-0-01-or-0-02-per-kilowatt-hour-within-15-years/.

Hambling, David. 2015. "Oil on Zap!" *New Scientist* 227(3034): 34–37.

Harvey, Abby L. 2017a. "Nation's First Offshore Wind Farm Releases Community from Decades of Diesel." *Power* 161(12): 24–25.

Harvey, Abby L. 2017b. "Anesco Celebrates Subsidy-Free Solar." *Power* 161(12): 26–27.

Hausfather, Zeke, and Glen P. Peters. 2020. "Emissions: The 'Business as Usual' Story Is Misleading." *Nature* 577: 618–620.

Helm, Dieter. 2017. *Burn Out: The Endgame for Fossil Fuels*. New Haven, CT: Yale University Press.

Houser, Trevor, Jason Bordoff, and Peter Marsters. 2017. "Can Coal Make a Comeback?" Columbia University Center on Global Energy Policy Report. New York: Columbia University.

Hughlett, Michael. 2020a. "Great River Closing Coal Plant." *Minneapolis Star Tribune*, May 8.

Hughlett, Michael. 2020b. "SMMPA to Make Shift from Coal by 2030." *Minneapolis Star Tribune*, February 6.

Ingersoll, A. 2015. "Why Is N. Dakota's Oil Land Stable?" *St. Paul Pioneer Press*, March 22.

Kennedy, Ryan. 2021. "Solar Deployed on Rooftops Could Match Annual U.S. Electricity Generation." https://pv-magazine-usa.com/2021/10/11/solar-deployed-on-rooftops-could-match-annual-u-s-electricity-generation.

Klein, Alice. 2017. "Can China Save the World?" *New Scientist* (3143): 20–21.

Krauss, Clifford. 2015. "Balance of Power Has Shifted in Oil Market." *San Antonio Express News*, April 24.

Larson, Aaron. 2019. "Competitive Advantages: Power Plant Wins with Low Fuel Costs, High Efficiency." *Power* 163(10): 28–29.

LePage, Michael. 2017. "The Green Revolution is Stalling." *New Scientist* 235(3137): 22–23.

Levorsen, Arville I. 1958. *Geology of Petroleum*. San Francisco: W. H. Freeman and Company.

Maddow, Rachel. 2019. *Blowout*. New York: Random House.

Malm, Andreas. 2016. *Fossil Capital*. London: Verso.

Mishnaevsky, Leon, Kim Branner, Helga N. Petersen, Justine Beauson, Malcolm McGugan, and Bent F. Sørensen. 2017. "Materials for Wind Turbine Blades: An Overview." *Materials* 10: 1285.

O'Sullivan, Meghan L. 2018. *Windfall*. New York: Simon and Schuster.

Outcalt, Chris. 2020. "Into the wind." *Wired* (4): 80–85.

Paylor, Adrian. 2017. "The Social-Economic Impact of Shale Gas Extraction: A Global Perspective." *Third World Quarterly* 38: 340–355.

Penn, I. 2020. "New Battle Lines Pit Natural Gas vs. Renewable Energy." *Minneapolis Star Tribune*, July 12.

Proctor, Darrell. 2017. "China's Renewables Strategy Shines in Massive Solar Park." *Power* 161(12): 28–29.

Redden, James. 2020. "Key Frac Ingredients Moving in Different Directions." *Exploration & Production* 93 (1):30–42.

Ritter, Stephen K. 2014. "A New Way of Fracking." *Chemical and Engineering News* 92(19): 31–33.

Rozenblat, Lazar. 2019. "Your Guide to Renewable Energy." http://www.renewable-energysources.com/.

Scheiber, Noam. 2021. "A Coal Miners Union Indicates It Will Accept a Switch to Renewable Energy in Exchange for Jobs." nytimes.com /2021/04/19/business/coal-miners-renewable-energy.html. April 19.

van Krevelen, Dirk W. 1993. *Coal: Typology – Physics – Chemistry – Constitution*. Amsterdam: Elsevier.

Wald, Lisa. 2017. "The Science of Earthquakes." U.S. Geological Survey. https://www.usgs.gov/natural-hazards/earthquake-hazards/science/science-earthquakes?qt-science_center_objects=0#qt-science_center_objects.

Webber, Michael E. 2015. "Coal." *The Washington Post National Weekly*. August 16.

Witt, S. 2020 "The Shining." *Wired*. (4): 90–94.

15
Climate

The facts that coal yields more carbon dioxide (CO_2) per unit of energy liberated than other fuels, that CO_2 levels in the atmosphere have been rising steadily since we started using coal on increasingly greater scales, and that the climate is changing, taken together, represent the second front in the war on coal. Some argue that we must stop burning coal almost immediately, or indeed stop using all fossil fuels. We explore the issues in this chapter.

In 1856, the American scientist Eunice Newton Foote first identified that CO_2 and water vapor have a role in how the atmosphere helps to retain heat on Earth (Schwartz 2020). The Irish physicist John Tyndall soon showed that the effect is due to absorption of infrared radiation (Perkowitz 2019). He also recognized that gases such as CO_2 and water vapor could influence the climate. Tyndall published his work in 1859.

The first calculations estimating the amount of increase in Earth's surface temperature were done in 1896, by Svante Arrhenius, whom we met in Chapter 7, though his model was very simple. Arrhenius reckoned that CO_2 from coal combustion could be a benefit, warding off another Ice Age and improving the climate. In 1896, the concentration of CO_2 in the atmosphere was 296 parts per million (ppm) (Economist 2013). It is currently 419 ppm, and rising (Daily CO_2 2022). There is a strong consensus within the scientific community that we are headed toward seriously unpleasant effects of climate.

The British turbine engineer Guy Callendar compiled and analyzed 50 years' worth of land-based temperature data from about 150 weather stations worldwide, a remarkable feat in the era before computers. His conclusions, published in 1938, showed that global temperatures had increased over that time.

In 1957, Roger Revelle and Hans Suess, at the Scripps Institution of Oceanography, stated that the emissions of CO_2 from fossil fuel combustion constituted a "large scale geophysical experiment" (Revelle and Suess 1957). In 1958, a measurement station was set up in conjunction with the Mauna Loa Observatory in Hawaii. Measurements of CO_2 in the upper atmosphere begun by Charles Keeling have continued to this day, representing the world's longest continuous set of data on CO_2 in the atmosphere. They

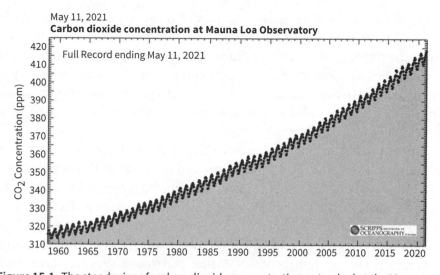

Figure 15.1 The steady rise of carbon dioxide concentration as tracked at the Mauna Loa observatory. This graph of carbon dioxide versus time is known as the *Keeling curve*. Courtesy of Scripps Institution of Oceanography.

show a steady rise in atmospheric CO_2., as in Figure 15.1. In 1967, Syukuro Manabe at Princeton University and Richard Wetherald, of the Geophysical Fluid Dynamics Laboratory, estimated that a global warming of 2.3°C would occur if the concentration of atmospheric CO_2 were to double (Manabe and Wetherald 1967), solidly in line with current estimates of an increase of 1.5–4.5°C (Lindsey 2014).

The Earth continuously receives energy from the sun, mostly in the visible and infrared regions of the spectrum. Heat energy always spontaneously flows from higher to lower temperatures. In keeping with this, Earth will try to transfer heat spontaneously into its frigid surroundings of outer space. During the night, objects that were heated during daylight hours will cool by radiating heat to outer space in the form of infrared radiation. Any physical system being studied can be described by an energy balance (Schobert 2014, 41):

$$Energy_{in} - Energy_{out} = Energy_{stored}.$$

The difference between the amount of energy put into a system and the amount of energy removed from it necessarily has to be stored up somehow in the system. The principal components of our atmosphere, nitrogen and oxygen, are essentially transparent to infrared radiation. However, the atmosphere also contains small amounts of other gases that absorb some infrared radiation. The average temperature of the Earth results, in part, from the trapping of infrared radiation in the atmosphere by some of the atmosphere's components. Because these components—CO_2 and water vapor being examples—act somewhat like the glass of a greenhouse, such infrared-absorbing gases are called *greenhouse gases*. Thus, the warming of Earth is sometimes called the *greenhouse effect*.

Infrared radiation does not become trapped permanently in molecules of greenhouse gases. The absorption of infrared radiation increases the energy of a molecule. At some point the molecule will release energy by reverting to its lower energy state. The energy released when the excited molecule relaxes is radiated equally in all directions. Some of that energy is radiated back to Earth's surface rather than escaping into outer space. For the Earth's energy balance, the $Energy_{out}$ term becomes smaller, and since $Energy_{in}$ (from the sun) is essentially constant, $Energy_{stored}$ necessarily increases. Macroscopically, we observe this effect as an increase in temperature.

If there were no greenhouse effect, the average temperature of Earth would still be described by an energy balance, but one unaffected by absorption of infrared radiation. In such a case, Earth would have an average temperature of $-18°C$. The warming of the planet as a result of the CO_2 and water vapor that occur naturally in the atmosphere is the **natural greenhouse effect**. It exists and occurs without human intervention.

The natural greenhouse effect is essential in keeping our planet habitable. Though water vapor and CO_2 represent only very small fractions of the total composition of the atmosphere—some 10,000–50,000 parts per million (ppm) and about 419 ppm, respectively—they provide significant contributions to the greenhouse effect. Any molecule that can absorb infrared radiation is potentially a greenhouse gas. Other greenhouse gases include methane, the main component of natural gas; nitrous oxide, a possible constituent of nitrogen oxides (NO_x); ozone; and chlorofluorocarbons (CFCs), formerly used in refrigeration and air conditioning.

The greenhouse effect is the warming of Earth's surface and atmosphere by absorption and reemission of infrared radiation by components of the atmosphere. The components of the atmosphere that do this are the greenhouse gases. There are two contributions to the greenhouse effect: the natural greenhouse effect, as we've seen, and an *anthropogenic greenhouse effect*, warming caused by additional quantities of greenhouse gases put into the atmosphere by human activities. It is now virtually universally accepted that the additional warming due to anthropogenic emissions of greenhouse gases is changing Earth's climate.

The effect of any greenhouse gas is gauged by its *global warming potential* (GWP), which takes into account the ability of a gas to absorb infrared radiation and the lifetime of that gas in the atmosphere (Anthony, Ferrett, and Bender 2003, 37). The GWP provides a way to compare the impact of different gases. The GWP of CO_2 is, by definition, 1. Over a 100-year time span, methane, the most common constituent of natural gas, has a GWP of 28–36 (EPA 2020). Why is the focus of concern on CO_2? A major reason is the much lower concentrations of the other gases. Gases other than CO_2 are commonly reported in parts per billion, a unit 1,000 times smaller than that used for reporting CO_2. CO_2 is the proverbial elephant in the room.

Evidence that the planet is warming, collected from many scientific disciplines and from many parts of the world, is unequivocal. So is evidence that the CO_2 concentration in the atmosphere is rising—and rising far more than could be attributed to natural phenomena. The temperature also is rising, as shown in Figure 15.2.

The expression "correlation is not causation" is important in this context. Correlation measures how closely related two things—such as sets of observations—are, period. A good correlation between *A* and *B* does not necessarily mean that *A* causes *B*, or vice versa. To establish whether there is a causation in the relationship between *A* and *B* requires us to probe more deeply.

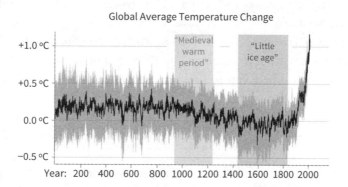

Figure 15.2 The global average temperature change during the last two millennia. Even noticeable changes that occurred during the Middle Ages and again later in the so-called Little Ice Age do not come close to comparing with the modern, rapid temperature change. The figure is known as the "hockey-stick graph," from the long, roughly flat portion followed by the sweeping curve at the end.
Courtesy of Ed Hawkins/Wikimedia Commons.

Most of the energy that we use derives in various ways from combustion. All common fuels contain carbon as a major constituent. Chapter 8 introduced the carbon–oxygen reaction:

$$C + O_2 \rightarrow CO_2.$$

Use of carbon-based fuels inevitably leads to the production and release of CO_2. To compare different fuels in terms of how much CO_2 is produced when they burn, a fair comparison should be made on the basis of a common amount of heat released. Coals will produce about 90–98 kilograms of CO_2 per gigajoule (kg CO_2/GJ) of energy, when calculated on a dry basis. The value depends on rank, with anthracites being the highest and lignites lowest. Woods will have greater CO_2 emissions per gigajoule, even on a dry basis, due to their low calorific values. Natural gas emits about 50 kg CO_2/GJ, and common liquid petroleum fuels about 68–70 kg CO_2/GJ.

In 2019, the CO_2 emitted by coal combustion amounted to 14.76 billion tonnes (Sönnichsen 2020). In the same year, anthropogenic CO_2 emissions from all sources were 36.8 billion tonnes (Harvey and Gronewold 2020). Coal combustion is responsible for 40% of anthropogenic CO_2 emissions. But still . . . is this correlation without causation?

Carbon exists as two stable isotopes, ^{12}C and ^{13}C. A third isotope, ^{14}C, is unstable, with a half-life of 5,700 years. Its concentration in the world amounts to about 1 part per trillion. The amount of ^{14}C in coals is zero—it has all decayed because of the geological ages of coals being in the tens to hundreds of millions of years. Even so, we will see shortly that ^{14}C has an important role in understanding the source of increased atmospheric CO_2.

An atom of ^{13}C is 8% heavier than an atom of ^{12}C on the atomic mass scale. In terms of their natural abundance, only 1 in 100 of all stable carbon atoms is ^{13}C. The chemical behavior of these two isotopes is virtually identical. But when we consider the rates of chemical reactions at a molecular level, often atoms of the heavier isotope react slightly more slowly than those of the lighter isotope (Moore and Pearson 1981, 367). This behavior is known as the *kinetic isotope effect*.

Coals—indeed all fossil fuels—derive from once-living organisms, in particular from plants. In photosynthesis the kinetic isotope effect favors a slight increase in ^{12}C relative to ^{13}C in glucose. The ratio of the amounts of ^{12}C and ^{13}C in glucose from photosynthesis will be higher than the ^{12}C:^{13}C ratio generally found in the world. Plants use the glucose produced in photosynthesis as a starting point to make all of the other compounds they need for their life processes. This slightly higher ^{12}C:^{13}C ratio will persist in plant components preserved in natural environments and experiencing diagenesis and catagenesis. Because of the kinetic isotope effect, coals have higher $^{12}C{:}^{13}C$ ratios as well.

The carbon in coals sooner or later winds up being returned to the atmosphere as CO_2. Because the coals have slightly higher $^{12}C{:}^{13}C$ ratios than the natural ratio of these isotopes, so too will the CO_2 produced when coals are burned. CO_2 emissions mix freely into the atmosphere. If billions of tonnes of CO_2 with higher $^{12}C{:}^{13}C$ ratios, coming from fossil fuel combustion, were to mix completely into the atmosphere, the inevitable outcome would be that samples of atmospheric CO_2 collected over a period of time should show increasing amounts of ^{12}C relative to ^{13}C. They do.

Because there is no ^{14}C in coals, the amount of $^{14}CO_2$ produced from coal combustion is zero. But there is naturally occurring $^{14}CO_2$ in the atmosphere, coming from other sources. When the atmosphere receives another 15 billion tonnes of CO_2 from coal combustion, and that amount of CO_2 has absolutely no ^{14}C, the effect should be to reduce the concentration of $^{14}CO_2$ in the atmosphere as a whole. That is exactly what is observed.

When billions of tonnes of CO_2 produced from coal combustion enter and fully mix with the atmosphere, CO_2 in the atmosphere will be found to be depleted in both ^{13}C and ^{14}C, but for very different reasons. In coals, the decreased ^{13}C relative to ^{12}C is a consequence of chemistry—the kinetic isotope effect. The complete absence of ^{14}C is a consequence of physics—the half-life of an unstable isotope. The depletion of these two isotopes in the CO_2 from coal combustion, when thoroughly mixed with the atmosphere, should produce a small but measurable change in the relative amounts of carbon isotopes in atmospheric CO_2: a bit less ^{13}C and ^{14}C and a bit more ^{12}C. These deductions are now amply confirmed by experimental measurements. Here lies the "smoking gun," the evidence that a major contribution to the increase in atmospheric CO_2 concentration comes from the human effect of fossil fuel combustion. These changes in isotopic composition of atmospheric CO_2 are known as the *Suess Effect*, in honor of the Austrian- American isotope chemist and physicist Hans Suess.

What, if anything, can we do about CO_2 emissions from coal utilization and their contribution to climate change? We can consider the two end-points first. The argument has been advanced that the world's burgeoning population faces many serious problems, of which hunger, diseases, and shortages of safe, clean water are only a few. There are limits to the world's manufacturing capacity and financial resources. Maybe we would collectively be better off to focus our limited abilities on making sure that everyone had enough to eat, had access to potable water, and had at least some level of medical care rather than worry about doing anything about climate change.

Given the increasing number of temperature spikes, droughts, and cata-strophic storms, the counterargument is that we need to stop burning fossil fuels—or at least stop using coal—right away. In fact, preferably tomorrow morning. No matter how well-intentioned and sincere are the arguments for a rapid phase-out of coal use, we simply don't have the capability to produce solar panels and wind turbines and to frack the world's shales in any reason-able time scale to replace quickly the energy derived from coal. Many people, if confronted with the option of keeping their lights on and having a refrig-erator versus producing a little less CO_2 from their personal use of electricity are very likely to opt for the former. If another billion people each bought one 60-watt light bulb and kept it on for only 4 hours per day, the world will have to add another 10 gigawatts of electricity-generating capacity, resulting in 50 million tonnes of additional CO_2 emission, assuming today's mix of energy sources for generating electricity.

Perhaps the best course is in the middle. Coal use for energy purposes is declining in most countries. Various approaches are being developed to re-duce CO_2 emissions and even to achieve "negative CO_2," by removing more CO_2 from the atmosphere than is emitted. The long-term future belongs to renewables (and possibly nuclear energy) because we can't operate forever on finite resources. But in the meantime. . . .

Whether to conserve energy or reduce emissions, always the best place to start is by increasing efficiency. In coal-fired power plants, increases in the efficiency of converting the chemical energy in coal to electrical energy supplied to consumers mean that we need less coal to produce the same amount of electricity. A one percentage point improvement in efficiency of a conventional pulverized-coal–fired plant leads to a 2–3% reduction in CO_2 emissions from that plant (World Coal Institute 2007). Current ultrasupercritical plants operate with efficiencies in the low 40% range. The next generation plants should achieve 45%, and the potential may be around 50% (World Coal Institute 2007). Present ultrasupercritical coal-fired power plants emit about 20% less CO_2 than the traditional subcritical plants for the same level of electricity output (Adams 2016).

What may seem to be a rather small increase in efficiency is applied against enormous tonnages of coal consumption. This can significantly reduce the amount of CO_2 that would be emitted or that would have to be captured and stored. If it were possible to convert the entire fleet of coal-fired power plants in the United States to ultrasupercritical generation—which would take a great deal of time and money—this seemingly small gain in efficiency would reduce CO_2 emissions by about a 100 million tonnes per year. But chances are that it would be faster, easier, and cheaper to go with natural gas and renewables instead of pursuing a major make-over of existing coal-fired power plants.

Efficiencies around 50% should be also be achievable in integrated gasification combined cycle (IGCC) plants (World Coal Institute 2007). Emissions should be about comparable to those from a natural-gas–fired plant (Goodell 2006). Also, because there is a higher concentration of CO_2 in the gas streams, it is easier—hence less expensive—to capture CO_2 at an IGCC plant than at a conventional pulverized-coal plant. IGCC plants will be more expensive to build than conventional coal-fired plants, by about one-third (EIA 2013).

In controlling CO_2 emissions, the focus has been on capturing CO_2 and then dealing with it, usually by finding ways to sequester it. Collectively, these strategies are known as **carbon capture and storage** (CCS).[1] The "and" in carbon capture and storage signals the need to deal with two separate issues: how best to capture the CO_2, and then, when we've done that, what to do with all that CO_2.

Coal flue gas contains about 12% CO_2. Most of the rest is nitrogen that came in with the air used in combustion. In any separation processes, the higher the concentration of the substance that we want to remove, the easier the separation. Enhancing the concentration of CO_2 in flue gas can be achieved by enriching the amount of oxygen used in combustion. An increase in the pro portion of oxygen necessarily results in there being less nitrogen. Since most of the oxygen is consumed in the boiler, the flue gas will have a higher concentration of CO_2 and less nitrogen than in a conventional combustion system. This strategy is referred to as **oxyfuel combustion**. This technology seems to be in the pilot-plant stage in several countries. The flue gas from oxyfuel combustion contains more than 80% CO_2 (NRCAN 2016). In any separation process, the higher the concentration of the substance we wish to remove, the easier it is. Oxyfuel combustion could be retrofitted to existing conventional coal-fired plants or could be used in new construction.

A standard approach to capturing CO_2 from flue gases relies on amines, a family of organic nitrogen-containing compounds. CO_2 dissolves in water to produce an acidic solution, commonly referred to as carbonic acid. Amines are bases. An acid-base reaction binds the CO_2 to amine molecules.

---

[1] There is potential confusion in this terminology. *Carbon*, in some elemental form, is not what is being captured—it's CO_2.

Depending on the method of contacting the flue gas with the amine solution, it's possible to capture about 90% of the CO_2 (Service 2016).

Amines used in CCS are too expensive to throw away. Liberating CO_2 from the capture solution allows the amines to be recycled. CO_2 can be released by heating the solution. Heat is often supplied by steam, a process called *steam stripping*. CO_2 released by steam stripping will be compressed and then either used or sequestered. CO_2 capture using amines is technologically mature (Kohl and Riesenfeld 1985, 31). The problem is in the economics. Steam consumed by stripper is not available for use in the turbines. As a result, the output of electricity is reduced by an estimated 20–30% (Mann 2014) parasitic load. A 1,000-megawatt (MW) plant would require about 300 MW of heat and electricity to operate the CO_2 capture system, resulting in an output of only 700 MW (Patel 2017).

The "S" in CCS indicates that CO_2 will somehow be stored to keep it out of the atmosphere, indeed, possibly stored forever. Oil is held underground in the pores of rocks. The important properties of reservoir rock are its porosity and permeability. Initially, reservoir pressure provides a natural driving force for the oil to migrate through the reservoir rocks to the well bore. After a portion of the oil has been extracted, the reservoir pressure has dropped to a point at which further natural migration of oil to the well is not feasible. A good portion of the oil is still in the reservoir, as much as 70% (Speight 1991, 146). We could recover more of that oil if we had a way to increase the pressure in the reservoir. Techniques for doing so are known as *enhanced oil recovery* (EOR).

Injecting compressed CO_2 into partially depleted oil reservoirs is already a proved and widely used technique for enhanced oil recovery. The intended benefit is to provide more oil. At the same time, much of the CO_2 pumped into such a rock formation becomes tightly held in the pores of the rock. The CO_2 used for EOR is a marketable commodity, so its sale for this application would generate revenues that could help offset the costs of the CO_2 capture system.

EOR projects using CO_2 produced from coal already exist. We have seen the CO_2 transport from the Dakota Gasification plant in North Dakota to the Weyburn and Midale oil fields in Saskatchewan (Griffiths 2006). Every tonne of CO_2 injected into the reservoir facilitates recovery of two to three barrels of oil[2] (Nijhuis 2014). The project successfully handles about 5,000 tonnes per day (Griffiths 2006). Taking 40 billion tonnes as the annual anthropogenic CO_2 emission means that we would need, worldwide, 22,000 projects the size of Weyburn.

---

[2] About 325–475 liters.

It should be possible to use coal seams for CO_2 storage. The concern is that a CO_2-laden coal seam will be permanently off-limits to mining, at least by any technology that we know of today. CO_2 will adsorb onto the surface of coals in preference to other gases, notably methane. Injecting CO_2 into a gassy coal seam provides enhanced *coalbed methane recovery* (ECBM) while sequestering the CO_2. A 6-year project in the southwestern United States injected about a quarter-million tonnes of CO_2, improving methane recovery and showing no signs of CO_2 in the methane (World Coal Institute 2007).

Implementing CCS increases the capital costs for a coal-fired power plant and reduces the efficiency, adding a performance penalty to the increased cost. Both of these issues will challenge the cost competitiveness of using coal in electricity generation. There is no off-setting financial incentive for sequestering the CO_2. In any case, we might have already moved past CCS, in the sense that cheap natural gas and a steady decline in costs of renewables are making new coal-fired generating capacity uneconomical.

The central problem in CCS is the gross mismatch of scale between our ability to make CO_2 relative to our ability to capture and sequester it. In 2019, the CCS projects then applied to coal-fired power plants had the capacity to capture 12 million tonnes of CO_2 per year (Cornot-Gandolphe 2019). Twelve million tonnes seem like a lot of CO_2. The problem is that, in 2018, CO_2 emissions from coal-fired power generation were 10 *billion* tonnes. Roughly, we produce a thousand times more CO_2 than we sequester. When other uses of coal are added, the total CO_2 emissions are 15 billion tonnes, almost equal to the value for all other fossil fuels combined. The plants on which the current CCS operations are installed have a combined generating capacity of 4 gigawatts. The International Energy Agency had projected that 20 gigawatts of coal-fired electricity generation would need to be fitted with CCS by 2040—less than 20 years away (Cornot-Gandolphe 2019). "The challenge to scale up the technology is enormous" (Cornot-Gandolphe 2019).

Instead of putting CO_2 into the ground and hoping that it stays there for eons, it might prove better if we did something useful with it. This broadens our horizons to technologies for CCUS: *carbon capture, utilization, and storage*. Being able to make materials or chemical products from CO_2 offers a double benefit. CO_2 that is usefully employed is CO_2 that does not escape to the atmosphere. Second, we can take advantage of CO_2 as a byproduct of another process and need not exploit fresh resources to get the CO_2. At current levels of industrial consumption of CO_2, these chemical applications would scarcely make a dent in total CO_2 emissions. Using captured CO_2 to supply all of the

CO_2 used in industry would account for about 1% of worldwide emissions (Bourzac 2017).

Some 200 million tonnes of urea are produced per year worldwide. The great majority is used in fertilizers, and much of the remainder in urea-formaldehyde resins. The modern industrial process for making urea starts with CO_2 and ammonia (Nijhuis 2014). Industrial production by this process amounts to about 100 million tonnes per year and consumes about that much CO_2 (Scott 2015).

Sodium carbonate, also known as soda ash and washing soda, is used in making glass, in water softening, as a food additive, and in many other applications. The predominant industrial process, the Solvay process (Mathur 1992), begins with reacting CO_2 with a solution of sodium chloride and ammonia. A soda ash plant in California is reported to use CO_2 captured from a coal-fired power plant using the amine scrubbing method (Ondrey 2014). Reacting sodium carbonate with additional CO_2 makes sodium bicarbonate, familiarly known and widely used as baking soda (Scott 2015). Both the carbonate and bicarbonate are produced in large quantities worldwide, amounting to about 52 million tonnes of the carbonate per year. Currently very little waste CO_2 is consumed in these processes, but it certainly could be in the future.

Sunfire, a company based in Dresden, uses renewable energy to produce hydrogen from the electrolysis of steam. Then, with waste CO_2, it is possible to run the *reverse* water–gas shift:

$$CO_2 + H_2 \rightarrow CO + H_2O.$$

With carbon monoxide made this way, and more hydrogen, we get right back to Fischer-Tropsch chemistry (Ondrey 2013). This is a potentially new approach to indirect coal liquefaction, capturing the CO_2 from coal combustion and converting it to liquid fuels. With CO_2-derived synthetic liquids, it would not be necessary to replace the existing liquid fuel infrastructure. In principle, though scarcely likely in practice, the synthetic liquids would be "CO_2 neutral" in the sense that all of the CO_2 produced when the liquids are burned as fuels would be balanced by the CO_2 used to make another equivalent amount of fuel.

Liquid hydrocarbons from CO_2 could be used for energy storage. We already have the technology for storing petroleum and its products in enormous quantities. When energy is needed, the liquid could be fed to a fuel cell for conversion to electricity. The energy density—the energy output per kilogram of material—is about 50 times higher for gasoline than it is for a lithium-ion rechargeable battery (Schlachter 2012). Using gasoline in a fuel cell will inevitably produce CO_2, but in a nearly pure stream that could be captured for reuse or sequestration.

CO_2 can also react with methane produce synthesis gas. The process, called dry reforming, has already been demonstrated in China (Tremblay 2017). Combined with methanol synthesis, dry reforming could be a route to methanol or useful chemicals from methanol, such as acetic acid.[3] As we have seen in Chapter 12, methanol can be used to make gasoline or to make dimethyl ether as a substitute diesel fuel. By using CO_2 as one of the feedstocks for making the original synthesis gas, the total carbon footprint for any of the downstream processes can be reduced relative to processes starting with coals or natural gas (Tullo 2016).

CCS alone may not be adequate to limit global warming to the level agreed upon in the 2015 Paris Agreement on climate change. It's argued that CCS needs to be augmented by reducing the amount of CO_2 in the atmosphere (i.e., removing more CO_2 than the amount emitted by human activities; Rosen 2018). The collection of strategies for doing this is called *negative emissions technologies* (NETs). The estimated burden on NETs is the capturing of somewhere between 100 billion and 1 trillion tonnes of CO_2 by the end of this century (Temple 2019). NETs do not intend to reduce the concentration of atmospheric CO_2 to zero. The intent is to remove a quantity of CO_2 from the atmosphere greater than the amount contributed by anthropogenic emissions.

Green plants capture CO_2 and sequester it so long as the plants are not harvested or cut down. A hectare of trees could capture 350 tonnes of carbon per year (Corkery and Wines 2016), equivalent to 1,280 tonnes of CO_2. Every tonne of coal burned would have to be offset by planting a certain number of trees. Though most trees grow slowly, the scheme could rely on faster-growing plants, which then would be collected and burned for their energy value, with the resulting CO_2 dealt with by CCS technologies (Rosen 2018).

The costs associated with CCS and with NET raise the question of how we get somebody to pay for them. Just as many technical approaches have been suggested, there are also many ideas of how to bring economic considerations into play.

---

[3] The primary industrial use of acetic acid is in the production of poly(vinyl acetate), an ingredient in many adhesives, as well as other vital commercial products such as paints. Very little acetic acid is commercially used directly as acetic acid.

Without putting some kind of price on CO_2 emissions, it will inevitably be cheaper to burn carbon-based fuels and let the CO_2 go up the stack rather than do something about it (Bourzac 2017). The market approach is a cap-and-trade system. Presumably it would be similar to the successful system for sulfur oxides (SO_x) emissions discussed in Chapter 9. *Offset credits* provide a system in which a company pays for CCS activities at some location other than their own site. For example, a coal-fired power plant constructed years ago might now be hemmed in by urbanization, the high cost of land making it very expensive to acquire additional space for retrofitting a CO_2 capture system. Instead, the company owning this plant could pay for a system to be installed at some other facility. One way or another, a company that does a poor job of reducing CO_2 will have to pay for it, while those that effectively reduce emissions will benefit. The market penalizes the big emitters.

The alternative approach is a *carbon tax*. This system would impose a tax on CO_2 emissions. In principle, the tax would be a mechanism to account for and recoup the costs associated with climate change. Various numbers have been suggested, one such being about $40 per tonne of carbon. If this value were assessed against carbon, burning 1 tonne of a dry, high-volatile C bituminous coal would incur a tax of $28 per tonne. Many power plants burn in excess of 10,000 tonnes of coal per day. If the tax were assessed against CO_2, then the tax would be about $100 per tonne of CO_2.[4] Taxes are thought to be simpler to implement than various cap-and-trade, market-based systems (Green 2017) but may face much stiffer political opposition in many countries.

What approach—if any—is ever adopted needs to take into account that, at the outset, there is an existing array of power plants, factories, vehicles, and other sources that cannot be replaced overnight. In other words, it is almost impossible to effect a nearly immediate change in emissions. The idea should be that, in the medium term, economic incentives and/or pressures will drive a change from high-CO_2 energy sources to ones that emit less CO_2 per unit of energy produced. This likely means a change from coal to natural gas. Any of the approaches discussed here have the greatest impact on coal. That results directly from the fact that coals produce more CO_2 per megajoule of energy than any other fuel. There is no getting around the rules of chemical stoichiometry, let alone trying to finagle the inexorable laws of arithmetic. And, in the long run, the economic approaches would be intended to provide the incentives to develop low-carbon, or no-carbon, energy sources—renewables.

---

[4] One unit mass of the element carbon (C) produces 3.67 units of CO_2. This is what accounts for the huge differences in numbers when we are talking about *carbon* as carbon or as CO_2.

Though the climate change focus is heavily on CO_2, it is not the only factor in the connection between coals and climate change. Combustion can add *aerosols* to the atmosphere. Aerosols are suspensions in air of solid particles or liquid droplets, usually smaller than 1 micrometer. Just as with CO_2 emissions, aerosols also can be natural or anthropogenic. Anthropogenic aerosols include particles of sulfate or nitrate compounds, which can derive from SO_x and NO_x emissions produced during combustion. They also include high-molecular-weight tarry organic compounds and soot, which can also come from combustion processes.

Aerosols scatter incoming solar radiation, deflecting a portion of it back into space. That is, the effect of aerosols is to reduce the *Energy*$_{in}$ term in the energy balance equation. With less energy getting to Earth's surface and lower atmosphere, aerosols will have a cooling effect. The worldwide cooling may amount to about 0.7°C (Mann 2013a). We should not, however, embrace a pro-aerosol policy as a means to counteract the warming effect of CO_2 and other greenhouse gases. Combustion-derived aerosols can have significant human health effects, especially affecting the respiratory tract. They can also alter circulation patterns in the atmosphere, among other things shifting the locations of abundant rainfall.

Burning coals as well as other fossil and biofuels with insufficient oxygen can lead to the formation of soot, tiny particles of unburned carbon possibly mixed with tars. Soot, when considered as an atmospheric pollutant, is also known as *black carbon*.[5] In the atmosphere, black carbon absorbs heat, contributing to global warming. Worldwide, black carbon is the second largest contributor to climate change (Mann 2013b).

Every energy source has technological disadvantages and disadvantages, economic advantages and disadvantages, and some effect on the environment. There is not a unique energy policy that provides the optimum balance among technology, economics, and environmental impact for all nations or all regions. Rather, nations need to find an energy policy that represents the best way forward among these often-conflicting concerns.

It is incontrovertible that anthracites and bituminous coals have the highest carbon contents among all the common fuels, that anthropogenic CO_2 emissions make a significant contribution to the rising amount of CO_2 in the atmosphere, and that Earth is warming. Abundant data have been gathered

---

[5] Though "soot" and "black carbon" are often used synonymously, some sources use "black carbon" more narrowly to refer to components of soot that are pure elemental carbon, not including tars and other aerosol particles.

around the world in laboratories and by field observations, by people of various nationalities and political leanings. Absolutely the most short-sighted and most wasteful thing we can do with coal is to burn it—even if there were no CO_2 issues. There are better options by far, as will be discussed in the next two chapters. But, as society considers the technology–economics–environment connections and what future there is for coal (if any), we should not lose sight of three key points:

The first key point is that CO_2 is not the only greenhouse gas. The characteristic that makes a substance a greenhouse gas is its ability to absorb infrared radiation as it is being emitted from Earth. We have met previously some of the other greenhouse gases—methane, nitrous oxide, ozone, chlorofluorocarbons—and discussed their global warming potentials.

The second key point is that human activity is not the only source of atmospheric CO_2. The natural aerobic decay of organisms releases CO_2. Oxidation on a much shorter time scale releases CO_2 from forest or prairie fires. Carbonate rocks exposed to high temperatures, as in volcanoes, will decompose to CO_2. Oceans represent a major source of CO_2, coming from the decomposition of marine animals and plants that have died and sunk to the floor.

And the third key point is that combustion is not the only human activity that releases CO_2. Ammonia and ammonium salts are widely used as fertilizers. Corn production in the Midwest of the United States uses about 170 kilograms of ammonia per hectare (Anthony et al. 2003, 51). Ammonia is made by direct reaction of nitrogen and hydrogen—the Haber-Bosch process. Hydrogen is commonly made from applying the carbon–steam reaction to natural gas, making synthesis gas, and then using the water–gas shift reaction to arrive at a mixture of hydrogen and CO_2. Without CCS, the CO_2 goes to the atmosphere. After ammonia application, the growing plants convert about 1–2% of the nitrogen into nitrous oxide, which itself is a greenhouse gas.

Hydrogen used for most other industrial applications is made in the same way. These additional applications include petroleum refining and petrochemical production, making other chemicals such as hydrochloric acid, and serving as a coolant in power-plant turbines. There are routes to hydrogen that do not produce CO_2 as a co-product, such as the electrolysis of water using electricity from renewable or nuclear generation. About half of the world's hydrogen production comes from the carbon–steam reaction, with natural gas as the source of the carbon.

Production of cement involves heating calcium carbonate, as in limestone for example, to produce calcium oxide. Commercial cement is about 66%

calcium oxide (Anthony et al. 2003, 51). Making calcium oxide by thermal decomposition of natural sources of calcium carbonate is inevitably accompanied by liberation of CO_2. About 0.8 tonnes of CO_2 are produced for every tonne of calcium oxide. In addition, CO_2 will also be produced in firing the kilns used to make the calcium oxide. Cement production could be responsible for about 10% of worldwide CO_2 emissions (Welland 2009, 240). If we stopped burning coal in power plants tomorrow and were able to replace the electricity using carbon-free sources, there are a few other things that we might still want to have around, such as steel, cement, and fertilizers. CCS will be valuable in those technologies as well.

Changes in land use, especially deforestation, cause two problems. First, growing plants remove CO_2 from the atmosphere through the process of photosynthesis. Their destruction diminishes CO_2 removal. And then, trees that have been deliberately cut down are either left to decay or are burned, either way producing CO_2. This double impact on CO_2 would be the equivalent of boxing's one-two punch. Restoring forests or establishing new ones on land that had previously been forested is a route toward negative carbon emissions that is comparatively less expensive than some of the other proposed strategies.

In the steel industry, blast furnace operation with coke produces CO_2 directly from coke combustion and from reaction of carbon monoxide with the iron ore. Additional CO_2 is produced during the coking operation. It is possible to reduce the carbon footprint by displacing some of the coke with anthracite, primarily because anthracite does not need to be run through a coke oven (Schobert and Schobert 2015). Beyond that step, there is interest in displacing coal entirely and going to hydrogen-based iron and steel production, using hydrogen produced by electrolysis with "non-carbon" electricity, but, if so, be prepared for a significant price hike in steel.

For about a decade, at least in the United States, it's been argued that there is a "war on coal." There is. Unfortunately, it is not the war that obdurate politicians and their supporters seem to think is going on. Coal is being squeezed from two directions at once.

Coming from the direction of economics, coal must deal with abundant and inexpensive natural gas made available by the shale revolution. For most applications, gas is a superior fuel to coal: double the calorific value per weight, easy to handle and control, instant on-off, essentially zero sulfur, and no ash. And now coal also has to deal with the steadily decreasing cost of renewables, particularly wind and solar. Not many years ago, homeowners who installed solar photovoltaic panels on their roofs were clearly marked as "solar freaks" or by terms even more pejorative. Not anymore.

Coming from the other direction are environmental issues, especially climate change. A quarter-century ago, the US National Research Council stated that, "Of all the environmental issues facing the future use of coal, none is as potentially far reaching as the worldwide concern over global climate change" (Longwell 1995). We can't get around the fact that coal produces more CO_2 per unit energy than any other fuel. There are technologies for CCS and NETs. Barring a tremendous breakthrough in CO_2 chemistry, perhaps accompanied by a great upsurge in demand for chemical products made using CO_2, dealing with the CO_2 emissions from coal combustion is going to be very expensive.

In 2015, renewable energy surpassed coal as the largest source of electricity generation globally (Economist 2016). A year later, worldwide investment in renewable energy exceeded that for fossil fuel power stations (Pearce 2016). This shift in investment is combined with existing coal-fired plants being retired at an unheard-of rate (Brahic 2017), a factor which is thought to contribute to a slowing in the rate of CO_2 emissions (Pearce 2015). In the United States, the 15% decrease in CO_2 over a 10-year period was linked to the displacement of coal by natural gas (Slezak 2013).

These trends, though they indicate a difficult time for the world coal industry, do not indicate that the use of coal will stop completely. Coke will still be important in metallurgy, especially in iron and steel. We will need steel to build devices that would be used in producing low-carbon energy. Coals remain abundant, relatively inexpensive, and distributed widely in many parts of the world. In those places where natural gas distribution systems don't exist, or where there is not yet infrastructure for renewables, coal might continue to be used. Coal could continue to be favored for small-scale, low-tech heating applications. In some geopolitical situations, domestic coal might be preferable to imported gas. In 2019, for example, Poland declined to join a European Union agreement establishing a goal of the EU becoming carbon-neutral by 2050 (Proctor 2019). Poland appears to be keenly interested in reducing its dependence on natural gas coming in from Russia. The alternative is to continue to rely on abundant Polish coal for electricity generation.

Furthermore, to burn coal is to waste a very valuable resource. Coals can be the raw materials for producing a wide range of useful chemicals. Coals also can be used to produce a variety of carbon-based products, such as activated carbon adsorbents and synthetic graphite. These opportunities will be explored in the next two chapters.

References

Adams, D. 2016. "Coal After Paris." *World Coal* 25(7): 10.

Anthony, Sharon, Tricia Ferrett, and Jade Bender. 2003. *What Should We Do About Global Warming*? New York: W. W. Norton.

Brahic, Catherine. 2017. "It's Not the End of the World. *New Scientist* 234 (3129): 22–23.

Bourzac, Katherine. 2017. "We Have the Technology." *Nature* 550: 566–569.

Corkery, M., and M. Wines. 2016. "A Curious Vision of Greener Coal." *Sunday Business, New York Times*, October 2.

Cornot-Gandolphe, Sylvie. 2019. "Carbon Capture, Storage and Utilization to the Rescue of Coal?" *Études de l'Ifri*. InstitutFrançais des Relations Internationales. https://www.ifri.org/en/publications/etudes-de-lifri/carbon-capture-storage-and-utilization-rescue-coal-global-perspectives.

Daily CO_2. 2021. https://www.co2.earth/daily-co2.

The Economist. 2013. "The Measure of Global Warming: Climate Change." *The Economist* 11: 85.

The Economist. 2016. "The Burning Question: Climate Change." *The Economist* 421(9017): 11.

EIA. 2013. "Updated Capital Cost Estimates for Utility Scale Electricity Generating Plants." U.E. Energy Information Administration. https://www.eia.gov/outlooks/capitalcost/.

EPA. 2020. "Understanding Global Warming Potentials." Environmental Protection Agency. https://www.epa.gov/ghgemissions/understanding-global-warming-potentials.

Goodell, Jeff. 2006. "Cooking the Climate with Coal." *Natural History* 115(4): 36–41.

Green, Jessica F. 2017. "Don't Link Carbon Markets." *Nature* 543: 484–486.

Griffiths, S. 2006. "Weyburn-Midale Project Update." *Greenhouse Issues* (81): 4–5.

Harvey, Chelsea, and Nathanial Gronewold, 2020. "CO_2 Emissions Will Break Another Record in 2019." Scientific American. https://www.scientificamerican.com/article/co2-emissions-will-break-another-record-in-2019/.

Kohl, Arthur, and Fred Riesenfeld. 1985. *Gas Purification*. Houston: Gulf Publishing Company.

Lindsey, Rebecca. 2014. "How Much Will Earth Warm If Carbon Dioxide Doubles Pre-Industrial Levels?" Climate.gov. https://www.climate.gov/news-features/climate-qa/how-much-will-earth-warm-if-carbon-dioxide-doubles-pre-industrial-levels.

Longwell, John P., ed. 1995. *Coal: Energy for the Future*, 62. Washington, DC: National Research Council/National Academies Press.

Manabe, Syukuro, and Richard T. Wetherald. 1967. "Thermal Equilibrium of the Atmosphere with a Given Distribution of Relative Humidity." *Journal of Atmospheric Sciences* 24, 241–259.

Mann, Charles C. 2013a. "What If We Never Run Out of Oil?" *The Atlantic* 311(4) 48–63.

Mann, Charles C. 2013b. "Black Magic." *Wired* 22(4): 73–81, 114–116.

Mann, Charles C. 2014. "Renewables Aren't Enough. Clean Coal is the Future." *Wired*. https://www.wired.com/2014/03/clean-coal/.

Mathur, Indresh. 1992. "Salt, Chlor-Alkali, and Related Heavy Chemicals." In: *Riegel's Handbook of Industrial Chemistry*, edited by James A. Kent, 408–441. New York: Van Nostrand Reinhold.

Moore, John W., and Ralph G. Pearson. 1981. *Kinetics and Mechanism*, 367. New York: John Wiley & Sons.

Nijhuis, Michelle. 2014. "Can Coal Ever Be Clean?" *National Geographic* 225 (4): 29–61.

NRCAN. 2016. "Near-Zero Emissions Oxy-Fuel Combustion." Natural Resources Canada. https://www.nrcan.gc.ca/our-natural-resources/energy-sources-distribution/clean-fossil-fuels/coal-co2-capture-storage/carbon-capture-storage/near-zero-emissions-oxy-fuel-combustion/4307.

Ondrey, Gerald. 2013. "CO_2 Utilization." *Chemical Engineering* 120 (7): 16–19.

Ondrey, Gerald. 2014. "CO_2 Gets Grounded." *Chemical Engineering* 121 (4): 21–23.

Patel, Sonal. 2017. "Capturing Carbon and Seizing Innovation: Petra Nova is POWER'S Plant of the Year." *Power* 161(8): 20–25.

Pearce, Fred. 2015. "Coal Bust Is Driving Down Emissions." *New Scientist* 225(3013): 16.

Pearce, Fred. 2016. "Hello, Cool World." *New Scientist* 229 (3061): 30–33.

Perkowitz, Sidney. 2019. "If Only a 19th-Century America Had Listened to a Woman Scientist." *Nautilus*, November 28. https://nautil.us/issue/78/atmospheres/if-only-19th_century-amer ica-had-listened-to-a-woman-scientist.

Proctor, Darryl. 2019. "Poland Pushing Back Against EU Goal to End Coal-Fired Generation." *Power* 163(11): 10.

Revelle, Roger, and Hans Suess. 1957. "Carbon Dioxide Exchange Between Atmosphere and Ocean and the Question of an Increase of Atmospheric CO_2 During the Past Decades." *Tellus* 9 (1): 18–27.

Rosen, Julia. 2018. "The Carbon Harvest." *Science* 359: 733–737.

Schlachter, Fred. 2012. "Has the Battery Bubble Burst?" *APS News* 21(8): 8.

Schobert, Harold H. 2014. *Energy and Society*. Boca Raton, FL: CRC Press.

Schobert, Harold H., and Nita S. Schobert. 2015. "Comparative Carbon Footprints of Metallurgical Coke and Anthracite for Blast Furnace and Electric Arc Furnace Use." Blaschak Anthracite Corporation. https://www.blaschakcoal.com/case-study-met-coke-vs-anthracite.

Schwartz, John. 2020. "Overlooked No More: Eunice Foote, Climate Scientist Lost to History." *New York Times*, April 27.

Scott, Alex. 2015. "Learning to Love CO_2." *Chemical and Engineering News* 93 (45): 10–16.

Service, Robert F. 2016. "Cost of Carbon Capture Drops, but Does Anyone Want It?" *Science* 354: 1362–1363.

Slezak, Michael. 2013. "At the Climate Coalface." *New Scientist* 220 (2947): 29.

Sönnichsen, N. 2020. "Coal Statistics & Facts." Statistica. https://www.statista.com/topics/1051/coal/.

Speight, James G. 1991. *The Chemistry and Technology of Petroleum*. New York: Marcel Dekker.

Temple, James. 2019. "Is Carbon Removal Crazy or Critical?" *MIT Technology Review* 122 (2): 28–32.

Tremblay, Jean-Francois. 2017. "'Made in China' Now Extends to Process Technology." *Chemical and Engineering News* 95 (42): 16–17.

Tullo, Alexander H. 2016. "Dry Reforming Puts CO_2 to Work." *Chemical and Engineering News* 94 (17): 30.

Welland, Michael. 2009. *Sand*. Berkeley: University of California Press.

World Coal Institute. 2007. *Coal Meeting the Climate Challenge*. Richmond, UK: World Coal Institute.

16
Chemicals

Our lives run on a very wide variety of chemical products, from pharmaceuticals and cosmetics to asphalt. Few of these needs could be met by relying solely on natural products. The organic chemical industry[1] relies quite heavily on raw materials derived from petroleum and natural gas. Some of the raw material demand of the chemical industry *could* be met by chemicals derived from coal. There are no technological barriers to doing this. Whether some of the raw material demand *should* be met by coal chemicals is a trickier question. There is not an answer that is universally applicable to all countries or all regions. Rather, the age-old response of "it depends" applies. It depends on the extent of indigenous reserves of oil and gas versus coal and on the regional geopolitical situation, especially regarding imports of these materials. The United States is currently nearly self-sufficient in oil production, thanks to the shale revolution, and though also richly endowed with coal reserves, at present there seems little incentive to consider coal as a raw material for the chemical industry. On the other hand, the idea of a coal-based chemical industry might be worth serious thought for coal-rich China and India, seeking to grow their economies rapidly, or for those countries that depend on Russian natural gas.

Industrial chemical products fall into three categories: commodity, specialty, and fine chemicals. *Commodity chemicals* are produced in large quantities and usually by two or more companies. They have essentially the same composition, properties, and structure, so are interchangeable regardless of the supplier. Commodity chemicals are of low cost. To keep the cost low, they need to be produced in continuous processes. *Specialty chemicals* might actually be a mixture of compounds but produced to meet a specific application. Specialty chemicals formulated by different suppliers might not be interchangeable. They tend to be more expensive than commodities and may have been produced in batch operations rather than in continuous-flow plants. *Fine chemicals* have very specific chemical structures, are produced in comparatively small quantities, and command high prices. They are produced in

---

[1] That is, excluding metals and alloys and excluding products such as cement made from industrial rocks or minerals.

batches. It's possible that fine chemicals could even be used as ingredients in specialty chemicals.

As raw materials for chemicals production, coals have two features that distinguish them from oil and gas. First, the molecular framework of most coals is based on aromatic carbon atoms. Natural gas and most petroleum derivatives are primarily aliphatic. This creates differences in the chemistry of coal and of oil and gas. Second, when a coal is used, the resulting ash needs to be dealt with. But the original minerals or the resulting ash can contain small, though possibly useful, amounts of valuable elements that might be worth extracting. This adds a component of inorganic chemistry to the possibilities for making chemical products from coals.

We have seen the strategies for producing synthetic liquid fuels from coals, both direct and indirect. Analogous choices relate to the present possibilities of making chemicals from coals. Direct routes lead to products that incorporate some of the molecular features of the parent coals. This is important because of the differences in molecular architectures between aromatic coals and the aliphatic petroleum and natural gas. Depending on the desired products, this distinction provides either an advantage for coal or an issue to be dealt with.[2] Indirect routes that completely destroy the molecular features of the coal generally lead to products similar to those already produced from petroleum or gas. Coals simply become an alternative raw material.

In 1856, August Wilhelm von Hofmann, the first director of the Royal College of Chemistry, suggested to one of his students, the 18-year-old William Henry Perkin, the possibility of synthesizing quinine from a compound extracted from tar that was a byproduct of coal carbonization to produce gas.

> I was in the laboratory of the German chemist Hoffmann. I was then eighteen. While working on an experiment, I failed, and was about to throw a certain black residue away when I thought it might be interesting. The solution of it resulted in a strangely beautiful color. You know the rest. (Perkin, as quoted in Garfield 2000)

Perkin's "strangely beautiful color" came to be called *mauve*, from the name for the French mallow flower.[3] The name was adopted in 1859, but Perkin, who had dropped out of college to go into business with his father and brother,

---

[2] The technology for making aliphatic products from aromatic starting materials is quite mature: hydrogenation. Many catalysts and reactor designs are available for doing this. Again, there is a similarity between chemicals production and coal liquefaction: at some point we need hydrogen. Coals could be the better choice when the desired products are aromatic chemicals.

[3] *Malva nicaeensis.*

already had a factory in production 2 years earlier. The opening of this factory was essentially the founding of the synthetic organic chemical industry.

Perkin's work soon led to the development of other dyes and chemical products from the components of coal tar. For a time, Britain led the world in the production of synthetic dyes, but, by 1914, it was the German chemical companies that had a near-monopoly on dye production. Perkin's son, Frederick, put his finger on the reason. In 1880s Britain, there was little support for education and research in the relevant areas of chemistry. For a company in the organic chemical industry to be able to expand, it needed a continuous source of well-trained chemists for both the development of synthetic routes to new dyes and the scaling up of the syntheses into industrial processes. Without the personnel to expand in the laboratories and pilot plants, a firm would, sooner or later, go out of business. The best-qualified chemists were German. Frederick Perkin acknowledged that the German chemists were good and, from a business perspective, gave good scientific value for the salaries they received. But a great many of them went home again after a stint in Britain. In coal chemistry today, China is the new Germany.

The decline of the coal-gas industry in many countries necessarily decreased the amount of tar produced. The primary source of tar became the byproduct recovery coke oven. The non-distillable remainder of the tar is a highly viscous material that solidifies at ordinary temperatures, known as *coal-tar pitch*. Pitch has become the dominant product of coal-tar distillation. An important use is its role as a binder in the production of synthetic graphite, a process discussed in the next chapter. The graphite electrodes used in steelmaking arc furnaces are made from two raw materials: petroleum coke and coal-tar pitch. Increasing use of electric furnace technology reduces the demand for metallurgical coke to feed blast furnaces. Reduced demand for metallurgical coke in turn reduces the production of coal-tar pitch, which causes shortages of a vital raw material needed to make the electrodes for the arc furnaces—an interesting embodiment of the law of unintended consequences.

The potential of coal-tar chemistry has been illustrated in the coal products tree discussed in textbooks from the first half of the twentieth century—with an example shown in Figure 16.1—and in such books as *The Treasures of Coal Tar* (Findlay 1917). Developing and making chemical products from coal tar was a glorious chapter in the history of coal chemistry and, indeed, of organic chemistry. Those days will not return.

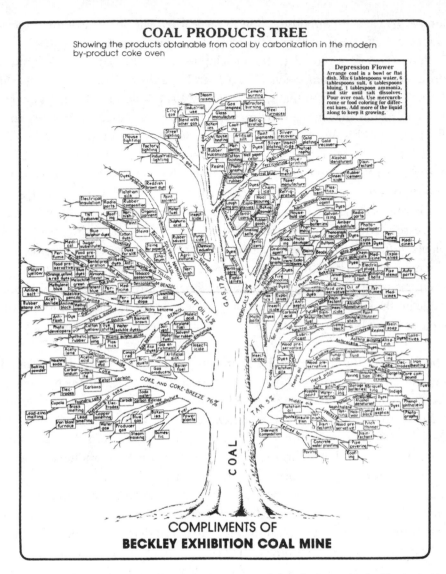

Figure 16.1 The importance of coal tar as a source of chemicals used to be illustrated by the coal products tree. Such a tree will not again be based on coal tar, but a great many of the products shown on this tree could be made from coals by new processes.
Courtesy of National Archives, Public domain.

Three issues affect the use of byproduct coal tar as a source of raw materials for the chemical industry. The first is the complexity of the mixture that is coal tar. Any sample of coal tar contains hundreds, perhaps thousands, of individual compounds. A 1945 compilation (Rhodes 1945), done before the advent of today's powerful analytical instruments, lists 348 compounds. Since the sum of all the components must always be 100%, in a sample containing hundreds of components, getting to a sum of 100 dictates that most of them will be present in amounts of less than 1%. Extracting one desired compound from this mixture offers significant challenges in separation processing. Likely it would be uneconomical.

To illustrate, current world production of metallurgical coke amounts to 640 million tonnes (Garside 2019). If all the coke were made in byproduct recovery ovens, there would be about 30 million tonnes of tar.[4] The dry tar might contain about 1% phenol[5]. In the hypothetical situation that all the coke ovens in the world were byproduct recovery ovens and that all the tar would be processed to recover pure phenol, the global phenol production would amount to about 300,000 tonnes. In the next few years, the global market for phenol is forecast to hit about 12 million tonnes (Wood 2018). Byproduct coal tar is not going to do it.

Even the limited supply of tar is shrinking. Because of environmental concerns, the trend is toward construction of "non-recovery" ovens (Chapter 10) with provision for burning the volatiles inside the oven. There is no tar production from such ovens.

From the first development of byproduct coke ovens until the ready availability of inexpensive petroleum after World War II, coal-tar chemicals were a major feedstock for the chemical industry. Then these "carbo-chemicals" steadily lost ground to petrochemicals. The fact that chemical byproducts might amount to only about 5% of the amount of coal fed to the ovens did not help. Steady progress in steelmaking resulted in reduced coke demand and production, which reduces production of coal-tar chemicals. Today, coal-tar chemicals represent only about 2% of total organic chemical production. However, they provide 15–25% of a mixture of benzene, toluene, and xylenes (BTX) and about 95% of polycyclic aromatic compounds (Song, Schobert, and Andrésen, 2005).

The challenge and opportunity of the future for chemicals derived from coal lies in developing entirely new processes for obtaining valuable chemical products, but ones that do not require high-temperature carbonization in coke ovens. Coals are potentially good raw materials for specialty and fine chemicals and for monomers to make high-performance polymers (Song et al. 2005).

---

[4] Similar to every other coal conversion process, the actual yields of coke, tar, and other products depend on the specific coal or blend of coals being coked and on the operating conditions of the coke oven. This example relies on typical results of 750 kilograms coke and 40 kilograms tar per tonne of coal.

[5] This figure considers phenol derivatives such as the cresols and xylenols as well.

A half-century ago, the great Indian coal scientist Adinath Lahiri, predicted that, "With progressive shrinking of the conventional markets for coal and coke, interest in the production of carbo-chemicals and liquid fuels from coal is being revived. . . . The preferred route now appears to be the once-discarded Bergius process" (Lahiri and Mazumdar 1967). There are certainly parallels between the early stages of development of coal fluidity and the direct liquefaction of coal (Neavel 1982). Direct liquefaction preserves some of the molecular structural features of the parent coal. So we might reasonably expect that liquids produced by such a process would, like coal tar, be a potential source of aromatic compounds, indeed many of the same kinds of compounds that could be obtained from tar. Many of the components of coal-derived liquids can be converted into valuable chemicals (Burgess and Schobert 2000).

The BTX liquids are sometimes present in relatively high concentrations in liquefaction products. Benzene is mainly converted to other chemicals used to make polystyrene, phenolic resins, and nylon. Most toluene is used to make more benzene, but it also serves as a precursor to polyurethane foam. Xylene gets us to poly(ethylene terephthalate) polymers, often known as PET, as in plastic beverage bottles.

Phenols are not acceptable components of liquid transportation fuels. These compounds could be separated and sold as a chemical product while the rest of the liquid product goes off to fuel markets (Burgess and Schobert 2000). Phenols from coal liquefaction could be converted to ethers that could serve as octane improvers in gasoline or as diesel fuel extenders (Burke et al. 2001).

Naphthalene is a valuable starting material in making plasticizers. Naphthalene can be converted to special kinds of pitch that can be used to make carbon fibers. The related methylnaphthalene, also prevalent in coal liquids, yields a pitch that has been converted to carbon foams, graphitized to graphite powder, or used itself as a high-temperature lubricant (Song et al. 2005). These special forms of carbon will be discussed in Chapter 17.

Complete hydrogenation of naphthalene results in decahydronaphthalene, usually known by its common name, decalin. Decalin occurs in two *stereoisomer* forms. Catalysts have been developed to favor production of one or the other of these forms (Schmitz, Bowers and Song 1996). *trans*-Decalin, as a component of aviation fuels, provides extremely good resistance to degradation at high temperatures (Yu and Eser 1998). It has a greater volumetric energy density than fuels from petroleum, translating into greater flying range. The other isomer, *cis*-decalin, can be converted to one of the monomers used to make Nylon-6,10 (Song et al. 2005).

The economics of direct coal liquefaction have never been competitive with petroleum, at least when not affected by wars or embargoes. The economics might change in a dual-product strategy, in which one stream was used to produce specialty or fine chemicals and other devoted to commodity fuels. It's also been suggested that a coal liquefaction plant could be dedicated to chemical production and integrated into an integrated gasification combined cycle (IGCC) plant, making the overall facility less costly than a stand-alone liquefaction plant (Burke et al. 2001). Integrated plants in which coals would be converted into products for at least two different markets, such as electricity and fuels or electricity and chemicals, provide opportunities for significant cost reduction relative to plants making only a single product (Longwell 1995, 106).

Use of coal liquids for chemicals faces the same problem as coal tar: the liquids contain hundreds of components. Solvent extraction could be used to "skim" some desired materials from a selected coal, with the residue remaining from extraction fed to a different process, such as gasification. Selectively extracting naphthalene and its derivatives from coals has been explored (Clifford et al. 2009), with their hydrogenation to decalins. Montan wax, solvent-extracted from lignites, has been used to make electrical insulators and—a long, long time ago—carbon paper and phonograph records (Schobert 1995, 655). The lignite remaining after wax has been extracted can be added to fertilizers as a mulch and to improve soil porosity.

The Fischer-Tropsch (F-T) process provides an extremely powerful and versatile route to an array of synthetic fuels and chemicals. As with liquid fuels, F-T is an indirect route to making chemicals. Except for the Sasol facility in South Africa, most of the chemical products made from synthesis gas nowadays start with natural gas. Sasol shows the way to start with coal.

F-T plants also produce large amounts of carbon dioxide (CO_2). Synthesis gas that has been shifted will have a higher concentration of CO_2 than is found in the flue gas of a combustor and will be easier to separate and capture. Conceptually, it would be possible to run an F-T plant with capture of the CO_2, with processes designed to produce chemicals from the F-T products and parallel processes to make chemicals from CO_2.

F-T waxes consist of linear chains of 20–40 carbon atoms. The German company Deurex markets such waxes. They happen to be located on Dr. Bergius Street in their home city. Chemically cracking waxes leads to compounds commonly known in industrial practice as *α-olefins*, feedstocks for detergents, plasticizers, and alcohols.

The Oxo synthesis, developed in Germany during the 1930s, is a route to the synthesis of alcohols. Nowadays all of the synthesis gas comes from oil or natural gas. The original interest was to make raw materials to produce detergents. Alcohols with fewer than 12 carbon atoms are used to form plasticizers.

Continued development of the F-T process has been spurred mainly by the chemical industry. Conversion of synthesis gas to ethylene, propylene, and butylenes offers a direct route to these very important monomers (Janardanarao 1990). Research groups in China have reported the development of new catalysts for this conversion (Rouhi 2015). Ethylene, the most important industrial organic chemical in the world, can be converted into commodity plastics such as polyethylene, poly(vinyl chloride), and poly(ethylene terephthalate), produced in enormous tonnages every year and practically ubiquitous in our daily lives. Propylene also is a starting point for many commodities, such polypropylene plastics. This approach is referred to as Fischer-Tropsch-to-olefins (FTO) technology.

More than half of total world methanol production is done in China, mostly via coal gasification (Tullo 2014). The development of methanol-to-olefins (MTO) technology, mainly being pursued in China, provides a route to ethylene, propylene, and related compounds from synthesis gas via methanol. Coal gasification combined with MTO gives a route to making olefins from coal. This provides incentives for countries such as China, with abundant coal reserves and steadily growing demand for consumer products, to develop MTO processes (Tremblay 2014). In China, coal-to-olefins plants are integrated facilities that convert coal into methanol and then to olefins. Ethylene glycol, used for making polyester fibers, is made from coal-based MTO technology, mainly using low-rank coals, and is estimated to have about 40% of the Chinese market (Tremblay 2014).

At the extreme temperatures of an electric arc furnace, several thousand degrees Celsius, calcium oxide reacts with anthracite or with coke to make calcium carbide. Calcium carbide then can be reacted with water to form acetylene. Acetylene was once used as a starting material for many important chemicals, including ethyl alcohol and acetic acid, as well as products such as plastics and synthetic rubber. In wartime Germany, when raw materials from conventional sources were becoming increasingly scarce, monthly production of calcium carbide reached 100,000 tonnes. Along with coal-tar chemicals, acetylene served as a significant foundation for the synthetic organic chemical industry. Both coal-derived acetylene and coal-tar chemicals steadily declined in importance with the widespread availability of relatively inexpensive petroleum and natural gas after World War II.

Concerns about high electricity usage and dealing with byproduct calcium hydroxide[6] raise the question of whether it is possible to make acetylene directly from coal, eliminating the carbide step. The French chemist Marcellin Berthelot observed that striking an arc between carbon electrodes in an atmosphere of hydrogen would produce some acetylene (Schobert 2014). The best conversions in the direct production of acetylene from coals are in the 40–50% range. Though direct production would be simpler than the route via calcium carbide, large-scale processes involving electric arcs will still be prodigious consumers of electricity.

Acetylene can be converted to a range of chemicals, such as vinyl chloride and vinyl acetate used to make polymers: poly(vinyl chloride), commonly known as PVC, and poly(vinyl acetate). PVC is produced in enormous quantities, about 44 million tonnes in 2018, and is commonly encountered in daily life, including as a material for making credit cards. Poly(vinyl acetate) is widely used in many kinds of glues and adhesives, even in some chewing gums.

At least 20 other compounds with present or potential commercial uses can be made from acetylene (Szmant 1989, 193). Many of the advances in using acetylene are due to Walter Reppe and colleagues at I. G. Farbenindustrie, in the years from 1925 to 1945. These include processes leading to acrylonitrile, a precursor material to carbon fibers, and to acrylate esters used to manufacture acrylics. Postwar investigation of the wartime German chemical industry identified "Reppe chemistry" as one of three areas from which the Allies could derive useful knowledge—along with production of synthetic fuels from coal and production of synthetic rubber (Szmant 1989, 261).

Technologies for producing acetylene from coal, and for then producing an extensive variety of commodity chemicals from acetylene, are well known. Many acetylene-to-commodity processes have been demonstrated at commercial scale somewhere in the world. There are no technical barriers to making chemicals from coal-based acetylene. The issues are primarily economic, especially the price competition with ethylene and the high energy costs for calcium carbide or direct acetylene production. The choice among coals, biomass, or petroleum for chemical production has to be made on a country-by-country basis according to the availability of raw materials and on financial and technical constraints. The future of coal-derived acetylene as a

---

[6] Produced when calcium carbide reacts with water to make acetylene. The calcium winds up as the hydroxide.

large-scale source of chemicals and materials depends on the efficiency of the electric-arc process or on the emergence of other significant breakthroughs in converting coal to acetylene. Ongoing work in China suggests continuing interest in improving the arc process.

Making chemicals from coals has been commercially feasible in the past and in some places still continues. Products include dyes, fertilizers, sweeteners, explosives, plasticizers, wood preservatives, antiseptics, detergents, and fragrances. Chemicals from coal can also be used to make carbon fibers, and polyester fibers, polystyrene, polyethylene, polycarbonates, nylon, and poly(vinyl chloride). This is just the short list. We can likely produce most of the chemical products and synthetic materials that we would expect to be available for modern life in some way by using coals as the starting point. About 60 years ago the English science writer J. Gordon Cook mentioned the possibility of "turning coal into clothes" (Cook 1967, 9). It can certainly be done, with nylon and polyesters at least.

For a chemical industry based on carbo-chemicals, the barriers are formidable but not insurmountable. Since its high point in 2014, coke production has been on a slow but steady decline. As we discussed in Chapter 10, new coke oven batteries are based on heat-recovery ovens, not byproduct ovens. And then there's CO_2: producing a tonne of coke makes about a tonne of CO_2 at the same time (Schobert and Schobert 2015).

Direct coal liquefaction plants could run on a dual-product strategy so that a process stream targeted for chemicals production would be a true co-product with a separate stream headed to fuels. The relative proportions of chemical and fuel products could be altered from time to time to respond to changing market demands. There are no technical barriers to running a direct liquefaction plant in a "chemicals-only" mode. Using hydrogen made by processes that have no significant CO_2 production, such as electrolysis of water using "non-carbon" sources of electricity, could significantly reduce the carbon footprint of a chemicals-plus-fuels liquefaction plant.

F-T processing is already proved to be a feasible route to chemical products. More versatility is gained from F-T methanol. CO_2 could be captured and then itself used to make chemicals. A chemical and fuel complex can be envisioned in which coal is converted into F-T chemicals, F-T fuels, methanol and its derivatives, and CO_2-based chemicals.

In considering a prospective future for chemicals from coal, it's useful to consider how chemicals are usually synthesized in the laboratory. A chemist selects as a starting material a compound for which the elemental composition and molecular structure are very accurately known. A reaction is selected for which, usually, the mechanism[7] is known. Ideally, the reaction would be one in which desired product is formed in high yield or produces only a small number of easily separable products. Can this approach ever be applied to coals? The best answer at present would be, "sort of."

The molecular structures of coals are not, and never will be, accurately known because there isn't one, in the sense that even large biochemical molecules have clearly defined structures. But we need to remember Andreas Osiander: "Nor is it necessary that these hypotheses [i.e., structures of coals] should be true, nor indeed even probable, but it is sufficient if they merely produce calculations which agree with the observations" (Osiander 1993). Continued probing of coals with a wide array of instrumental techniques and the continued development of powerful computer-based structural modeling provides an ever-increasing base of useful knowledge.

The way in which many coal reactions proceed (i.e., their mechanisms) is reasonably worked out thanks to extensive fundamental research in coal chemistry in laboratories worldwide, much of it done during the last quarter of the previous century. The mechanisms of radical reactions are quite well known. However, we should not fail to consider that there are alternative ways of breaking bonds. It might prove better to design a reaction following some other pathway. It is very common in coal chemistry to do bond-breaking with thermal energy.[8] There might be advantages in such options as ultrasonic energy, microwaves, or extremely fast laser pulses. A challenge for the future is to develop controlled means to break desired bonds selectively.

The biggest challenge is to find reactions that lead to a small number of easily separable products. Given the complexity of composition and the variability among coals, this may well prove impossible, which then comes back to the notion of designing process plants around a dual- or multi-product strategy. It may not be a good idea to try to drive a coal to 90% conversion to liquid products if those products need considerable downstream processing to meet market specifications. Perhaps it would be better to skim a much lower percentage of light liquids for making fuels or chemicals and convert the heavy material to something else, such as pitch.

---

[7] A reaction mechanism is the detailed description, usually in a series of elementary reaction steps, of what occurs in every stage of the transformation of the reactant to the product—what bonds break or form and when, and the relative rates of the various steps.

[8] This is embodied in what is facetiously known as the First Law of Experimental Chemistry: when in doubt, heat the hell out of it.

We cannot write off the future for specialty and chemicals derived from coal. The challenge and opportunity lie in developing entirely new processes for producing valuable chemical products, but ones that do not require coke ovens, calcium carbide, F-T reactors, or coal liquefaction. A worthy challenge would be to develop routes to chemicals from coal following the philosophy of green chemistry.

The intent of *green chemistry*, as originally defined by the US Environmental Protection Agency, is "To promote innovative chemical technologies that reduce or eliminate the use or generation of hazardous substances in the design, manufacture, and use of chemical products" (Lancaster 2016). The practice of green chemistry is based on 12 tenets (Lancaster 2016, 5), of which several have immediate relevance to the development of forward-looking processes for producing chemicals from coals. One is the concept of atom economy, using synthetic methods that maximize the incorporation of all materials used in the process into the final product. Second, wherever possible, synthetic methods should be designed to use and generate substances that have little, preferably no, harm to people or the environment. Third, solvents or separation agents should be made unnecessary whenever possible and innocuous when used. Fourth, energy requirements of chemical processes should be minimized. Green processes are thought likely to be more cost effective (Anastas and Warner 1998). A laboratory program exploring the green chemistry approaches to conversion of coals could pay off handsomely. "Make your mistakes in test tubes, make your profits in vats" (Gilbert, as quoted in Reynolds 1993).

Atom economy ideally should be 100% (Manahan 2011, 28). Given the heterogeneity of coals, that does not seem attainable. A more likely scenario would be most of the coal winding up as product, with any byproducts having uses of their own.

Coals do not meet the green chemistry goal of using renewable resources rather than resources that can be depleted. However, despite the consumption of vast amounts of coal worldwide for more than two centuries, a very large trove remains—about a trillion tonnes (BP 2019). We need sustained support for further advances in coal chemistry, coal petrography, and coal structural modeling. The field of coal preparation has opportunities for developing processes for concentrating desired macerals. Frederick Perkin has pointed out what happens as a consequence of a period of neglect of coal science.

We've seen in Chapter 9 that ash has several possible uses, an important one being in the production of cement. Only a few inorganic constituents of coals occur in significant proportions. Many of the elements that are present only at a fraction of a percent, or even at a parts-per-million (ppm) level, have valuable commercial applications. This makes it worth considering the possible inorganic chemical products from coal. Concentrations of potentially valuable trace elements in coals vary from one coal to another and certainly from one element to another, but, as an example for comparison, the mean values of a few trace elements in coals[9] of the western United States are 11 ppm of cerium, 2.5 ppm of gallium, and 1.2 ppm of uranium.

The chemical family often called *rare earth elements* (REE) represents a remarkable misnomer. The rare earth elements are not rare and are not earths. In the nineteenth century, oxides of elements were called "earths." The rare earths are usually found in nature as phosphates, silicates, or carbonates, not oxides. In terms of average abundance in Earth's crust, many of the rare earths are at least as abundant as copper and lead and much more than silver, metals still actively mined and used today. The rare earths include the family of 15 elements from lanthanum through lutetium. All but one (promethium) occur in nature. Sometimes it is convenient to include yttrium and scandium among the rare earths because the chemical behavior and properties of these two elements are very similar to those of the lanthanum family.

REEs are used in many applications in consumer goods and in many items of hardware used in defense. China currently holds a near-monopoly on REEs, accounting for about 95% of the world market. If China were to embargo REE exports other countries, there would be a desperate scramble for alternative sources. This accounts for increasing interest in seeking such alternative sources, even if their REE concentrations are well below those in ores that are currently exploited.

Though many commercially important metallic elements occur at low average concentrations in the Earth's crust—indeed lower than those of the REEs—various geological processes have produced localized deposits in which their concentrations are enhanced many-fold, resulting in useful ores. This has not happened with the rare earths. They commonly occur as mixtures at comparatively low concentrations in many kinds of rocks. The great similarity of chemical behavior among the rare earths results in their all tending to be found together in the same rocks.

This means it's necessary to develop mineral processing schemes to produce a beneficiated product having enhanced concentrations of the REEs.

[9] There is potential for inadvertent confusion when checking the literature on trace elements in coals. Some sources report values as ppm in the *coal*, usually referred to as a "whole-coal basis," whereas other sources might report values as ppm in the *ash*. It's always important to check to see how the values are reported in a particular publication. For a coal having an ash yield of 10%, trace element concentrations in ash will be an order of magnitude higher than those on a whole-coal basis.

Extraction processes need to be able to obtain the REEs from relatively un-reactive rocks such as granite, even after the rocks have been beneficiated. Separation methods have to yield the 16 pure elements (i.e., including scandium and yttrium) from a process stream in which they all occur together and in the face of their being very similar to one another chemically.

In spite of the ingrained, and inaccurate, usage of the adjective "rare," the REEs pervade our daily lives. Members of the rare earth family—mainly neodymium, cerium, and lanthanum—are used extensively in consumer electronics such as disk drives, earphones, speakers, and television screens; in turbine-driven electric generators, as in wind farms; in laser applications such as laser pointers; in fiber optics; and in catalytic converters in vehicle exhaust systems. Scandium alloys with aluminum have applications in aerospace and in various kinds of sporting equipment. These few examples only touch on the many other applications of rare earths (Krishnamurthy and Gupta 2016, 43).

Although the concentrations of total and individual REEs in coal ashes are lower than in current commercial ores, ashes nevertheless could be an attractive source. Compared to the effort and costs of mining a hard-rock REE ore, coal ash is "free" because the coal from which it came was mined and consumed for a different purpose. Besides, ash will be produced in combustion or gasification processes anyway, whether we want it or not. Global production of coal ash is about a billion tonnes per year (Harris, Heidrich, and Feyerborn 2019).

Using coals as the raw material to extract REEs would require the handling of prodigious quantities of materials to obtain a few kilos of a selected element. Processing coals for the sole purpose of recovering REEs is not feasible. Considering ash as an REE source is a much better choice since the elements would already be concentrated by about an order of magnitude compared to the parent coal. Still, economic feasibility will require a processing scheme in which a coal is first used in some other application—perhaps more than one other application—and rare earth recovery is considered a value-added by-product obtained from the ash. Possibly coal preparation operations could be modified to provide a rare earth–rich refuse stream.

Many methods for obtaining rare earths begin with leaching the ore in strong acids (Erickson 2018). Water flowing through mines in which the coals have high levels of pyritic sulfur becomes highly acidic. Lime is used to raise the pH. A sludgy precipitate forms, primarily consisting of compounds of iron, manganese, and aluminum, like the "yellow boy" discussed in Chapter 6. The precipitate also contains the REEs, at a concentration about 2,600 times higher than their concentration in the water, and which can be treated to obtain a material that is 90% rare earths. A possibly easier way would be to extract the REEs directly from the acid mine drainage, by-passing the precipitate and its treatment (Song et al. 2020).[10]

Some lignites contain sufficient uranium to be economically attractive as sources of uranium. Uranium was discovered in North Dakota lignites in 1948 (Schobert 1995, 24), as the Cold War was beginning. One year later, the Soviet Union exploded its first atomic bomb. The Cold War and the "Red Scare" of the late 1940s and '50s led to considerable interest in finding and exploiting domestic sources of uranium in the United States.

Lignites of the Sentinel Butte Formation contain about 850 ppm of uranium.[11] This is close to the concentration in a low-grade ore body, which would be 1,000 ppm (Ulmer-Scholler 2019), and much larger than the average concentration in Earth's crust of 2.8 ppm. Recovering uranium directly from the lignite using processes of the 1950s was not feasible, because lignite cannot be ground in the same equipment as the sandstone ores. Instead, the lignite was first burned and then the recovered ash could be mixed with sandstones for processing. From 1962 to 1967, about 270 tonnes of U_3O_8 were produced from North Dakota lignite.

Gallium and germanium are frequently considered together as potentially valuable products from ash. Both are used in making semiconductor devices. Many years ago small amounts of gallium and germanium were obtained from coal ashes (Greenwood and Earnshaw 1984, 245). Though they occur in ash in concentrations measured in ppm, their high costs[12] could repay the effort involved in recovering them from ash.

Gallium is found in nature associated with bauxite, the principal ore of aluminum, but also with sulfide minerals. Most gallium used today is a by-product of the aluminum industry; the raw gallium is obtained by electrolysis and further purified to ultra-high grades.

Gallium in the form of gallium nitride or gallium arsenide is used in integrated circuits. Roughly half of worldwide gallium consumption goes into cellphone components. Increasing demand for gallium is driven by the evolution of 4G and 5G smartphone technology, which requires an order of magnitude more gallium than earlier generations.

---

[10] This pertains only to REE extraction. Regardless of what happens to the REEs, the acid mine drainage still has to be treated to raise the pH to an acceptable level and to remove the iron, aluminum, and manganese sludge.

[11] Equivalent to 0.1% uranium expressed as the oxide U_3O_8.

[12] In 2018, prices for high-purity gallium were in the range of $10 per gram.

During World War II, germanium was recovered from power plant ash in the northeast of England. Some coals with 8,000 ppm germanium in the ash have been reported, but most germanium is recovered as a byproduct of zinc smelting. It has uses in fiber optics and in the glass formulations used in high-quality lenses for wide-angle photography and microscopy. Alloys of germanium with silicon are important in semiconductors. Germanium is also incorporated into photovoltaic cells.

The moisture held in soils is actually a solution of soluble inorganic constituents, such as calcium and magnesium. REEs are more likely to dissolve in a small volume of water, as in the water in soils; this preferential accumulation is called *fractional leaching* (Goldschmidt 1950). REEs can be concentrated relative to the more common elements. The water— more accurately, the solution—in soil is taken up by the roots of plants. As water evaporates from plants, the dissolved elements are left behind. Ashes of oak leaves might contain more than 10% of manganese oxide (Goldschmidt 1950).

Plants can be found growing on the residues of coal cleaning operations. We could even be proactive and encourage their growth by seeding and fertilizing such accumulations. By fractional leaching, these plants might have concentrated some valuable trace elements. Harvesting and ashing plants and then extracting trace elements from the ashes might be easier than trying to deal with extracting the coal cleaning residues.

This approach would use living plants to do much of the work of extracting elements from coals. This approach is known as *phytomining*, the prefix *phyto-* denoting the role of plants. Phytomining offers opportunities for recovering valuable elements from coal wastes (Williams 2019). Burning plants to recover ash will certainly produce CO_2, but, at least in principle, that amount of CO_2 would be removed from the atmosphere by replanting in the following year. The "miners" are renewable.

Coals should be viewed as a hydrocarbon source having multiple prospective uses, all of which are deserving of serious consideration as prospective uses for this resource (Schobert 1998). Coal utilization today is dominated by combustion—not only of the coal itself, but also combustion of coal products such as coke and synthetic fuels. If some useful byproducts happen to be made along the way, they represent just a small added bonus. Combustion applications of coals will dominate in the near-term and may remain in use for decades, but to ignore now the potential for alternative uses is only to short-change ourselves in the future (Schobert and Song 1996).

What coals should we select as raw materials for making chemicals? Most chemical products of interest have higher hydrogen-to-carbon ratios than do most coals. We are probably best off working with coals at the lower end of the rank range—the lignites to high- or medium-volatile bituminous—with relatively small aromatic ring systems and a relatively higher amount of aliphatic carbon. We should also think about looking at coals that might have high concentrations of liptinites. We will save the low-volatile bituminous coals and the anthracites for Chapter 17.

Our hydrocarbon resources, including coals and the other fossil fuels, are valuable raw materials for making a wide range of the chemical products used in everyday life and, as we'll see, for making carbon products as well. Dimitri Mendeleev, in addition to devising the Periodic Table, was also involved in petroleum chemistry. He recognized the wastefulness of using hydrocarbons as fuels instead of, in his case, making petrochemicals. He remarked that using petroleum simply as a fuel would be equivalent to "firing up a kitchen stove with bank notes" (Moore, Stanitski, and Jurs 2007, 197). One could say the same about using coals.

Likely for centuries to come, human society will still require carbon-based raw materials that can be turned into organic chemicals and carbon products. Coals can have a role in fulfilling such raw materials needs. Opportunities abound for developing green chemistry approaches for making chemicals and carbon products from coals, especially as other coal technologies, including combustion for power generation, continue to decline.

There is great potential for coals as a source of specialty chemicals and of carbon materials as well. This potential derives from the special chemical characteristics of coals: high carbon content, significant aromaticity, and the prevalence of polycyclic aromatic structures. But we must be realistic: the old ways, meaning the use of byproduct coke ovens for chemicals from coal tar, will never come back. New ideas and new strategies are needed. We need to rethink coal.

References

Anastas, Paul T., and John C. Warner. 1998. *Green Chemistry Theory and Practice.* Oxford: Oxford University Press.

BP. 2019. "Statistical Review of World Energy." https://www.bp.com/en/global/corporate/energy-economics/statistical-review-of-world-energy.html.

Burgess, Caroline E., and Harold H. Schobert. 2000. "Direct Liquefaction for Production of High Yields of Feedstocks for Specialty Chemicals or Thermally Stable Jet Fuels." *Fuel Processing Technology* 64: 57–72.

Burke, Frank P., Susan D. Brandes, Duane C. McCoy, Richard A. Winschel, David Gray, and Glen Tomlinson. 2001. *DOE Direct Liquefaction Process Development Campaign of the Late Twentieth Century: Topical Report.* U.S. Department of Energy Report. No. DOE/PC 93054-94.

Clifford, Caroline E. B., Harold H. Schobert, Josefa Griffith, and Leslie R. Rudnick. 2009. "Solvent Extraction of Bituminous Coals Using Light Cycle Oil: Characterization of Diaromatic Products in Liquids." *Energy and Fuels* 23: 4553–4561.

Cook, J. Gordon. 1967. *Virus in the Cell.* New York: Dial Press.

Erickson, Britt. 2018. "Rare-Earth Recovery." *Chemical and Engineering News* 96 (28): 28–33.

Findlay, Alexander. 1917. *The Treasures of Coal Tar.* London: George Allen & Unwin Ltd.

Garfield, Simon. 2000. *Mauve.* New York: W. W. Norton and Company.

Garside, Michelle. 2019. "Global Coke Production 1993–2018." Statistica. https://www.statista.com/statistics/267891/global-coke-production-since-1993/.

Goldschmidt, Victor M. 1950. "Occurrence of Rare Elements in Coal Ashes." In: *Progress in Coal Science*, edited by Donald H. Bangham, 238–247. New York: Interscience Publishers.

Greenwood, N. N., and A. Earnshaw. 1984. *Chemistry of the Elements.* Oxford: Pergamon Press.

Harris, David, Craig Heidrich, and Joachim Feyerborn. 2019. "Global Aspects on Coal Combustion Products." Coaltrans. https://www.coaltrans.com/insights/article/global-aspects-on-coal-combustion-products.

Janardanarao, Mulpuri. 1990. "Direct Catalytic Conversion of Synthesis Gas to Lower Olefins." *Industrial and Engineering Chemistry Research* 29: 1735–1753.

Krishnamurthy, Nagaiyar, and Chiranjib Kumar Gupta. 2016. *Extractive Metallurgy of Rare Earths.* Boca Raton, FL: CRC Press.

Lahiri, Adinath, and B. Mazumdar. 1967. "Studies on Dehydrogenation of Coal and Structural Implications." In: *Symposium on the Science and Technology of Coal*, edited by T. E. Warren, 175. Ottawa: Mines Branch, Department of Energy, Mines and Resources.

Lancaster, Mike. 2016. *Green Chemistry: An Introductory Text.* Cambridge: Royal Society of Chemistry.

Longwell, John P. 1995. *Coal: Energy for the Future.* Washington, DC: National Academy Press.

Manahan, Stanley E. 2011. *Green Chemistry and the Ten Commandments of Sustainability.* Columbia, MO: ChemChar Research.

Moore, John W., Conrad L. Stanitski, and Peter C. Jurs. 2007. *Chemistry: The Molecular Science, Volume 1.* 197. Pacific Grove, CA: Brooks/Cole.

Neavel, Richard C. 1982. "Coal Plasticity Mechanism: Inferences from Liquefaction Studies." In *Coal Science*, edited by Martin L. Gorbaty, John W. Larsen, and Irving Wender, vol. 1, 1–19. New York: Academic Press.

Osiander, Andreas. 1993. "To the Reader on the Hypotheses in This Work." Preface to Copernicus, N. *De RevolutionibusOrbiumCoelestium.* Translated by A. M. Duncan. Norwalk, CT: Easton Press.

Reynolds, Francis D. 1993. *Crackpot or Genius?* Chicago: Chicago Review Press.

Rhodes, E. O. 1945. "The Chemical Nature of Coal Tar." In *Chemistry of Coal Utilization*, edited by Homer H. Lowry, 1293. New York: John Wiley & Sons.

Rouhi, A. Maureen. 2015. "China's Coal-to-Chemicals Success." *Chemical and Engineering News* 93 (34): 30–31.

Schmitz, Andrew, Grainne Bowers, and Chunshan Song. 1996. "Shape-Selective Hydrogenation of Naphthalene over Zeolite-Supported Pt and Pd Catalysts." *Catalysis Today* 31: 45–56.

Schobert, Harold H. 1995. *Lignites of North America.* Amsterdam: Elsevier: Amsterdam.

Schobert, Harold H. 1998. "Coal Conversion Chemistry as a Problem in Organic Synthesis." In *Proceedings, 6th Japan-China Symposium on Coal and C1 Chemistry*, 358–361.

Schobert, Harold H. 2014. "Production of Acetylene and Acetylene-Based Chemicals from Coal." *Chemical Reviews* 114: 1743–60.

Schobert, Harold H., and Nita S. Schobert. 2015. "Comparative Carbon Footprints of Metallurgical Coke and Anthracite for Blast Furnace and Electric Arc Furnace Use." Blaschak Anthracite Corporation. https://www.blaschakcoal.com/case-study-met-coke-vs-anthracite.

Schobert, Harold H., and Chunshan Song. 1996. "Carbochemicals: Toward a New Coal Chemistry." In *Proceedings, 7th Australian Coal Science Conference*, edited by Phil Waring, 29–36. Melbourne.

Song, Chunshan, Harold H. Schobert, and John M. Andrésen. 2005. "Premium Carbon Products and Organic Chemicals from Coal." International Energy Agency Report No. CCC/98, 2005.

Song, Guanrong, X. Wang, Carlos Romero, H. Chen, Z. Yao, A. Kaziunas, R. Schlake, M. Anand, Thomas Lowe, J. Gregory Driscoll, Boyd Kreglow, Harold H. Schobert, and Jonas Baltrusaitis. 2020. "Extraction of Selected Rare Earth Elements from Anthracite Acid Mine Drainage Using Supercritical CO2 Via Coagulation and Complexation." *Journal of Rare Earths*. doi.org/10.1016/j.jre.2020.02.007.

Szmant, H. Harry. 1989. *Organic Building Blocks of the Chemical Industry*. New York: John Wiley & Sons.

Tremblay, Jean-Francois. 2014. "China's Feedstock Revolution." *Chemical and Engineering News* 92 (17): 18–19.

Tullo, Alexander H. 2014. "Methanol for Sale." *Chemical and Engineering News* 92 (32) 12–13.

Ulmer-Scholler, Dana S. 2019. "Uranium: Where Is It Found?" New Mexico Bureau of Geology & Mineral Resources. https://geoinfo.nmt.edu/resources/uranium/where.html.

Williams, Orla. 2019. "Opportunities for Critical Raw Material Production from Biomass and Coal Waste Resources." *17th International Conference on Coal Science and Technology*. Krakow, Poland.

Wood, Laura. 2018. "Global Phenols Market Analysis and Outlook 2017–2023." BusinessWire. https://www.businesswire.com/news/home/20181126005423/en/Global-Phenols-Market-Analysis-Outlook-2017-2023-Applications-in-the-Field-of-Medicine-Has-Been-Growing-With-the-Highest-Rate-of-About-6.15-Per-Year---ResearchAndMarkets.com.

Yu, Jian, and Semih Eser. 1998. "Thermal Decomposition of Jet Fuel Model Compounds under Near-Critical and Supercritical Conditions: 2. Decalin and Tetralin." *Industrial and Engineering Chemistry Research* 37: 4601–4608.

17
Carbons

In a sense, coals *are* carbon materials, especially those at the upper end of the rank range. The direct use of coals as solid materials is not significant. Coals are highly variable in composition and properties. It would be likely impossible for engineers to design applications for a material that does not have consistent properties from one batch to the next. The electrical, thermal, and mechanical properties of coals have not been explored or related to composition nearly to the extent that their chemical behavior has been, nor to the extent that such materials as polymers or alloys have been. These concerns are largely overcome by using coals as raw materials to make carbon products of consistent composition and properties.

When considering the potential of coals as raw materials, two other factors come into play. Many carbon materials enjoy steadily growing market demand, a very different situation compared to the decline in demand for coal to be consumed in power plants. Many carbon materials also command prices significantly higher than current market prices of coals, by two or three orders of magnitude. Making and selling such materials, even if only as a byproduct stream, represents a significant revenue source.

Attractive targets for making carbon materials from coals are activated carbons and graphites. The technology for producing these materials from coals is known. Activated carbons are already made commercially from coals. There have been promising pilot trials of graphitizing anthracites. Activated carbons and graphites have large, existing markets that will continue growing. The best-quality products are priced at dollars per kilo, whereas coals sell for dollars per tonne.

Activated carbons are adsorptive materials. They have been produced by treatment of a carbonaceous solid with reactive gases, sometimes with the addition of chemical reagents, under carbonizing conditions. The treatment aims to obtain a substance with adsorption properties good enough to be used in the purification of liquids or gases. All activated carbons work by adsorbing the compound or compounds sought to be removed. Most activated carbons are isotropic solids having disordered structures.

Adsorption occurs at the carbon surface. Many activated carbons have surface areas greater than 1,000 square meters per gram. Activated carbons are used in such important processes as purifying drinking water and removing mercury from power plant flue gases. A monograph published 80 years ago lists 134 products treated with activated carbon (Hassler 1941, 154). Today such a list would go on for multiple pages.

Different coals, processed under identical conditions, will yield activated carbons having different physical and chemical properties. The same coal processed under different conditions can also yield carbons of very different properties. Given a selected coal, a combination of activation procedures and conditioning can often tailor a product for a desired application.

Coals are not the only raw materials for making activated carbons. Desirable raw materials are of low cost, generally high carbon content, and low levels of inorganic impurities. Though it is difficult for some coals to be cost competitive with such alternatives as olive pits and coconut shells, coals offer the opportunity of making high-quality activated carbons. The overriding issues are the properties of the activated carbon and its suitability for the intended use.

Two strategies—physical activation and chemical activation—are used to make activated carbons. Both rely on chemical reactions. The difference lies in the activating agents and how they are applied. In physical activation, carbonization to remove volatiles is followed by a second step of partial gasification. The gasification does the activation, opening or creating porosity in the carbon and increasing the internal surface area. The carbonization and activation steps could be done simultaneously. Air can also be used as an activating agent, but it can be difficult to produce a desired distribution of pore sizes.

Chemical activation reacts the raw material with strong acids, such as phosphoric acid, or with strong bases, such as potassium hydroxide. The added reagent decomposes part of the molecular architecture of the feedstock, creating or opening internal porosity. Washing the reagent out of the now-activated char at the end of the carbonization step leaves the activated carbon. Chemical activation operates at lower activation temperatures, needs less activation time, and gives higher yields of carbon with higher apparent surface areas. However, most of the reagents used for chemical activation are strong acids or bases, potentially hazardous and corrosive. The subsequent washing step needed to remove the activating reagent adds a complication to the process. Whichever activation strategy is employed, the objective is to make a porous carbon material having surface area, porosity, and adsorption characteristics suitable for the intended applications.

The chemical nature of the carbon surface can also affect adsorption. Depending on the intended use of the carbon, there might be reason to modify the surface composition after the activation process. If so, a final processing step is used to condition the carbon surface by appropriate reaction. Various chemical treatments will modify the surface of a carbon by introducing oxygen, sulfur, or nitrogen *functional groups*. The acidity or basicity of these groups affects the kinds and amounts of species that will adsorb on the surface.

When coals are carbonized to make activated carbons, many of the oxygen, nitrogen, and sulfur atoms will be driven off in gaseous molecules, along with some of the small aliphatic structures. Larger pieces of the molecular framework will also come off, later to condense as tars.

At carbonization temperatures, the requirements for internal redistribution of hydrogen atoms are partially satisfied by dehydrogenative polymerization, aromatic rings joining to produce larger and larger aromatic "sheets." Some of the aromatic rings may contain atoms of nitrogen or oxygen, making the sheets non-planar. As char forms, these aromatic sheets begin to come together into stacks. The stacks are of irregular size and randomly oriented, which means that there can be spaces or crevices between them. There might be some crosslinking between the aromatic sheets. This provides an incipient porous structure. The free spaces can become filled with tar or even with amorphous, very carbon-rich solids.

Steam or carbon dioxide (CO_2) used to do the activation is most likely to react with the amorphous, disordered carbon and with the condensed tars because the condensed tars and disordered carbon are less stable chemically than the stacks of aromatic sheets.[1] There will certainly be some reaction of the activating gas with the solid carbon. Removal of the tar and disordered carbon via the carbon–steam or Boudouard reaction clears out the blocked pores, giving a porous solid carbon.

The weight of activated carbon produced will be less than the weight of coal used, sometimes significantly less. The loss in weight is referred to as the burn-off. It's important to keep the burn-off as low as possible, consistent with obtaining an activated carbon of the desired quality. Any material burned off and not winding up in the product is lost money. On an industrial scale, the production of activated carbon represents a compromise between the development of the adsorption properties—such as porosity—and the yield of activated carbon. Where does the burned-off carbon go? In activation with steam or CO_2, carbon will be lost from the solid as carbon monoxide (CO). The CO

---

[1] This argument is based on thermodynamic stability. Graphite is the thermodynamically stable form of carbon under ordinary conditions. The char is not truly graphite, but the stacking and alignment of aromatic layers is getting toward a graphite-like structure. Compared to the much more disordered structures in condensed tar or amorphous carbon, char is more stable and less likely to react.

might eventually be converted to CO_2. With activation in air, carbon really is burned off and forms CO_2 directly.

Carbonization and activation do not remove mineral matter. The ash yield will be higher than that of the parent coal.[2] The inorganic components of the coal could have a catalytic effect on the carbon–steam or Boudouard reaction, enhancing their rates. Since catalysts function by changing the pathways of reactions, mineral-matter catalysis can also affect the ways in which pores are created and evolve during activation (Rodríguez-Reinoso 1997). Some of the ash particles will likely be exposed on the surfaces of the activated carbon, where they might affect the adsorption processes because of their own interactions with the materials being adsorbed.

The porosity and pore-size distribution of the carbon product depend not only on the coal, but also on the type of activation method and the extent of activation or burn-off. There is great versatility in manufacturing families of activated carbons in which the pore characteristics and surface areas vary greatly and can be tailored to specific adsorption requirements.

The global market for activated carbons from all sources is about 2.5 million tonnes and is worth approximately $5 billion. That represents a cost for activated carbon of about $2,000 per tonne. That's a pretty good mark-up compared to a coal price of, say, $100 per tonne. In 2017, coal-based carbons had the major share of the activated carbon market (Cision 2019).

Fred Cannon and his colleagues at Penn State showed that it was possible to exploit the water originally in coals, along with the waste heat in the exhaust gases of a foundry cupola, to activate coals (Huang, Wang, and Cannon 2009). Not surprisingly, lignites work best in this scheme because they normally have the highest moisture contents of any rank of coal and are fairly reactive coals. Possibly other ranks might be activated to comparable extents if additional water were added to the coal or if the process could be adapted to facilities that have both waste heat and waste steam. Even if the homemade activated carbon were not comparable to commercial products, it should be able to be produced at relatively low cost because water comes in with the coal whether we want it or not, and the waste heat and waste steam needed are "free." The activated carbon needs only to be good enough to do the clean-up or pollution-prevention job needed at that particular site.

---

[2] This is because burn-off only affects the carbonaceous portion of the coal. As an example, 100 grams of a dry coal of 10% ash yield would contain 10 grams of minerals and 90 grams of carbonaceous material. If physical activation resulted in a 50% burn-off, the activated carbon would consist of 10 grams of minerals and 45 grams of carbonaceous material, equivalent to an 18% ash yield. (In this example we have ignored the distinction between mineral matter and ash.)

A vast array of industrial processes generate waste heat and CO_2. Many also have waste steam on hand. If not, coal-bed methane or other gaseous fuels might be available. Coals have water in them. With some clever engineering, and possibly some modest retrofitting of plant, there could be ample opportunities for producing relatively inexpensive, good-enough activated carbons from coal.

In low-NO_x burners, the benefit of reduced NO_x is accompanied by a reduction in the efficiency of combustion. Reduced efficiency is seen in an increased amount of partially burned char particles in the ash, particles commonly referred to as "unburned carbon." When the carbon content of ash rises above 6%, that ash is usually not acceptable as an ingredient in making cement. The carbon-laden ash can be retained at the power-plant site in holding ponds or sent off to a landfill. Neither is an appealing choice. Separating the unburned carbon from the ash allows the ash to be marketed to the cement industry, but also gives us another source of carbon. It is, in a respect, "free" because it will be produced as part of the boiler operation. The only added cost would be separating it from the ash. Unburned carbon already is a char. Unburned carbons from combustion of high-volatile bituminous coals can be activated with steam, with the best cases exceeding 1,000 square meters per gram of surface area (Lu, Maroto-Valer, and Schobert 2008).

Potentially the biggest-ever application of activated carbons has a direct linkage to coal technology: mercury capture. Carbons from activation of bituminous coal provide better mercury adsorption than those made from agricultural or forestry raw materials (Rodriguez, Contrino, and Mazyck 2020). Ash yield is not a critical factor in determining mercury adsorption capacity (Satterfield 2017). Microporous carbons have large surface areas for mercury adsorption (Office of Mercury Management 2018).

Because mercuric sulfide is stable and highly insoluble, sulfided activated carbons have been used for capturing mercury compounds (Reddy 2014, 157). Elemental mercury is poorly adsorbed on most carbons, so it needs to be oxidized for effective capture. Activated carbons treated with bromine or bromide salts work well in this application (Reisch 2015). Brominated activated carbons provide about 90% capture of mercury (Reisch 2015). Their effectiveness means that less carbon is required, thus reducing cost. But even with carbons treated with bromine or other agents, mercury capture remains a potentially huge market for activated carbons, especially in countries such as China, which still use coal extensively in power generation (Reisch 2014).

Ordinary sieves are made of woven wire mesh or perforated plates. If we made the aperture sizes small enough, it would be possible to separate molecules of different sizes. We would have a *molecular sieve*. Carbon molecular sieves can be made to use for separating molecules. The analogy with sieves of macroscopic size is not perfect (Spencer 1967). Pores in a molecular sieve are unlikely to be all of a perfectly regular geometric shape, as in an ordinary sieve. Some pores have slit-shaped entrances. Furthermore, not all molecules are of the same shape, let alone of the same size. Benzene is flat, methane is spherical. In a macroscopic sieve we don't expect interactions between the material being sieved and the material from which the sieves are made. In a molecular sieve, the ultrafine pores,[3] of molecular dimensions, are essentially the very narrow spaces in the molecular framework of the solid carbon. Molecules passing into the sieve can interact in various ways with the walls of the pores, thus having an effect on the separation process.

Development of carbon molecular sieves and the recognition that they can be made from coals comes from the work of, among others, Philip Walker at Penn State and Denis Spencer at the British Coal Utilisation Research Association. The molecular sieve market is currently dominated by synthetic zeolites, but carbon-based materials can offer several advantages. Carbon materials are less expensive. They have less affinity for water, which is adsorbed by zeolites and requires that the zeolite molecular sieve be preconditioned before being put into service. They are better for adsorbing oxygen. For service in acidic media, carbons are acid-resistant, whereas zeolites are not.

As raw materials for molecular sieve carbons, anthracites have advantages. First, *native* anthracites already have a high degree of microporosity (Patel and Walker 1973). Second, in high-temperature applications bituminous coals can soften, losing much or all of their original porosity. Anthracites don't do this. The pore openings in anthracite are very small, so that only a few kinds of molecules can enter the pores of native anthracite, and the adsorption process is slow. Only a small amount of activation is needed to enlarge the pore openings to create a greatly enhanced rate of penetration by such molecules as methane (Patel, Nandi, and Walker 1972). Very likely, native anthracites or activated anthracites with carefully tailored pore size distributions could be used as molecular sieves in other separation applications.

Carbon molecular sieves can be used in air separation processes. Chapter 11 mentioned the need for producing pure oxygen to be used in oxygen-blown gasifiers. An air separation plant relying on carbon molecular sieves would

---

[3] Typically, the pores for molecular sieving are on the order of 0.3–0.5 nanometers, and generally less than 1 nanometer diameter (Ceçen 2014).

be cheaper, smaller, and less energy-intensive than traditional cryogenic separations. Carbon molecular sieves can also separate methane from CO_2. This is useful in the purification of biogas or landfill gas. Biogas is easily produced by the anaerobic digestion of human or animal waste. Anaerobic decay of organic matter produces CO_2 and methane. The mixture of CO_2 and CH_4 is an adequate fuel, but nearly pure methane would be much better, no different chemically from natural gas. Removing the CO_2 would provide a gas stream of very high CO_2 content for sequestration.

Butane, a useful fuel gas and refrigerant, can be used to produce fiberglass-reinforced plastics, with a wide range of engineering uses. Isobutane can be used to make high-octane rating components of gasoline, which are also valuable products but quite different from plastics. Being able to separate isobutane from butane can be a help regardless of whether one is making gasoline or plastics. A very slightly activated anthracite—only about a 7% burn-off—is able to do this (Patel et al. 1972).

Because graphite represents the ultimate end product of catagenesis, it tends to be found with rocks that have experienced a high degree of metamorphosis. Examples would be rocks altered by contact with magmas or by tectonic stresses in the surrounding rocks (Nover, Stoll, and von Der Gönna 2005). Graphite occurs naturally in many parts of the world, including Siberia, Mexico, Madagascar, Sri Lanka, Quebec and Ontario in Canada, and the state of New York in the United States.

Graphite has a highly anisotropic structure with planes of hexagonal rings of aromatic carbon atoms, the planes being flat and aligned vertically with a well-defined spacing, shown in Figure 17.1. In a turbostratic carbon, the layers are not exactly parallel and do not have an orderly stacking as occurs in graphite. Many of these "planes" are folded, bent, or crumpled. The alignment may also be twisted. Graphite is the thermodynamically stable form of carbon. Heating turbostratic carbons to sufficiently high temperatures can cause them to convert to graphite. This implies close parallels between coalification processes in nature and laboratory carbonization reactions (Marsh 1997). Both create graphitic material from the original carbonaceous feedstocks.

Planes of carbon atoms in hexagonal array are characterized by very strong bonds between the carbon atoms. These individual planes are called graphene layers, a term that we will revisit later. Between the planes, only relatively weak interactions provide the bonding from one plane to another. As a result, when graphite experiences a mechanical force in the direction of the planes, they slide across each other quite easily. This behavior gives natural graphite one of its major applications, as a lubricant. Powdered graphite makes a superb

GRAPHITE CRYSTAL MODEL - ATOMIC CRYSTAL

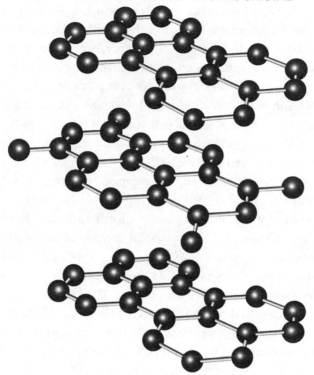

Figure 17.1 The structure of graphite. Each plane represents a sheet of carbon atoms in hexagonal aromatic rings. The strong bonds between the carbon atoms and the very weak interactions between the planes account for the highly anisotropic nature of graphite.
Courtesy of Erebor Mountain/Shutterstock.com.

lubricant for high-temperature applications, where lubricating oil would vaporize or char. Graphite also works well as a lubricant at very low temperatures, when oils would be too viscous to allow parts to slide against one another.

Natural graphites can contain impurities—such as iron, silicon, and sulfur—and possible structural defects that could limit their use in some commercial applications. Making synthetic graphites gives us control over their purity and structures. Graphites have a very wide range of commercial applications (Schobert and Song 2002), which include electrodes in arc furnaces for metallurgical operations and anodes in lithium-ion batteries. Because graphite is the thermodynamically stable form of carbon, it should be possible to find ways of converting various carbonaceous raw materials into graphite.

Agnès Oberlin, at the Laboratoire Marcel Mathieu in France, developed the concept of the *basic structural unit* (BSU), consisting of two to four stacked layers of polycyclic aromatic structures. How the BSUs are arranged in the original raw material and how, or if, they can subsequently be fitted together into a graphitic structure has a significant role in the course of graphitization. She, and her many collaborators over the years, developed insights through the use of electron microscopy (Oberlin 1989). She and her colleagues attributed graphitization to the preferred orientation of pores, with the orientation resulting from the tectonic stresses experienced during up-ranking of the coal to anthracite (Oberlin and Terriere 1975).

Graphitization can be catalyzed. There appear to be two pathways, possibly slight variations on the same thing. In one case, disordered, non-graphitic carbon dissolves into melts of metals to form carbides. When elemental carbon precipitates from the melt phase, it comes out as the thermodynamically stable product, graphite (Fitzer and Kegel 1968). Alternatively, quartz, which occurs in virtually every coal, can react with carbon to produce silicon carbide at temperatures higher than 1,600°C. This reaction, using anthracite or metallurgical coke as the carbon source, has been used for a century to produce silicon carbide, sold under such trade names as Carborundum and used as a refractory and an abrasive. At much higher temperatures, silicon carbide decomposes back to the elements. The literature explicitly warns that, "if that temperature [for the formation of silicon carbide] is exceeded, metallic silicon will be volatilized from the formed Carborundum and graphite will be left behind" (Landis 1920). It is thought that silicon will likely react first with metastable disordered carbon,[4] but, when the silicon carbide decomposes, it yields the stable form of carbon, graphite (Pappano and Schobert 2009). In this connection it can be noted that natural graphites can be found closed associated with quartz or other silica-bearing minerals, as Lyell discovered with the plumbaginous anthracite in Massachusetts.

Graphitization converts a raw material—petroleum coke, anthracite, or other forms of carbon—into the well-ordered structure of graphite. Graphitization requires temperatures in the vicinity of 2,800–3,000°C. Reaction conditions and the forming process allow manipulating the sizes of individual graphite crystallites and how they are oriented, giving products having the desired macroscopic properties.

---

[4] A metastable substance can remain in the condition or state that it is in, even though it is theoretically unstable. Specifically, the stable form of carbon at these conditions is graphite. Structurally disordered carbon is unstable relative to graphite but exists in its disordered state until something triggers its transformation to the stable graphite.

Making a synthetic graphite product requires two raw materials: a filler and a binder. The filler supplies most of the necessary carbon and provides the bulk of the article to be made. The binder holds the particles of filler together through the forming process. The preferred fillers in the early days of the synthetic graphite industry were anthracite, metallurgical coke, and bituminous coals of low ash yield (Liggett 1964). Currently the principal filler material is petroleum coke. The binder is often coal-tar pitch.

As the item made from the filler-binder mixture is heated to graphitization temperatures, sulfur, oxygen, nitrogen, and then hydrogen begin to be driven off, starting around 1,300°C. Above about 1,800°C, turbostratic carbon begins to align and start forming the graphitic structure. Above 2,000°C, most atoms other than carbon have been driven off. Alignment of turbostratic carbon continues, increasing as the temperature passes 2,200°C. The development of a fully graphitic structure is usually complete at 3,000°C.

Anthracites should be excellent fillers. Comparing Hirsch's anthracite structure with a structure of graphite drawn in the same style suggests this, as in Figure 17.2. On a dry, ash-free basis they are typically 90–95% carbon. More so than any other coals, they are already carbon materials. More than 95% of the carbon atoms in anthracites are present in large, polycyclic aromatic ring systems. These systems are not flat and neatly stacked, as they are in a graphite crystal, but they are clearly on the way to becoming graphite and can do so with a little help from us (Pappano et al. 2004).

A major industrial process for making synthetic graphite was discovered by the American chemist Edward Acheson, in the 1890s. He first worked for Thomas Edison, developing carbon materials for filaments in incandescent lighting. Acheson's key discovery was mentioned above: allowing a furnace to

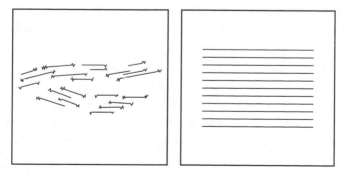

Figure 17.2 Anthracite is not graphite, but it's getting close. Some thermal energy put into the Hirsch anthracitic structure on the left would eliminate any disordered carbon and align the aromatic planes into the graphite structure, shown edge-on on the right. Artwork by Lindsay Findley, from the author's sketch.

get too hot forms graphite instead of Carborundum. This laid the foundations of the synthetic graphite industry. Commercialization of synthetic graphite, sometimes called Acheson graphite, began in the late 1890s (Marsh 1997). Natural graphite was almost completely displaced. Though petroleum coke is currently the preferred carbon source for most of the world's synthetic graphite, high-rank coals could serve as raw materials.

The use of graphite electrodes in arc furnaces for steelmaking is an extraordinary triumph of carbon material science. An electrode can be some 8 meters long. At the tip, the temperature might reach 3,500°C. In a bath of molten steel and slag, the temperature is 1,500–2,000°C. At the other end, a few meters away, the temperature is close to ambient. The carbon maintains a temperature differential of perhaps 2,000 degrees over a distance of a few meters. The electrode is clamped at the top and must maintain sufficient mechanical strength to support its own weight while sustaining this temperature difference. The electrode has to resist splitting due to thermal shock. Thermal expansion must be low to keep the hot electrode from expanding to a point at which it cannot be withdrawn through the roof of the furnace. When the white-hot electrodes are pulled from the furnace, they must withstand oxidation by exposure to air and resist catching fire. As the philosopher Thomas Hobbes said in a different context, a graphite electrode has "a life that is nasty, brutish, and short" (Hobbes 2012).

For a great many of us, the world suddenly shifted in 1991, though we probably didn't realize it at the time. Sony introduced the first practical, commercial lithium-ion battery. Since then, these batteries have allowed billions of people to make use of electronic devices such as smartphones, computers and tablets, and power tools. The world now has more cellphones than people (Murphy 2019). Larger versions are used in power supplies, electric and hybrid vehicles, and for energy storage. Lithium-ion batteries have almost completely wiped out the competing technologies of metal hydride batteries and nickel-cadmium (NiCad) batteries.

The graphite structure, with strong carbon-to-carbon bonds in a given layer and weak interlayer interactions, provides the possibility of incorporating atoms of other elements in spaces between the layers. This process is called *intercalation*, and the products are known as graphite intercalation compounds (Chung 2019, 48).

In a lithium-ion battery, lithium atoms are stored in the anode, for which graphite is a popular material. When we use the battery, lithium atoms release their electrons, becoming lithium ions. The lithium ions leave the anode and pass through the electrolyte to the cathode, where they recombine with the electrons. While this ionic process is happening inside the battery, the electrons released at the anode travel through an external circuit, providing energy to operate the electronic device. The electrons complete the circuit by returning to the cathode, where they reduce lithium ions back to lithium.

Aside from its ability to intercalate lithium, graphite has other attractive properties for this application. Low density[5] means that it contributes little to the weight of a battery, good for portable personal devices and vehicular applications. Graphite also has good thermal and electric conductivities, is chemically resistant to any compound likely to be chosen as an electrolyte, and it is not toxic.[6]

Synthetic graphite for this application can be made from coals (Novák, Goers, and Spahr 2010). The first synthetic graphite for lithium-ion batteries was made from coal tar. Coal-tar pitch contains a wide variety of compounds whose molecules are built up of numerous aromatic rings. In the liquid state molten pitch is isotropic. Heating isotropic molten pitch to 400–500°C causes some of the molecules to react, forming larger, disk-shaped aromatic structures that begin to orient and align. The liquid is no longer isotropic. The material that forms from this thermal treatment is called *mesophase*. Carbonization of bituminous coals proceeds in much the same way. We can compare Hirsch's idea of the "liquid structure" with a similar concept of a nematic liquid crystal (Gray 1962, 85), as in Figure 17.3. Holding pitch at a temperature below T_r allows the mesophase spheres to grow. The solidified spheres, *mesocarbon microbeads*, can be collected by dissolving away the rest of the pitch. The mesocarbon microbeads can be graphitized to make a material suitable for intercalation with lithium.

The great difference between inter- and intra-layer bonding is responsible for the pronounced anisotropy in the properties of graphite. Given the relatively weak interlayer bonding, it seems that there should be some way to take graphite apart, layer by layer. In 2004, Andre Geim and Konstantin Novoselov, at the University of Manchester, found an elegantly low-tech method for isolating graphene layers from graphite. It involved pressing a piece of sticky tape onto a sample of graphite oriented so that the tape would be lying parallel to the direction of the planes, pressing against a single plane. Carefully peeling the tape away from the graphite brings with it a layer of graphene (Hazen 2019,

[5] The density of pure graphite is 2.25 grams per cubic centimeter. Comparing equal volumes, graphite is slightly more than twice as heavy as water, but only about one-fourth the weight of steel and one-fifth the weight of lead.

[6] This last point was an issue with nickel-cadmium (NiCad) batteries. Many cadmium compounds cause human health or environmental problems. NiCad batteries put into landfills or incinerators could emit cadmium or its compounds to the environment.

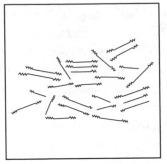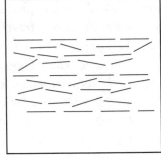

Figure 17.3 The Hirsch liquid structure of bituminous coals on the left can be transformed, with heating, into a mesophase liquid, sketched on the right. The liquid is the precursor to coke, to mesocarbon microbeads, and other carbon products.
Artwork by Lindsay Findley, from the author's sketch.

148). This discovery opened the way into fertile new territory in materials science, the study of graphene and related materials. Geim and Novoselov were awarded the 2010 Nobel Prize in Physics.

Graphene quickly became recognized for its extraordinary properties. Graphene is transparent, is a good conductor of heat and electricity, and has a tensile strength at least two orders of magnitude higher than the strongest steel wire. A true single layer of graphene would be one atom thick and would exist only in two dimensions.

Procedures are being developed for making graphene from coals. Some involve a sequence of wet-chemical processing steps (Wu et al. 2013). A process with fewer steps involves heating by passing an electric current through a material, a phenomenon known as Joule heating (Jasi 2020). In this case, a pulse of 200 volts heats coals to about 2,700°C in milliseconds. It's estimated that anthracite would provide yields of 80–90% and a product that was 99% pure (IEA 2020).

Numerous applications of graphene are constantly being considered and tested. A very small collection of examples includes biosensors for detection of disease (Leotaud 2019), materials for desalination of water (Cohen-Tanugi and Grossman 2012), fertilizer additives for slow release of plant nutrients (Kabiri et al. 2017), additives for high-strength cement (Ho et al. 2020), components of longer-lived fuel cells (Angel et al. 2020), fabric that can't be penetrated by mosquitos (Castilho et al. 2019), hair dyes (Patel 2018), and solar sails for spacecraft (European Space Agency 2020). Graphene is a candidate for developing new solar cell technologies.

Renewable energy generation from wind or solar sources requires either a back-up energy system or supporting technology for energy storage. Capacitors provide an option for the latter. Capacitors consist of two pieces of electrically conducting material, often in the shape of plates, separated by a nonconductor. The ability of capacitors to store electrical charge depends on the surface area of the plates and on the distance between them. The larger the surface area and the smaller the distance between the plates, the more charge can be stored. Graphene is an excellent conductor of electricity, exceptionally strong and very thin. These properties can be exploited to make plates with large surface areas and that can be tightly packed together into a small space. The low density of graphene means that the device is lightweight. Such devices can store a considerable amount of energy. Better than capacitors, they are *supercapacitors*. To increase even further the available surface area of the conducting material, graphene sheets can be folded, storing a large amount of electrical charge in a small volume. Better than super-, these devices are *ultracapacitors*. Super- and ultracapacitors could provide the energy storage needed for local, renewable energy systems (Johnson and Meany 2018, 145).

Graphene is not soluble in water, so graphene coatings could provide surface protection for items that would be affected by water. Graphene is flexible, providing the potential of developing wearable electronic devices or medical sensors based on graphene. The thermal and electrical properties of graphene lead to its potential of replacing silicon in various electronic applications. A good designer could take advantage of water resistance, flexibility, and electronic properties of graphene to come up with wearable devices that could be even be used while swimming or bathing (Hazen 2019, 150).

The development of incandescent electric lighting by Thomas Edison has almost become folklore. In 1879, Edison finally found that a piece of carbonized cotton thread—carbonized cellulose—glowed for about 40 hours in a vacuum. Though carbonized cellulose fibers were supplanted by metal filaments, they represent not only the beginning of practical incandescent lighting but also the first major application of carbon fibers.

From that start in 1879, the top 10 countries producing carbon fibers—led by the United States, Japan, and China—now make about 140,000 tonnes of material annually. Figure 17.4 shows carbon fibers being produced. Carbon fibers have high strength, low density, resistance to many chemicals, and good electrical and thermal conductivities (Song, Schobert, and Andrésen 2005). These properties change only little at elevated temperatures. Their comparatively high cost may hold back some commercial applications. The properties of carbon fibers have made them particularly useful in aerospace applications (Vohler 1986), where sometimes cost is not as important as performance.

A composite material is one made of two or more individual materials. The components of a composite are usually selected to have very different

Figure 17.4 Fibers are moving into a high-temperature furnace, where they will be converted into carbon fiber product.
Courtesy of US Department of Energy, Oak Ridge National Laboratory.

properties, with the idea that the composite material would have attributes that are superior to those of either of its components. Carbon fibers are used to produce composites with metals, polymers, cement, and with other carbon forms. Carbon fiber-reinforced polymers (CFRPs) are important in aerospace and defense applications. The Boeing 787 airplane is made with 35 tonnes of CFRPs, which contain 23 tonnes of carbon fibers (Chung 2019, 245).

For about the past half-century, wind turbine blades have been made from composites. While many blades are made using fiberglass composites, carbon fibers are a promising technology (Mishnaevsky et al. 2017). Compared to the traditional glass fibers, carbon fibers allow lighter, thinner, and stiffer blades, although they are not so resistant to damage. Because of their stiffness and strength, CFRPs are steadily increasing in use in wind turbines (Groh 2013).

The currently preferred raw material for most carbon fibers is polyacrylonitrile (PAN). PAN itself comes from acrylonitrile, made by the gas-phase reaction of propylene, ammonia, and air. It's possible to make propylene via the Fischer-Tropsch-to-olefins reaction, using coal-derived synthesis gas. Hydrogen for the synthesis of ammonia can come from the carbon–steam reaction followed by the water–gas shift. Coal-derived PAN is entirely feasible.

Carbon fibers can be made from a variety of materials, including coal-tar pitch.[7] Mesophase pitch fibers have better properties than those made without

---

[7] The word "pitch" refers in general to a black, viscous residue, usually from distillation. In carbon fiber technology, it could mean pitches derived from petroleum, coal tar, or other sources, or even synthetic

the preliminary step of generating mesophase. Mesophase pitch is an important precursor for high-performance fibers (Song et al. 2005). Carbon fibers made from pitches have higher densities and conductivities than do PAN fibers and undergo less expansion as the temperature coefficient is raised. On the other hand, PAN fibers are stronger when subjected to tensile forces (Suárez-Ruiz and Crelling 2008, 210). Very-high-performance fibers are produced from mesophase pitch.

Both pitch and mesophase pitch are solids at ambient temperatures. Fibers are made from either material by melt-spinning. Pulverized mesophase pitch is melted and forced through a nozzle, known in the business as a spinneret. Passing through the spinneret further aligns molecules in the mesophase. Fibers are drawn as the melt leaves the spinneret. Heating in air or oxygen stabilizes the fibers—it makes them no longer susceptible to melting. The stabilized fibers are carbonized in an inert atmosphere. Nitrogen and hydrogen are driven off. If the objective is to produce a high-strength fiber, carbonization will be followed by graphitization.

Active carbon fibers are usually produced by physical activation via either the carbon–steam reaction or the Boudouard reaction. Pitch-derived fibers have been used in this application (Song et al. 2005). Active carbon fibers have numerous applications that include water treatment, air separation, recovery of solvent vapors, and storage of methane on natural-gas-powered vehicles.

Fullerenes are molecules of pure carbon bonded into a closed shell, or mesh, made of a single layer of carbon atoms enclosing a hollow interior. The first example to be discovered consists of a sphere of 60 carbon atoms arrayed in pentagons and hexagons, much like the surface of a soccer ball. This structure resembled the geodesic dome buildings designed by the American architect Buckminster Fuller. This material, C_{60}, came to be termed *buckminsterfullerene*, informally called "buckyballs." Similar materials, such as C_{70}, were soon discovered. The term "fullerenes" refers to the family of such materials.

A research team at Exxon found evidence for C_{60} and C_{70} among many products of vaporizing graphite with a powerful laser (Rohlfing, Cox, and Kaldor 1984). A collaboration of Harold Kroto, of Sussex University, with the research team of Richard Smalley at Rice University verified these results, but then discovered how to make C_{60} as the single product (Kroto et al. 1985). Kroto, Smalley, and their colleague Robert Curl were awarded the 1996 Nobel Prize in Chemistry for their work.

pitches specially prepared from pure aromatic hydrocarbons. The main problem in making pitch-based carbon fibers is controlling the properties of the pitch itself.

Fullerenes can be incorporated into polymer-fullerene composites to make organic solar cells (Mathur, Singh, and Pande 2017, 68). Such cells are light and flexible, offering possible applications that would be impractical with rigid silicon-based cells, such as mounting them on the contours of vehicles. Fullerene-based materials have other renewable-energy applications, such as in hydrogen–oxygen fuel cells and improving the quality of insulation on electrical cables to allow them to carry higher voltages.

Fullerenes can be made from all ranks of coals, from brown coal to semianthracite (Pang, Vassallo, and Wilson 1992). The most useful approach seems to be to carbonize the original coal sample, then use the carbonized material to produce carbon electrodes for an electric-arc discharge (Pang et al. 1992). With a given set of arc conditions, the yield of fullerenes depends on the rank and ash yield of the specific coal being tested. An alternative route to fullerenes comes from heating graphite above 3,000°C in a solar furnace. A solar furnace has minimal energy costs compared to electric furnaces. If it can be done with graphite, it could be possible with anthracite.

Fullerenes can be imagined as if a flat graphene sheet were altered by converting some of its carbon hexagons into pentagons, allowing the sheet to curl up on itself and form hollow, caged structures such as buckminsterfullerene. A graphene sheet can also be rolled up to make a hollow, seamless cylinder. Such structures, called *carbon nanotubes,* were discovered by Sumio Iijima at the NEC Corporation in Japan, and, independently, by Donald Bethune at IBM in San Jose, California.

The original structures consist of a single roll of graphene, as in Figure 17.5. Nanotubes also exist in assemblies in which there is one tube inside another, as if one had pieces of pipe of several diameters, inserting one into another, but taking care they were aligned along a common axis. These two kinds of nanotubes are referred to as single-wall nanotubes (SWNTs) and multi-wall nanotubes (MWNTs). On an atomic scale, nanotubes are very long and very thin, with length-to-diameter ratios of 1,000 or more (Mathur et al. 2017, 75)—in some cases said to exceed 100 million to 1. They can be thought of as being one-dimensional (1-D) materials.

SWNTs have good absorption of light through the ultraviolet-visible-infrared regions of the spectrum, which makes them attractive candidates for solar panels. Light can be absorbed effectively in a thin layer of material. The less material that's needed, the lower the cost. Nanotubes have high electrical conductivities combined with low density. Furthermore, the hollow cores of nanotubes are well suited for the intercalation of lithium ions. They should also be attractive for use in supercapacitors (Sharon and Sharon 2010, 120).

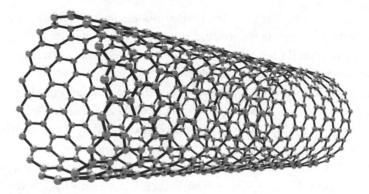

Figure 17.5 The structure of a single-wall carbon nanotube.
Courtesy of Evannovostro/Shutterstock.com.

Using hydrogen as an energy source faces the problem of how to store it in practical amounts, especially for relatively small-scale applications, as in vehicles. An option is to adsorb hydrogen on a low-density material that has a high adsorption capacity. Bundles of SWNTs can adsorb to about 5–10% of their weight of hydrogen. MWNTs should also work in this application.

Carbon nanotubes can be made from coals by the arc-discharge method (Sharma, Sharma, and Sharma 2015). There is the caution that "significant challenges remain in scale-up, manufacturability, and quality control" (Song et al. 2005).

Quantum dots, discovered less than 20 years ago, are nanoparticles, usually less than 10 nanometers in diameter. The diameter of a normal-sized human hair is about 60,000–80,000 nanometers. These particles are, in terms of geometry, literally dots. Quantum dots are zero-dimensional, or 0-D, materials. Thus, with carbon, we can have three-, two-, one-, and zero-dimensional materials: graphite, graphene, nanotubes, and quantum dots.[8] When exposed to ultraviolet light, quantum dots fluoresce, often emitting visible light. The color of the emitted light depends on the size and composition of the dots. The fabrication process can tailor quantum dots to emit a desired color. Their fluorescence and size, on the same order as some proteins or nucleic acids, make quantum dots very useful in medical imaging. It should be possible to image and study single cancer cells (Sharon and Sharon 2010, 365).

In 2004, Walter Scrivens and his colleagues at the University of South Carolina discovered carbon quantum dots while in the process of purifying

---

[8] In the late 1940s, there was a report of a four-dimensional organic compound, thiotimoline, which had interesting solubility behavior (Asimov 1948). This work does not seem to have been repeated.

nanotubes (Xu et al. 2004). Later, *carbon quantum dots* (CQDs) were made from powdered graphite by eroding the surface with a laser (Wang and Hu 2014). The potential uses of carbon dots extend beyond medical applications to solar cells, light-emitting diodes (LEDs), and catalysts for producing hydrogen or reducing CO_2 (Wang et al. 2019). CQDs could be used in supercapacitors (Wang and Aiguo 2014). CQDs are able to emit very pure colors over long periods of time, providing improved displays for televisions (Bailey 2019).

Being able to make CQDs from graphite suggests the possibility of making them from coals. In 2013, James Tour and colleagues at Rice University produced CQDs from bituminous coal, coke, and anthracite (Ye et al. 2013). It should be easier to make CQDs from coals than from graphite because it is easier chemically to cut out small regions of ordered structures from the relatively disordered frameworks of coals than to do so from the highly ordered structure of graphite. These workers obtained yields of about 20%. Consistent with the variation of coal structure with rank, CQDs from bituminous coal were about 2–4 nanometers wide. Those from coke were about double this size, with some graphite-like stacking. This supports the notion that coke formation involves dehydrogenative polymerization of aromatic units and, coming from a mesophase, should have some stacking of aromatic units. CQDs from anthracite were much bigger, up to 40 nanometers, and showed stacked structures. Subsequently, James Hower at the University of Kentucky Center for Applied Energy Research and co-workers in India have shown the possibility of making CQDs from Pennsylvania anthracites and from a high-volatile bituminous coal from southeastern Kentucky (Saikia et al. 2019).

Jeffrey Grossman, Nicola Ferralis, and colleagues at MIT have pointed out that, given the variety of compositions of coals collectively, with relatively simple processing coals could provide carbon materials encompassing a range of electrical and optical properties (Keller, Ferralis, and Grossman 2016). Potential applications include, among others, battery anodes, conductive films, and solar cells. Often the properties of synthetic carbons can be tailored to meet the requirements of a specific application. All carbon materials have a wide range of potential uses, only a few examples of which have been discussed here. Coal-derived carbons have applications that include supercapacitors, anodes, and hydrogen adsorbers for energy storage; carbon fibers for wind turbine blades; solar cells; 3D-printable power sources; susceptors and trays for production of silicon-based electronic components; molecular sieves for taking CO_2 out of biogas; and soil conditioners that could help in the growth of biomass. All of these applications directly relate to the steadily growing renewable energy industry.

An American folk saying from the 1930s: "If you can't beat 'em, join 'em." If you can't defeat your opponent, you might as well work alongside them or do what they do. Coal's position in electricity generation is steadily losing ground to wind and solar. It seems likely to keep losing ground. So we might consider this scenario: improved solar cells made with coal-derived graphene. Carbon fiber blades for wind turbines from coal-derived pitch. Neodymium, extracted from coal ash, to make the powerful magnets in wind turbines. Wind and solar systems feeding electricity into improved batteries and into supercapacitors of coal-derived carbons. Most of the technology for doing all this exists. It may need some tweaking and system optimization, but it will all work. Coal does not need to roll over and play dead. We need to rethink coal to recognize where the real opportunities are and rise up to meet them. Why not let the coal industry build the renewables industry?

References

Angel, Gyen Ming A., Noramalina Mansor, Rhodri Jervis, Zahra Rana, Chris Gibbs, Andrew Seel, Alexander F. R. Kilpatrick, Paul R. Shearing, Christopher A. Howard, Dan J. L. Brett, and Patrick L. Cullen. 2020. "Realising the Electrochemical Stability of Graphene: Scalable Synthesis of an Ultra-Durable Platinum Catalyst to the Oxygen Reduction Reaction." *Nanoscale* 12: 16113–16122.

Asimov, Isaac. 1948. "The Endochronic Properties of Resublimated Thiotimoline." *Astounding Science Fiction* 41(1): 120–125.

Bailey, Mary P. 2019. "A Bright Future for Quantum Dots." *Chemical Engineering* 126 (4): 14–17.

Castilho, Cintia J., Dong Li, Muchun Liu, Yue Liu, Huajian Gao, and Robert H. Hurt. 2019. "Mosquito Bite Prevention through Graphene Barrier Layers." *Proceedings of the National Academy of Sciences* 116: 18304–18309.

Çeçen, Ferhan. 2014. "Activated Carbon." In *Kirk-Othmer Encyclopedia of Chemical Technology*, edited by Claudia Ley, 4: 1–34. New York: John Wiley & Sons.

Chung, Deborah D. L. 2019. *Carbon Materials*, 21–89. Singapore: World Scientific.

Cision. 2019. "Activated Carbon Market - Global Industry Analysis, Value, Share, Growth, Trends, and Forecast 2018–2026." *Cision PR Newswire*, February 26. https://www.prnewswire.com/news-releases/activated-carbon-market---global-industry-analysis-value-share-growth-trends-and-forecast-2018---2026-300790778.html

Cohen-Tanugi, David, and Jeffrey C. Grossman. 2012. "Water Desalinating Across Nanoporous Graphene." *Nano Letters* 12: 3602–3608.

European Space Agency. 2020. "One Atom Thick Graphene Light Sail Could Speed Journey to Other Star Systems." SciTechDaily. https://scitechdaily.com/one-atom-thick-graphene-light-sail-could-speed-journey-to-other-star-systems/ May 19, 2020.

Fitzer, E., and B. Kegel. 1968. "Reaktionen Von Kohlenstoff Gesattigter Vanadium Carbide Schmelze Mit Ungeordnetem Kohlenstoff." *Carbon* 6: 433–446.

Gray, George W. 1962. *Molecular Structure and the Properties of Liquid Crystals*. London: Academic Press.

Groh, Rainer. 2013. "Composite Materials and Renewables: Wind energy." Aerospace Engineering. https://aerospaceengineeringblog.com/composite-materials-wind-energy/.

Hassler, John W. 1941. *Active Carbon: The Modern Purifier.* New York: Githens-Sohl Corporation.

Hazen, Robert M. 2019. *Symphony in C.* New York: W. W. Norton.

Ho, Van Dac, Ching-Tai Ng, Campbell J. Coghlan, Andy Goodwin, Craig McGuckin, Togay Ozbakkaloglu, and Dusan Losic. 2020. "Electrochemically Produced Graphene with Ultra Large Particles Enhances Mechanical Properties of Portland Cement Mortar." *Construction and Building Materials* 234: 117403.

Hobbes, Thomas. *Leviathan.* 2012. London: The Folio Society.

Huang, He, Yujue Wang, and F. S. Cannon. 2009. "Pore Structure Development of In-Situ Pyrolyzed Coals for Pollution Prevention in Iron Foundries." *Fuel Processing Technology* 90: 1183–1191.

IEA. 2020. "Green Process Converts Almost Any Carbon Source into Graphene." International Energy Agency Clean Coal Centre. https://www.iea-coal.org/green-process-converts-almost-any-carbon-source-into-graphene/.

Jasi, Amanda. 2020. "Green Process Converts Almost Any Carbon Source into Graphene." Chemical Engineer. https://www.thechemicalengineer.com/news/green-process-converts-almost-any-carbon-source-into-graphene/.

Johnson, Les, and Joseph E. Meany. 2018. *Graphene.* Amherst, NY: Prometheus Books.

Kabiri, Shervin, Fien Degryse, Diana N. H. Tran, Rodrigo C. da Silva, Mike McLaughlin, and Dusan Losic. 2017. "Graphene Oxide: a New Carrier for Slow Release of Plant Micronutrients." *Applied Materials and Interfaces* 9: 43325–43335.

Keller, Brent D., Nicola Ferralis, and Jeffrey C. Grossman. 2016. "Rethinking Coal: Thin Films of Solution Processed Natural Carbon Nanoparticles for Electronic Devices." *Nano Letters* 16: 2951–2957.

Kroto, Harold W., James R. Heath, Sean C. O'Brien, Robert F. Curl, and Richard E. Smalley. 1985. "C 60: Buckminsterfullerene." *Nature* 318(6042): 162–163.

Landis, Walter S. 1920. "Electrochemical Industries." In: *Industrial Chemistry*, edited by Allen Rogers, 281–303. New York: D. Van Nostrand.

Leotaud, Valentina R. 2019. "Gold and Graphene Now Used in Biosensors to Detect Diseases." Mining.com. https://www.mining.com/gold-graphene-now-used-biosensors-detect-diseases/.

Liggett, L. M. 1964. "Carbon: Baked and Graphitized Products, Manufacture." *Kirk-Othmer Encyclopedia of Chemical Technology*, vol. 4. New York: John Wiley & Sons.

Lu, Zhe, Mercedes M. Maroto-Valer, and Harold H. Schobert. 2008. "Role of Active Sites in the Steam Activation of High Unburned Carbon Fly Ashes." *Fuel* 87: 2598–2605.

Marsh, Harry. 1997. "Carbon Materials: An Overview of Carbon Artifacts." In: *Introduction to Carbon Technologies*, edited by Harry Marsh, Edward A. Heintz, and Francisco Rodríguez-Reinoso, 1–34. Alicante: University of Alicante.

Mathur, Rahesh Behari, Bhanu Pratap Singh, and Shailaja Pande. 2017. *Carbon Nanomaterials.* Boca Raton, FL: CRC Press.

Mishnaevsky, Leon, Kim Branner, Helga N. Petersen, Justine Beauson, Malcolm McGugan, and Brent F. Sørensen. 2017. "Materials for Wind Turbine Blades: An Overview." *Materials* 10: 1285.

Murphy, Michael. 2019. "Cellphones Now Outnumber the World's Population." Quartz. https://qz.com/1608103/there-are-now-more-cellphones-than-people-in-the-world/.

Novák, Petr, Dietrich Goers, and Michael E. Spahr. 2010. "Carbon Materials in Lithium-Ion Batteries." In: *Carbons for Electrochemical Energy Storage and Conversion Systems*, edited by François Béguin and Elżbieta Frąckowiak, 263–328. Boca Raton, FL: CRC Press.

Nover, Georg, Johannes B. Stoll, and Jutta von Der Gönna. 2005. "Promotion of Graphite Formation by Tectonic Stress: A Laboratory Experiment." *Geophysical Journal International* 160: 1059–1067.

Oberlin, Agnes. 1989. "High-Resolution TEM Studies of Carbonization and Graphitization." In: *Chemistry and Physics of Carbon,* vol. 22, edited by Peter A. Thrower, 1–143. New York: Marcel Dekker.

Oberlin, Agnes, and G. Terriere. 1975. "Graphitization Studies of Anthracite by High Resolution Electron Microscopy." *Carbon* 13: 367–376.

Office of Mercury Management. 2018. "Removal of Mercury from Flue Gas Using Activated Carbon." Mercury Technology Bulletin. Series 006. Tokyo: Environmental Health Department, Ministry of the Environment (Japan).

Ōtani, S., and A. Ōya. 1986. "Progress of Pitch-Based Carbon Fiber in Japan." In: *Petroleum-Derived Carbons,* edited by J. D. Bacha, J. W. Newman, and J. L. White, 323–334. Washington, DC: American Chemical Society.

Pang, Louis S. K., Anthony M. Vassallo, and Michael A. Wilson.1992. "Fullerenes from Coal: A Self-Consistent Preparation and Purification Process." *Energy and Fuels* 6: 176–179.

Pappano, Peter J., Frank Rusinko, Harold H. Schobert, and David P. Struble. 2004. "Dependence of Physical Properties of Isostatically Molded Graphites on Crystalline Height." *Carbon* 42: 3007–3009.

Pappano, Peter J., and Harold H. Schobert. 2009. "Effect of Natural Mineral Inclusions on the Graphitizability of a Pennsylvania Anthracite." *Energy and Fuels,* 23: 422–428.

Patel, Prachi. 2018. "The Quirkier Uses of Graphene." *Scientific American* 318 (6): 20–24.

Patel, Raman L., S. P. Nandi, and Philip L. Walker. 1972. "Molecular Sieve Characteristics of Slightly Activated Anthracite." *Fuel* 51: 47–51.

Patel, Raman L., and Philip L. Walker. 1973. "Preparation of Molecular Sieve Materials from Anthracite." Office of Coal Research, Research and Development Report No. 61.

Reddy, P. Jayarama. 2014. *Clean Coal Technologies for Power Generation.* Boca Raton, FL: CRC Press.

Reisch, Marc S. 2014. "Activated for Growth." *Chemical and Engineering News* 92 (43): 18–19.

Reisch, Marc S. 2015. "Bromine Bails Out Big Power Plants." *Chemical and Engineering News* 93 (11): 17–19.

Rodriguez, Regina, Domenic Contrino, and David W. Mazyck. 2020. "Role of Activated Carbon Precursor in Mercury Removal." *Industrial and Engineering Chemistry* 59: 17740–17747.

Rodríguez-Reinoso, Francisco. 1997. "Activated Carbon: Structure, Characterization, Preparation and Applications." In: *Introduction to Carbon Technologies,* edited by Harry Marsh, Edward A. Heintz, and Francisco Rodríguez-Reinoso, 35–102. Alicante: University of Alicante.

Rohlfing, Eric A., Donald M. Cox, and Andrew Kaldor. 1984. "Production and Characterization of Supersonic Carbon Cluster Beams." *Journal of Chemical Physics* 81 (7): 3322–3330.

Saikia, Monikankana, James C. Hower, Tonkeswar Das, Trisharani Dutta, and Binoy K. Saikia. 2019. "Feasibility Study of Preparation of Carbon Quantum Dots from Pennsylvania Anthracite and Kentucky Bituminous Coals." *Fuel* 243: 433–440.

Satterfield, John. 2017. "Effective Mercury Sorbents." *International Cement Review* 4:1–4.

Schobert, Harold H., and Chunshan Song. 2002. "Chemicals and Materials from Coal in the 21st Century." *Fuel* 81: 15–32.

Sharma, Ritu, Anup K. Sharma, and Varshali Sharma. 2015." Synthesis of Carbon Nanotubes by Arc Discharge and Chemical Vapor Deposition Method with Analysis of Its Morphology, Dispersion, and Functionalization Characteristics." *Cogent Engineering,* 2. https://doi.org/ 10.1080/23311916.2015.1094017.

Sharon, Maheshwar, and Madhuri Sharon. 2010. *Carbon Nanoforms and Applications.* New York: McGraw-Hill.

Song, Chunshan, Harold H. Schobert, and John M. Andrésen. 2005. "Premium Carbon Products and Organic Chemicals from Coal." International Energy Agency Clean Coal Centre Report No. CCC/98.

Spencer, Denis H. T. 1967. "The Use of Molecular Probes in the Characterization of Carbonaceous Materials." In: *Porous Carbon Solids*, edited by R. L. Bond, 87–154. London: Academic Press.

Suárez-Ruiz, Isabel, and John C. Crelling. 2008. *Applied Coal Petrology*. Amsterdam: Elsevier.

Vohler, Oskar J. 1986. "Carbon and Graphite in Future Markets." *Erdöl & Kohle Erdgas Petrochemie* 39: 561–567.

Wang, Xiao, Yongqiang Feng, Peipei Dong, and Jianfeng Huang. 2019. "A Mini-Review on Carbon Quantum Dots: Preparation, Properties, and Electrocatalytic Applications." *Frontiers in Chemistry*. doi.org/10.3389/fchem2019.00671.

Wang, Youfu, and Aiguo Hu. 2014. "Carbon Quantum Dots: Synthesis, Properties, and Applications." *Journal of Materials Chemistry C* 2: 6921–6939.

Wu, Yingpeng, Yanfeng Ma, Yan Wang, Lu Huang, Na Li, Tengfei Zhang, Yi Zhang, Xiangjian Wan, Yi. Huang, and Yongcheng Chen. 2013. "Efficient and Large Scale Synthesis of Graphene from Coal and Its Film Electrical Properties Studies." *Journal of Nanoscience and Nanotechnology* 13: 929–932.

Xu, Xiayou, Robert Ray, Yunlong Gu, Harry J. Ploehn, Latha Gearheart, Kyle Raker, and Walter A. Scrivens. 2004. "Electrophoretic Analysis and Purification of Fluorescent Single-Walled Carbon Nanotube Fragments." *Journal of the American Chemical Association* 126: 12736–12737.

Ye, Ruquan, Changsheng Xiang, Jian Lin, Zhiwei Peng, Kewei Huang, Zheng Yan, Nathan P. Cook, Errol L. G. Samuel, Chih-Chau Hwang, Gedeng Ruan, Gabriel Ceriotti, Abdul-Rahman O. Raji, Angel A. Martí, and James M. Tour. 2013. "Coal as an Abundant Source of Graphene Quantum Dots." *Nature Communications* 4, Article no. 3943.

18
W(h)ither?

The odd-looking word in the title was coined by Geoff Dolbear, who has made many contributions to petroleum science and technology. It's used here to express two questions at once: Where is the coal industry going—*whither*—or will it *wither*—dry up, shrivel, and blow away?

Let's begin by considering the risky business of making predictions. We can circle back to Chapter 1 and the expression, "It's difficult to make predictions, especially about the future," which, as we saw, has been attributed to sources as diverse as Yogi Berra and Niels Bohr.[1] Regardless of who actually said it, it's true. Anyone tempted to make authoritative predictions might first try this thought experiment: Suppose that in 2015, or 2018, or even as recently as 2019, you had been asked to say what you expected to be doing in 2020. Almost every one of us would have had an answer that differed considerably from the real events of 2020. Very few of us had heard the word *covid*. Likely some may not have been familiar with the word *pandemic*. And then....

But we must also remember that,

> If we have learned one thing from the history of invention and discovery, it is that, in the long run—and often in the short one—the most daring prophecies seem laughingly conservative. (Clarke, as quoted in Shermer 2015, 397)

Arthur C. Clarke, the person who offered this observation, had some relevant experience. In 1951, he made a seemingly wild prediction that there might someday be communications satellites in orbit around Earth (Clarke 1951, 152). Nowadays we take them for granted.

Not only can it be difficult to make predictions about the future, it's also dangerous to make extrapolations from recent trends.

> [T]his seems no more accurate than to conclude that because we observe a cannon-ball has traversed a mile in a minute, therefore in an hour it will be sixty miles off, and in the course of ages that it will reach the fixed stars. (Jenkin, as quoted in Gould 1991, 348)

---

[1] A hall-of-fame catcher for the New York Yankees and a Nobel Laureate in Physics, respectively.

As Lincoln Moses, the first head of the US Energy Information Administration, has said, "There are no facts about the future" (O'Sullivan 2018, 51).

There are no substantial technical barriers to the conventional use of coals—the applications discussed in Chapters 8 through 13. We know how to build and operate coal-fired power plants, steadily increasing their efficiency. Carbon consumption is usually in the high 90% range. Coke plants represent mature technology. Improved operations are reducing the coke rate in blast furnaces. Gasification technology is amply demonstrated by Sasol and Dakota Gasification. Downstream conversion of synthesis gas to substitute natural gas or to liquid products is well established. Direct coal liquefaction has been commercialized in at least one plant in China. Of course, we can always work to improve efficiencies and conversions in all coal technologies, but certainly we know how to build plants and run them. Yet the coal industry is in trouble in many parts of the world.

Several factors have combined to establish the present situation with coal. These factors are not primarily technological issues, though carbon capture and storage methods that are effective on very large scales yet inexpensive would represent a very desirable development. We need to rethink coal because coal faces a perfect storm of economic, environmental, and political or societal issues. Let's consider some of the storm winds.

Coal mines are being closed. The cost of mining is going up because in many parts of the world the easiest-to-mine seams have been worked out. This is combined with the increasing efficiency of power plant operation providing the ability to generate the same amount of electricity with less coal. Because electricity generation is, in most countries, by far the largest market for coal, when this market does not grow, there is no alternative growth opportunity for coal that might have gone into the power plants.

In many places, surface mining is replacing deep shaft mining. In the United States, coal produced from shaft mines in Kentucky costs three times as much as the coal produced from surface mines in Wyoming (Hansen 2016). Also, mining itself has become more automated, especially underground mining. One miner can produce about three times as much coal per shift as could be done 40 years ago. Even if there were to be a significant uptick in coal demand, the greatly increased productivity per miner means that fewer miners are needed now and fewer will be needed in the future. Continued improvements in mining productivity will continue to reduce employment even if coal use remains steady. They will reduce employment even faster in a climate of shrinking coal demand. It's been said, albeit with some hyperbole, that, "Deep mining is a dead duck" (Hansen 2016). If the duck is not exactly dead yet, there is the increasing likelihood of deep mining with robots (Ge 2018). Not only would this development require no human miners at all underground, those few workers on the surface would need skill sets totally different from those of the traditional coal miner.

In most coal-using countries, coal-fired power plants account for a significant majority of the demand for coal, up to 90% in some countries. When plants close, the coal that they consumed has no other ready market to absorb what would have been consumed in those plants. A coal-fired power plant that burns 10,000 tonnes of coal per day and operates about 90% of the time would by itself consume 3.3 million tonnes of coal per year. There's no place else to sell that coal, and there is no reason to mine coal that can't be sold. "It seems the writing is on the wall for thermal coal" (Arden 2018).

A further economic difficulty comes from some very major financial institutions no longer being willing to fund coal projects—or indeed projects involving any kind of fossil fuels. As one example, in early 2020, JPMorgan Chase, a very important lender to fossil fuel projects worldwide, announced that it would stop lending to coal firms (Mufson and Grandoni 2020). Not only that, this lender will not refinance loans on existing coal plants or provide money to any companies that obtain the majority of their revenue from coal. Existing loans will be phased out by 2024.

Despite this litany of bad news, coal does have an advantage relative to natural gas and petroleum in that it is comparatively simple to extract, especially with surface mining. Coal extraction does not require platforms or pads, drilling rigs, drill strings, and piping—all of the hardware that we discussed in Chapter 14. If the cost of coal were to become too low for profitable extraction, a mining company could consider just stopping production and starting again when economic conditions are more favorable.

For a long stretch of years, coal enjoyed the position of being the cheapest available fuel in terms of dollars per megajoule of energy. Now there is serious competition. Thanks to the shale revolution, those countries able to take advantage of shale gas find that gas is now the cheapest fuel. Gas seems to be capturing the market share that coal is losing. It's been suggested that "Much of the damage to the US coal sector is irreparable" (Doyle 2017).

Nevertheless, there is public resistance to fracking and, in some places, outright bans on this practice. Four states in the United States, as well as parts of others in the Delaware River Basin, have banned fracking. At least one state in Australia and one province in Canada have done likewise. Countries such as France, Bulgaria, and Germany have enacted bans, others have put moratoria on fracking. While the regulatory outlook worldwide is mixed, in addition we should not forget that, even with fracking technology, the reserves of natural gas are necessarily finite. We cannot sustain infinite growth on finite resources.

On April 21, 2017, Britain went a full day without using coal to generate electricity for the first time since the 1880s (Dartnell 2019). Not only is shale gas a strong competitor, both wind and solar are also taking business from coal. This is happening even though both need to be combined with energy storage methods or back-up generating capacity. Fossil fuel plants on standby to back up solar or wind will very likely be natural gas plants, not coal. Installation costs for solar and wind have been dropping and continue to drop. Energy producers in many nations are installing wind farms and solar installations of increasingly larger sizes. In the United States in particular, the trend to renewables is supplemented by an "ocean" of natural gas from fracking (Kellman 2016).

The trajectory of public opinion and policy seems likely to drive continued growth in renewables (Rowland 2016). This factor is combined with an increasing awareness of and concern for global warming and its effects. Evidence is clear that anthropogenic emissions of greenhouse gases, including carbon dioxide (CO_2) from fossil fuel combustion, are a factor in global warming. This is combined with the incontrovertible fact that coals produce more CO_2 per megajoule of energy than petroleum or natural gas.

It is necessary to recognize that coal has a negative public perception in many parts of the world. Anyone aspiring to financial independence would benefit from establishing a mechanism that would allow him or her to collect one dollar for every time the expression "coal is a dirty fuel" has been uttered or has appeared in print someplace. Of course it is a dirty fuel, if we, the public, via our elected officials and appointed regulatory boards, allow coal projects to be built and operated according to the standards of the 1970s, or, worse, of the 1920s. Effective emission control technologies for sulfur oxides (SO_x), nitrogen oxides (NO_x), and particulates exist. We know how to build them and make them work. However, much remains to be done on carbon capture and storage (CCS), especially getting it to very large scales.

Nevertheless, there is the negative public perception that will be difficult, maybe impossible, to dispel. As a small personal example, anthracites, because of their combustion characteristics, are superb fuels to fire pizza ovens. Nita and I have collected abundant empirical evidence that pizza made in such ovens is excellent. Anthracite provides a hot, short flame, with no tars and no smoke. But on recommending the restaurant to a family member, the immediate response was, "Oh, yuk! Coal!"

For many years, sulfur was considered to be the "bad guy" in coal because of the contributions of SO_x to environmental problems related to acid rain. Now sulfur's role seems to have been taken by mercury. Coal-burning power plants do indeed contribute 40% of annual mercury emissions to the environment (Guffey 2015). These concerns very likely have a role in some consumers' decisions to switch to cleaner energy sources such as renewables or natural gas—no sulfur, no ash. The positive environmental perceptions of renewables come on top of the increasing economic advantages of these sources.

If the foregoing factors were not enough, we must also consider that we are getting better at using all kinds of energy more efficiently. We've seen this a few times in the previous chapters, just in the context of the heat rate in boilers and the coke rate in blast furnaces. As those technologies continue to become more efficient, we can generate the same amount of electricity or hot metal with less coal, or we can make more of those commodities with the same amount of coal that we use now. It is hard for coal to carve out any growth in that situation.

For the coal industry, an additional complication makes things still worse. Many of the developed countries of the world are decreasing the *energy intensity* of economic growth. That is, less and less energy is required to achieve a particular level of growth in the economy. A hundred workers in a steel mill consume much more energy than a hundred workers seated at computer terminals. And most of us, regardless of occupations or financial means, are also contributing simply by being more conscious of energy efficiency at home, from air conditioning to various efficient appliances to LED lighting. If the growth in energy demand is leveling off thanks to greater efficiencies, essentially the "energy pie" is staying about the same size. The trouble for coal is that natural gas, wind, and solar are all cutting themselves increasingly larger slices.

We will introduce two points here, both of which we will return to in more detail later in the chapter. First, not all of the issues and concerns that have been discussed in this section apply to all countries, at least not to all countries at the same time. China and India seem likely to be reliant on use of coal on massive scales for decades to come. Second, we should not neglect the "if you can't beat 'em, join 'em" principle. We have seen examples—by no means a complete catalog—in Chapter 17 of how carbon materials derived from coals have many possible roles in contributing to renewable energy technology. There lies a new and growing market for coals.

In Chapter 6, we mentioned the "canary in the coal mine" and how this expression has become an idiom used far beyond the realm of coal mining, referring to something—CO_2 levels in the atmosphere, for example—that serves as an early indicator of adverse conditions, some impending greater trouble or outright danger. Never mind the canary in the coal mine—what we should be concerned about is the black swan in the coal mine (Figure 18.1).

So-called *black swan events* are very rare and unpredictable, but nevertheless have severe impacts. They can occur in fields as seemingly diverse as finance, science and technology, and international relations. The assassination of Archduke Franz Ferdinand of Austria was a black swan event. Its severe impact was called World War I. An additional characteristic of black swan events is that people will look at them with hindsight and claim that they can be rationalized or, indeed, were obvious—after the fact. Black swan events have been discussed eloquently by the Lebanese-American polymath Nassim Nicholas Taleb in a series of books (Taleb 2007). I've seen two black swans affect coal technology. The first was the OPEC oil embargo of 1973, which had an immense effect on reviving coal research and development worldwide. The second was the development of practical fracking techniques, about a decade ago, for accessing vast amounts of gas and oil in shales. This one seems pretty effectively to have ended interest in coal in some—but not all—countries. By

Figure 18.1 Never mind the canary in the coal mine. Watch out for the black swan in the coal mine—the very rare, unpredictable event that has a severe impact.
Courtesy of Tratong/Shutterstock.com.

their nature, we can't predict when the next "coal black swan" will appear or what its impact will be.

World population is steadily growing. After a very strange year ushered in by another black swan, the COVID-19 pandemic, forecasts seem to be that the global economy should be growing through the next few years. So should the global demand for energy. There is a good, albeit not perfect, correlation between per capita energy use and per capita gross domestic product (Schobert 2014, 6). The US Energy Information Administration has predicted a 50% increase in world energy usage by 2050 (EIA 2019). Its seems likely that countries with both large and burgeoning populations and large coal reserves, which are also enjoying economic development, will be relying on coal for power plant use for decades simply to meet their expectations for better standards of living (Arden 2018). This includes the BRICS countries—Brazil, Russia, India, China, and South Africa. Though some countries are seeing significant gains for shale gas and renewables at the expense of coal, there is really no "one size fits all" answer for satisfying the energy needs for economic development.

In China, coal is certainly much cheaper for electricity development than natural gas (O'Sullivan 2017, 223). Barring a change that brings a much bigger role for gas in China's energy economy, that country will likely continue to rely heavily on coal. Despite the phenomenal gains that China has made in its economy and standard of living, a large fraction of the population—perhaps a fifth to a quarter—is still in need of better quality housing, highways, sanitation, and educational opportunities (Mann 2014). It's difficult to see how to get those folks out of dire poverty without using coal to supply much of the needed energy.

In the latter half of the 2010s, India was second only to China in terms of installing solar projects. Several huge projects are rated at 2–5 gigawatts, costing in the billions of dollars. The problem is that the Indian gross domestic product could expand by a factor of five in the next two decades (Temple 2019). Like China, India has several hundred million people in serious poverty. In parts of India, air conditioning is becoming a public health necessity. Air conditioners run on electricity. Ideally, India would need to install 550 gigawatts of new electricity capacity by about 2030 to address the growth in demand. If India were to choose to obtain this new capacity entirely from renewables, meeting that target is highly unlikely (Temple 2019). Coal will remain essential.

The situation of renewables vis-à-vis coal in India is a reminder that there is not enough manufacturing capacity and not enough financial resources to displace coal from a major role in the projected growth of energy needs. There exists, worldwide, an immense investment in infrastructure for coal utilization, most notably in electricity-generating plants. The world has neither the money nor the factories and workshops to retrofit all of these plants to handle other fuels, let alone replace the complete plant with a new one in a short span of years. A complete turnover of infrastructure would take a very long time (Fallows 2010). Renewables are virtually inevitable in the long term. Until then, coal for electricity generation will be regarded as an option, especially in the developing economies, for decades (Arden 2018). If economic and practical strategies are developed for CO_2 sequestration that can be deployed on a scale coming close to matching our rate of CO_2 production, it's possible that such a development could sustain coal in electricity generation in developing countries. Some countries may have abundant coal reserves but lack indigenous petroleum or natural gas. For them, it might make good economic sense to rely on domestic coal rather than on imported gas or oil.

The fact that there is a struggle among coal, natural gas, renewables, and possibly nuclear is a signal that, regardless of the outcome, there are alternatives to fossil fuels for electricity generation that are actually in practical use. But electricity generation is not the only use of coal. There are industrial processes for which, at present, there are no good alternatives to coal. The steel industry is one; cement is another.

The worldwide steel industry continues to need large quantities of metallurgical coals for coke production. This is still the case despite inroads being made by electric furnace processes. Many of the deposits of good-quality coking coals have already been consumed or are quickly on the way to depletion. Coke producers and coal petrographers have been learning how to blend coals so that, even if no single coal is ideal, the blend might have good coking qualities. It is highly likely that metallurgical coals will be in demand for decades to come (Arden 2018). They will almost certainly sell at increasingly premium prices compared to coals used for steam generation in power plants.

Cement production has large energy requirements. About 4 billion tonnes of cement are produced annually. Half of that is in China. Cement production starts with heating limestone to produce lime (calcium oxide) and drive off CO_2. Lime is then reacted with various silicate and aluminate minerals at high temperatures. Coal still has a crucial role in cement production due to the requirements for large amounts of heat. Two hundred kilograms of coal are needed produce a tonne of cement (Arden 2018). It takes 300–400 kilograms of cement to make 1 cubic meter of concrete (Arden 2018). Burning large amounts of coal to supply process heat, combined with the calcining of limestone, means that the cement industry is also a prodigious emitter of CO_2. It has been estimated if, somehow, the world's cement industry could be counted as a country, the "country of cement" would be the third-largest emitter of CO_2 in the world, trailing only the United States and China (Watts 2019).

Unfortunately, there are also countries that must make a decision between using indigenous coal or relying on other energy resources imported from neighbors who might be capricious or downright hostile at times. As this chapter is being edited, Russia has launched a brutal invasion into Ukraine, a country richly endowed with coal. It seems reasonable that someone in the energy ministries of such potentially vulnerable countries might well employ his or her time learning about coal gasification as a replacement for imported natural gas.

As this book is written, we are getting close to welcoming our eight-billionth person to our planet.[2] All of us together use energy, and we use a lot of it. Collectively, we use enough energy in the course of a year to consume the equivalent of almost 9 million tonnes of oil. The equivalent amount of electricity is about 10,000 terawatt-hours (TWh) per year. In terms of the more familiar unit that most of us consume or pay for every month, a TWh is a billion kilowatt-hours. If this consumption were converted to an annual rate of using power, the global consumption by all of us is 12 terawatts (TW), which is 12,000,000,000 kilowatts. By the year 2100, we collectively will use 18 TW of power (Gibbs 2017).

Sadly, the state of our world is such that a very large fraction of humanity lacks access to some of the crucial necessities of life: clean water, adequate nutrition, shelter, at least enough education for literacy, basic health care. Estimates vary, but roughly somewhere between a third and half of the world's population is not able to obtain at least one of these services and often does not have access to more than one. Suppose we were to try to lift all of the world's population to a level such that any person, any place, could reasonably expect to have safe and warm shelter, drinking water, enough food, and access to a school and a clinic.[3] This does not mean providing everyone with the re-sources to use energy on the profligate scale at which we use it in the United

---

[2] As this paragraph was being edited, the count was 7,943,484,888. It can be a frightening thing to watch the "worldometer" (https://www.worldometers.info/world-population/) even for a few minutes.

[3] Clearly, the supposition ignores details of the mechanism by which this immense international human-itarian effort could be carried out. Even coming to agreement on how to achieve it, let alone actually doing it, is likely a vastly more difficult task than solving the scientific and engineering problems.

States, but rather to a level of a modest standard of living that would at least assure health, nutrition, and safety. Doing only this will require about 12 more TW, if we could achieve it in a reasonable time frame. Where are we going to get the additional 12 TW?

During my career, I have been fortunate to meet executives and senior managers from many energy industries and government agencies. When the topic comes up of addressing future energy needs of a rapidly growing world population, all of them, regardless of personal professional interests, have expressed the same sentiment in almost exactly the same words: "We're going to need everything." That is, every energy resource we have available will have to be mobilized to meet the future energy needs and expectations of the ever-growing world population.

In mobilizing the global energy resources, we need to keep three things in mind: first, every energy source has some technical advantages and some technical disadvantages. Second, every energy source has some impact on the environment, and some may even have advantages or benefits for the environment, especially vis-à-vis other sources. And third, every energy source has some economic advantages and some economic disincentives. These facts mean that all of us, not just professional workers in energy science and engineering, have to try to understand how to navigate between the technical advantages and disadvantages, the environmental impacts and benefits, and the economic incentives and disincentives to arrive at a mixture of energy sources, for wherever we happen to live, that provide a reasonable balance among reliability, affordability, and impact on the environment. There is no solution to energy needs that is optimum in every place at every time.

John Hofmeister, former president of Shell Oil Company and current head of the US-based organization Citizens for Affordable Energy, has described the solutions to future energy needs in terms of the "four mores" (Hofmeister 2012).

More energy from all sources, certainly including all of those discussed throughout this book: coal, shale gas, and renewables. This is just another way of saying that we're going to need everything.

More efficiency in producing and using energy. The better the efficiency of energy conversion, the more useful things—such as producing electricity—that can be done for a given amount of energy. Or, from the other perspective, we can use less energy to produce the same quantity of useful work as now. We see this in improvements in the heat rate in power plant boilers and in the efficiency gains in ultrasupercritical plants. More electricity for less coal. We are also reminded that, "It's always better to use energy rationally than to come up with new ways of using it" (Martin 2018).

More environmental protection. This "more" applies to every aspect of our environment, whether land, or air, or water, and certainly includes other living organisms. The greatest concern at present is global climate change, now recognized as an environmental issue with significant anthropogenic contribution and as something requiring action. This recognition is afforded by the overwhelming majority of the scientific community and a steadily growing percentage of non-scientists.

More infrastructure. These are the technologies that transport or transmit energy from wherever it happened to be produced—the electricity-generating station, the fracking pad, or wherever—to the places where it is needed for use.

In many countries, coal is likely to continue to continue on a downward trajectory, caught between the market forces of gas and renewables on one side and climate change on the other side. Continued improvements in mining productivity, coal-to-busbar efficiencies in power plants, and reduced coke rates in blast furnaces will give coal an extra push downward. Coal is not going to disappear completely any time soon, but it seems very unlikely to regain the prominence it held even a decade ago, let alone in its glory years.

Perhaps the key point to rethinking coal is to recognize that the absolute worst thing we can do with coal is to dig it up and burn it. Coal is a valuable resource for producing chemicals and carbon materials. Chapters 16 and 17 discussed this, but provide only examples of what might be done, not an encyclopedic list. It took a long time to produce the substances we recognize as coals. Converting all those useful carbon structures and carbon–carbon bonds straightaway back into CO_2 is a great waste, especially when there are other viable, possibly better, ways of getting heat.

In the field of chemicals from coals, we will never go back to the great days of coal-tar chemistry. The biggest roadblock is simply that there's not enough. The scale of tar production from byproduct recovery coke ovens, even if we were to retain that technology, cannot possibly keep up with the demands for chemical products. A first focus should be on those chemicals that could be made from coals more readily than they are from petroleum, especially aromatic chemicals. The opportunity to apply at least some of the principles of green chemistry to coal chemistry seems to be almost completely neglected so far. With coals, much more so than with petroleum, we can also consider the inorganic components. There is the potential for recovery of rare earths and elements that have applications in electronics, such as gallium and germanium.

The expansion of coals into carbon materials markets means first working toward increasing the opportunities in activated carbon and graphite applications. There is also a tremendous opportunity working toward building the renewable energy infrastructure with carbon, especially carbon materials from coals: carbon fiber wind turbine blades, electrical carbons for energy storage in batteries or supercapacitors, carbon molecular sieves, graphene and carbon quantum dots. Flexible solar cells are one of many potential applications.

An approach that should help coal both in energy markets and in the emerging chemicals and materials markets is to focus strongly on a dual-product or even a multi-product strategy for coal utilization. For much of my career I benefitted greatly from federally funded research projects. In that era, it seemed that much of the funding was along technology lines. That is, a person working on liquefaction was encouraged to push for conversions to liquids of close to 90%. Meanwhile, a colleague working on gasification was expected to convert all of the coal (sometimes the very same coal!) to gas. But there may be situations in which such an approach doesn't make sense. Possibly a better strategy would be to take off liquid products that would be comparatively easy to convert to chemicals or distillate fuels and then carbonize or gasify the rest. We must also recognize that, even with a reorientation to more emphasis on coal conversion to chemicals and carbon materials, it will continue to be necessary to have a clear focus on environmental issues, including mined land reclamation, emissions, and CO_2.

Well, then, what can be done? I suggest three things:

Put people first. It seems certain that the workforce in traditional mining, especially deep underground mining, will continue to shrink. This is not the fault of the people who are doing the mining. They should not be consigned to the rubbish bin. Those people who have spent their lives in underground mining, especially in the bad old days, to provide for their families are heroes. An op-ed article by the novelist Tawni O'Dell following the 2010 Upper Big Branch mine disaster very eloquently shows miners as "heroes, victims, martyrs, fools" (O'Dell 2010). Active plans are needed to retrain persons whose jobs have been displaced, for reinvestment in traditional mining towns to create new job opportunities, and for educational opportunities for young people in those towns.

There's a need to pull together the human and financial resources in traditional coal-mining regions to attract new opportunities for jobs and economic growth (Houser, Bordoff, and Marsters 2017). As with technology, different approaches work for different situations. Two examples from the United States: in 2018, Wyoming, a major coal-producing state, mandated that all public school students in all grades receive instruction in computer science (Goldstein 2019). As the news article put it, "Coal country [is] turning into code country" (Goldstein 2019). On a more local scale, the area around Tazewell, in the coal country of southwestern Virginia, is transforming to focus on its cultural and natural offerings, such as music festivals, regional crafts, and abundant opportunities for many kinds of outdoor recreation (Furchgott 2018).

> [T]he responsible response from policymakers is to be honest about these facts—about the causes of coal's decline and unlikeliness of its resurgence—rather than offer false hope that the glory days can be revived. . . . we must redouble efforts like those above to rebuild these communities, as well as fulfill pension obligations and pay back the debt we owe to workers and families who spent generations, often at the expense of their own health and well-being, providing the energy that powered a good part of the [world's][4] economy. (Houser et al. 2017)

Learn. We are not anywhere close to knowing as much about coals—and their variability—as we need to know. The great Dutch coal scientist Dirk van Krevelen reminds us that, "difficulties of a major nature in technical practice nearly always originate from unexpected peculiarities in the compositions of the materials used. This may be the composition on a macroscopic, microscopic or molecular level. These unexpected and mostly unobtrusive peculiarities can only be evaluated by the coal scientist" (van Krevelen 1993, 873).

We can continue to make strides in coal preparation to produce maceral concentrates that might offer better chances to have a prepared material well-suited for the production of chemicals or carbons. Scientists well-trained in coal petrography are vital. It might prove possible to concentrate mineral streams further, for example to obtain a rare earth concentrate. Even with Andreas Osiander's and George Box's admonitions in mind, we need to continue modeling the molecular frameworks of coals. These give us the mental maps that we need, so that we can plan how to remove, neatly and cleanly, the chemical structures we want from the molecular frameworks of coals. We need to continue to explore reaction chemistry, particularly getting energy into the coal particles without having to heat up everything in sight, and employing green chemistry where we can. We certainly need more focus on the solid-state physics of coals, or perhaps we might say on the materials science of coals. How can we best take advantage of the physical and mechanical properties of coals? And we need to keep in mind that certainly the high-rank

---

[4] I have changed the wording of the original, which referred to the American economy, but in fact these comments can apply to most coal-producing countries worldwide.

coals are carbon materials. Turning again to van Krevelen, "As coal scientists our strive [sic.] is: the desire to understand and the feeling of happiness and satisfaction when we do understand" (van Krevelen 1993, 874).

Be in the game for the long haul. Let's recall the years, sometimes decades, of patient, dedicated work by some of those who we met in previous chapters: Friedrich Bergius, Fritz Haber, Carl Bosch, George Mitchell. Many similar stories could be told. Scientific breakthroughs and technological developments can't be purchased online. They don't spring *de novo* from the soil—think of what human and material resources were already in place worldwide to develop the covid vaccine so quickly (Rhodes 2021). We need trained, dedicated people in well-equipped laboratories and pilot plants, and we need to support them for as long as it takes.

> Supporting applied and not-yet-applied research is not just smart, but a social imperative. . . . to enable and encourage the full cycle of scientific innovation . . . it is more productive to think of developing a solid portfolio of research in much the same way as we approach well-managed financial resources. Such a balanced portfolio would contain predictable and stable short-term investments, as well as long-term bets that are intrinsically more risky but can potentially earn off-the-scale rewards. (Flexner 2017, 32)

Is there a long haul for coal? I think so, but I do not see it in conventional combustion applications. I think it is going to have to be in organic chemicals and carbon materials, with careful attention to mining at the front end and capturing or eliminating CO_2 at the back end. To me, the best overall view of the future of coal comes from a cat named mehitabel:[5]

there s a dance or two
in the old dame yet

—Marquis (1950)

---

[5] mehitabel and her amanuensis archy are the marvelous creations of Don Marquis. mehitabel fancied herself to be the reincarnation of Cleopatra. archy was a free-verse poet reincarnated as a cockroach who recorded mehitabel's story on a manual typewriter by diving head-first onto the keys. archy was not able to work the shift key nor to reach the punctuation keys. That's why archy's writings are devoid of uppercase letters and punctuation.

References

Arden, H. 2018. "The Future of Thermal Coal in Power Generation." *World Coal* 27(7): 58–64.

Clarke, Arthur C. 1951. *The Exploration of Space*. New York: Harper & Brothers.

Dartnell, Lewis. 2019. *Origins*. New York: Basic Books.

Doyle, S. 2017. "USA: The Coal Hard Truth." *World Coal* 26(8): 12–15.

EIA. 2019. "EIA Projects Nearly 50% Increase in World Energy Usage by 2050, Led by Asia." Today in Energy. U.S. Energy Information Administration. https://www.eia.gov/todayinene rgy/detail.php?id=41433.

Fallows, James. 2010. "Dirty Coal, Clean Future." *The Atlantic* 306(8): 64–78.

Flexner, Abraham. 2017. *The Usefulness of Useless Knowledge*. Princeton, NJ: Princeton University Press.

Furchgott, Roy. 2018. "The Coal Is Gone: It's Time to Get Creative." *New York Times*, October 7.

Ge, Shirong. 2018. "Robotic Mining Technology for Underground Mines." International Pittsburgh Coal Conference, Xuzhou, China. October 16, 2018.

Gibbs, W. Wayt. 2017. "How Much Energy Will the World Need?" Anthropocene. https://www.anthropocenemagazine.org/howmuchenergy/.

Goldstein, Dana. 2019. "Coal Country Turning into Code Country." *Minneapolis Star Tribune*, September 8.

Stephen Jay Gould. 1991. *Bully for Brontosaurus*. New York: W. W. Norton.

Guffey, Roger. 2015. "Don't Fall for the 'Truthiness' About an Obama 'War on Coal.'" *Lexington Herald-Leader*, April 23.

Hansen, Peter A. 2016. "Coal: A Twisted Future." *Trains* 76(3): 40–47.

Hofmeister, John. 2012. "The Promise and Reality of Energy Independence." *Hydrocarbon Processing* 91(2) 42–44.

Houser, Trevor, Jason Bordoff, and Peter Marsters. 2017. "Can Coal Make a Comeback?" Columbia University Center on Global Energy Policy Report.

Kellman, J. 2016. "Illinois' Coal Country Teeters; Next President May Tip Balance." *Mankato Free Press*, October 3.

Mann, Charles C. 2014. "Black Magic." *Wired* 22(4): 73–81, 114–116.

Marquis, Don. 1950. *The Song of Mehitabel*. Garden City, NY: Doubleday.

Martin, Glen. 2018. "Power Shift." *Utne Reader* 198: 6–13.

Mufson, S., and D. Grandoni. 2020. "JPMorgan Pulling Back from Coal Companies, Arctic Drillers." *Minneapolis Star Tribune*, February 26.

Rowland, J. 2016. "Light at the End of the Tunnel?" *World Coal* 25(9): 17–23.

O'Dell, Tawni. 2010. "Heroes, Victims, Martyrs, Fools." *Pittsburgh Post-Gazette*, April 11.

O'Sullivan, Meghan L. 2018. *Windfall*. New York: Simon and Schuster.

Rhodes, John. 2021. *How to Make a Vaccine*. Chicago: University of Chicago Press.

Schobert, Harold H. 2014. *Energy and Society*. Boca Raton, FL: CRC Press.

Shermer, Michael. 2015. *The Moral Arc*. New York: Henry Holt and Company.

Taleb, Nassim Nicholas. 2007. *The Black Swan*. New York: Random House.

Temple, James. 2019. "India's Dilemma." *MIT Technology Review* 122(3): 14–19.

van Krevelen, Dirk W. 1993. *Coal: Typology – Physics – Chemistry – Constitution*. Amsterdam: Elsevier.

Watts, Jonathan. 2019. "Concrete: The Most Destructive Material on Earth." *The Guardian*, February 25. https://www.theguardian.com/cities/2019/feb/25/concrete-the-most-destruct ive-material-on-earth.

Appendix

1 Conversion of Coal Analyses to Different Analytical Bases

Converting the analytical value of some property of interest, which we can simply call X, from an as-received or as-analyzed basis to a moisture-free (mf) basis can be done as

$$X_{mf} = \frac{100 \times X_{ar}}{\left(100 - M_{ar}\right)},$$

where M denotes moisture, and the subscripts ar indicate as-received basis and mf indicate moisture-free basis.

Converting to a moisture-and-ash-free (maf) basis becomes

$$X_{maf} = \frac{100 \times X_{ar}}{\left[100 - \left(M_{ar} + A_{ar}\right)\right]}.$$

Here the symbol A indicates the ash yield and subscript maf denotes a "moisture-and-ash-free" basis.

If the value for mineral matter has been determined, or has been calculated using a formula such as the Parr formula (below), the conversion to a moisture-and-mineral-matter-free basis is done by

$$X_{mmmf} = \frac{100 \times X_{ar}}{\left[100 - \left(M_{ar} + MM_{ar}\right)\right]},$$

where now MM represents mineral matter.

The ASTM system for classification by rank requires the calorific value to be known on a moist, but mineral-matter-free basis, which can be found from

$$X_{mf} = \frac{100 \times X_{ar}}{\left(100 - MM_{ar}\right)}.$$

There are two useful tips for anyone wanting or needing to try these calculations. First, if a whole suite of data, such as the proximate or ultimate analysis, are being converted for a particular sample, an easy and quick check on the calculations can be done by remembering that analytical data always have to sum to 100 (or, allowing for possible round-off errors, say 99.5–100.5). If not, then something went awry in the arithmetic. Second, anyone doing many of these conversions, or needing to do them often, will find it convenient to enter these formulas into a spreadsheet or even a programmable hand calculator.

2 The Parr Formula for Calculation of Mineral Matter Content

Ash is relatively easy to prepare and weigh. Ash is the product of various chemical reactions and phase changes affecting the minerals and other inorganic constituents that were originally present in the coal sample. Ash and mineral matter are not the same. Until the advent of modern computer-interfaced analytical instruments about 30 years ago, it was extremely difficult to separate and weigh the minerals present in a sample. Over the years scientists in several countries have developed formulas to calculate a weight of mineral matter from more easily acquired analytical data. The formula developed by Samuel Parr of the University of Illinois, and commonly known as the *Parr formula* is one that has achieved fairly wide acceptance. The formula is

$$\text{Mineral matter} = 1.08 \times [\text{Ash} + 0.55 \text{ x Total sulfur}]$$

The mineral matter will be calculated on the same basis—as-received or mf—as used for the values of ash and total sulfur.

Mineral matter can also be determined by petrographic analysis, if such data are available.

3 Calculation of the Atomic Hydrogen-to-Carbon Ratio

Almost always, information on coal composition will be available on a weight basis (i.e., percent by weight). Conversion to an atomic basis requires that the weight percent data be divided by the atomic weights of hydrogen and carbon, which are 1 and 12, respectively. Since division by 1 is trivial, the hydrogen-to-carbon atomic ratio can be calculated from

$$\text{H/C atomic} = \frac{\text{H} \times 12}{\text{C}},$$

where the symbols represent the weight percents of hydrogen and carbon. The hydrogen and carbon data can be used on any basis—as-received, mf, or maf. The result will be the same.

4 Information on some Principal Macerals, Their Likely Origin, and Properties

This is not a comprehensive listing of all known macerals. It indicates the three principal maceral groups—vitrinite, exinite, and inertinite—some of their properties, a few of the macerals in each group, and the type of plant components from which those macerals originated. In humic coals of Carboniferous age, the vitrinites dominate.

The vitrinite group macerals have a shiny, almost glass-like appearance. Their name derives from the word "vitreous," meaning *glass-like*. Chemically, they are the most oxygen-rich of the three maceral groups. Hydrogen and aromatic carbon are moderate. Important macerals include

Collinite: Derived from gels of humic acids or humic-like materials
Tellinite: Derived from wood, bark, or layers immediately below the bark

Exinite group macerals have resinous or waxy appearances. The name derives from the outer layers of pollen and spores, which are called *exines*. Exinites are the most hydrogen-rich and least aromatic of the maceral groups. Important macerals in this group include

Alginite: Derived from algae
Cutinite: From leaf cuticles (i.e., the waxy surface film on leaves)
Resinite: Primarily from resins, such as are used to seal wounds
Sporinite: Derived from spores, such as from fungi

The inertinite group macerals derive their name from their relative unreactivity in coking and in other reactions. However, this pertains to northern-hemisphere Carboniferous coals. Permian coals from the southern hemisphere, such as South Africa, have some macerals in the inertinite group that are reactive. Intertinites have a dull, charcoal-like appearance and may represent the carbonized remains of plants from ancient forest fires. Of the three maceral groups, inertinites are the most carbon-rich and tend to be highly aromatic. This groups includes

Fusinite: From carbonized wood or woody tissues
Macrinite: Undifferentiated detritus, with grains larger than 10 microns
Micrinite: Similar to macrinite but for grains being smaller than 10 microns
Sclerotinite: From hardened tissues (sclerotia) used for food storage by fungi
Semifusinite: Carbonized woody material

5 Common Minerals in Coals

As with the principal macerals, the list provided here is not intended to be a catalog of all the minerals ever reported to have been observed in a coal sample at some place, some time. Only relatively few minerals are relatively common in most coals. There are probably a hundred more that might be characterized as rare or occurring in trace amounts. The list is intended to indicate the major minerals because in most instances the nature of ash will be determined by the reactions and phase changes (e.g., melting) of these minerals and the interactions among them at the temperature and time during which the coal is being burned or gasified. Furthermore, the elemental composition of ash will be determined by the compositions of the minerals and their relative proportions in the original coal sample. We can see from this list that the elements that can be expected in all coal ashes are silicon, aluminum, iron, calcium, magnesium, potassium, and sodium.

The common minerals are the following:

Carbonates
Ankerite: $Ca(Fe,Mg)(CO_3)_2$
Calcite: $CaCO_3$
Dolomite: $CaMg(CO_3)_2$
Siderite: $FeCO_3$

Chlorides (common in a few, but not all, coals)
Halite: $NaCl$
Sylvite: KCl

Clays

Illite: $(K,H_3O)(Al,Mg,Fe)_2(Si,Al)_4O_{10}[(OH)_2,(H_2O)]$

Kaolinite: $Al_2(OH)_4Si_2O_5$

Montmorillonite: $(Na,Ca)_{0.33}(Al,Mg)_2(Si_4O_{10})(OH)_2 \cdot nH_2O$

Hydroxides

Limonite: $FeO(OH) \cdot nH_2O$

Oxides

Quartz: SiO_2

Sulfides

Pyrite: FeS_2

Glossary

Acid deposition. The most generic term for accumulation of acidic substances from the atmosphere onto surfaces or into natural waters, encompassing rain, sleet, snow, hail, and dry acidic solids.

Acid gases. Hydrogen sulfide and carbon dioxide, as they are found in the untreated product of coal gasification. Both dissolve in water to form weak acids.

Acid precipitation. A more inclusive term for acid rain that recognizes that other forms of precipitation—snow, sleet, or hail—can also be acidic due to nitrogen and sulfur oxides in the atmosphere.

Acid rain. Rainfall that is more acidic (i.e., with a pH less than 5.6) than rainfall in a completely pristine, non-polluted environment.

Aerobic. Processes that take place in or require air (oxygen). In the context of coalification, this term refers to bacteria that require oxygen for their metabolism.

Air-blown. A type of coal gasifier in which the oxygen needed is supplied by using air rather than pure oxygen.

Afterdamp. A mixture of gases accumulating in underground coal mines that consists of the products of a fire or explosion. Afterdamp is deficient in oxygen and may contain carbon monoxide.

Aliphatic. A class of hydrogen-rich organic compounds in which molecular structures can have either chains or various-sized rings of carbon atoms.

Anaerobic. Processes that do not require or take place in air or oxygen.

Angiosperm. Plants with enclosed seeds, the dominant form of plants today.

Anthracosis. A disease of the lungs caused by long-term inhalation of coal dust. Commonly known as "black lung disease."

Apparent density. The density of a porous solid as measured by immersion in a liquid. The numerical value can vary depending on the specific liquid chosen for the measurement and the ability of that liquid to penetrate the pores in the solid.

Aromatic. The class of carbon-rich organic compounds in which the carbon atoms are (with rare exceptions) arranged in hexagonal rings.

Aromaticity. In coals, the fraction of the total number of carbon atoms present in aromatic structures.

Ash. The non-combustible inorganic residue remaining after the combustion of coal, consisting of a mixture of various minerals.

Autogenous heating. The heating and possible ignition of stored coal, without human intervention, due to such causes as slow oxidation by air or release of heat of wetting.

Availability. In the context of power plants, the percentage of time that the unit is actually operating. A unit that is operating 900 hours out of every 1,000 would have an availability of 90%.

Baghouse. A device use to remove fine particles of solids from the flue gases of a boiler, consisting of fabric with pores small enough to trap the particles but large enough to allow easy passage of the gases.

Bark coal. A type of coal, found mainly in China, in which a major constituent is coalified tree bark.

Biochemical phase. The first phase of coal formation, marked by bacterial and fungal attack on accumulated plant debris.

Biomarker. A chemical compound found in coals (or in petroleum) that is identical to compounds found in living organisms or for which the structure is clearly derived from components of living organisms.

Bitumen. A viscous mixture of mostly high-molecular-weight hydrocarbons formed in natural processes or during pyrolysis of carbonaceous substances.

Black carbon. Narrowly, particles in the atmosphere that are pure carbon. More broadly used to include soot, condensed coal tars, and possibly inorganic particulate mixed with the carbonaceous material.

Blackdamp. An accumulation of high concentrations of carbon dioxide in an underground coal mine. This gas is not poisonous but is dangerous because it deprives the miners of oxygen.

Boiler. In a power plant, the device in which steam is produced from water. In coal-fired power plants, the furnace and boiler are one combined unit.

Bottom ash. In the boiler of a coal-fired power plant, the ash that is of large enough particle size to fall to the bottom of the boiler, as distinct from fly ash.

Bucket-wheel excavator. A machine used in the surface mining of coals, using a wheel having a series of digging buckets attached to its rim; as the wheel rotates, the buckets cut and scoop up the coal.

Bulk density. The weight of a solid that will fill a volume of given size. For any solid, the bulk density measurement depends on the particle size of the solid since more small particles can be packed into a given volume than large particles.

Burden. In a blast furnace, the mixture of iron ore, coke, and limestone flux.

Byproduct recovery coke oven. A device used in the production of metallurgical coke from bituminous coals, configured so that gases, oils, and tars driven off the coals during coking can be collected for other uses.

Caking coals. Bituminous coals that, upon heating, soften and agglomerate into a single solid mass.

Calorific intensity. The rate at which heat is liberated during the combustion of a coal.

Calorific value. The heat released upon combustion of coals (or other solid fuels), expressed as megajoules per kilogram, kilocalories per kilogram, or British thermal units per pound.

Capacity factor. For any type of electricity generation, the ratio of the electricity actually generated (e.g., over a year) to the capacity of the plant at maximum output.

Carbon capture and storage. The various techniques and processes used to capture carbon dioxide from flue gases or other sources and then to sequester the CO_2 so that it cannot be emitted to the atmosphere.

Carbonization. A thermal process in which coals or other organic materials are converted to carbon-rich solids, such as in the production of metallurgical coke.

Catagenesis. The phase of coal formation in which the principal agents of change are the temperature gradients and tectonic pressures inside Earth's crust.

Char. A carbonaceous solid formed in a thermal process but in which the original raw material did not pass through an intermediate fluid phase.

Chemical fractionation. A set of procedures for investigating how the inorganic constituents are incorporated in coals, consisting of sequential leaching or extraction of the sample with reagents that remove ion-exchangeable cations, coordinated or chelated ions, elements present in acid-soluble minerals, and elements present in silicate or aluminosilicate minerals.

Chemisorption. An adsorption process in which chemical interactions (usually weak) help bind the adsorbed species to the surface.

Coal ball. A petrified aggregation of plant material, impregnated with minerals such as calcite and pyrite, found in coal seams.

Coalbed methane. Methane that occurs naturally in coal seams, formed either by bacterial action on the coal or thermal processes during catagenesis. It is potentially a useful fuel resource.

Coal beneficiation. A general term for any operations that improve the quality of coals.

Coal cleaning. Processes for reducing the amount of sulfur, minerals, and extraneous rock in coals.

Coalified. Plant matter that has been converted to coal.

Coal oil. A liquid fuel obtained from heating coals in the absence of air to drive off and collect volatiles. Sometimes informally but incorrectly called "kerosene."

Coal preparation. An inclusive term for the various operations intended to make coals ready for utilization by removing moisture and sulfur and reducing ash yield.

Coal washing. Those operations of coal cleaning that use water or aqueous slurries solutions of various reagents.

Coke. A carbonaceous solid formed in a thermal process, in which the original raw material passed through an intermediate fluid phase in the course of forming the final solid product.

Coke breeze. Pieces of metallurgical coke that are too small to be used in a blast furnace but which might serve as fuel in other applications.

Coke rate. In a blast furnace, the quantity of coke required to produce 1 tonne of hot metal from the smelting of iron ore.

Coking coals. The subset of bituminous coals that pass through a fluid phase upon heating and which then resolidify to a strong, hard, porous, high-carbon solid suitable for use in a blast furnace.

Cooling tower. Large structures with open interiors into which the heated water from a power-plant condenser can be sprayed to allow it to cool to near-ambient temperature before being discharged into the environment.

Crosslink. In true polymers and, it is thought, in many coals, one or more atoms that connect polymer chains to each other via covalent bonds.

Crosslink density. In polymers and coals, a measure of the number of covalently bonded crosslinks in a given length of the polymer chain.

Damps. A collective name for various gases or mixtures of gases that accumulate in underground coal mines. All are dangerous in some fashion. The word is an Anglicization of the German *dampf*, meaning vapor.

Dehydrogenative polymerization. A chemical process in which two aromatic molecules fuse to form one larger structure via the loss of hydrogen atoms from the periphery of reacting molecules and the formation of new carbon–carbon bonds between them.

Delocalized. Electrons in a molecule that are not confined to the immediate region of an electron pair covalent bond, but rather that can move throughout the bonding system of the molecule. Some electrons in aromatic molecules or structures are delocalized.

Detrital mineral. Minerals in coals that originated elsewhere and were transported into the coal-forming environment by wind or water.

Dewatering. Any of several operations applied to remove loosely held moisture in coal, as after coal washing processes.

Diagenesis. The phase of coal formation in which accumulated organic matter is transformed into kerogens, primarily by the action of bacteria and fungi.

Direct liquefaction. Various processes that convert coals into high yields of liquids without requiring the intermediate step of forming synthesis gas.

District heating. A form of energy utilization in which hot water produced in a power plant is piped to homes, industries, or businesses near the plant to be used for heating, thereby displacing alternative energy sources that would otherwise have been needed to provide the heat.

Dragline. A type of equipment used in the surface mining of coal that operates by pulling or dragging a large bucket across the surface to be removed.

Drift. An entryway or tunnel driven into a coal seam to gain access.

Dry-bottom. A gasifier boiler from which ash is removed as a solid, in contrast to allowing it to melt to a slag.

Dry scrubber. A device used to capture sulfur oxides from flue gases, relying on the reaction of SO_x with a pulverized solid rather than a water-based slurry of lime or limestone.

Dunkelflauten. A German term meaning "dark doldrums," and referring to a problem encountered with a combined solar–wind energy system, in which there is insufficient wind at night to compensate completely for the solar energy not being generated in the dark.

Dusting. A practice used in underground coal mines in which a powder of a non-combustible, non-silicate mineral such as limestone is spread in the mine to suppress coal-dust explosions.

Ebullated-bed reactor. A reactor design for effective conversion of three-phase mixtures, with a high flow rate that gives the appearance of the reactor contents boiling.

Electromagnetic induction. The generation of an electric current in a conductor by a changing magnetic field, which may be obtained by moving a conductor through a magnetic field or by moving a magnetic field relative to the conductor.

Electrostatic precipitator. A device used to remove fly ash and other particulate matter from flue gases, operating by imposing a strong electric field which gives the particles an electric charge and causes them to be attracted to an oppositely charged electrode.

Endothermic. A process, such as a chemical reaction, that absorbs heat as it proceeds.

Energy intensity. In the context of economic growth, the amount of energy required to achieve a given level of growth.

Enhanced oil recovery. Any of several techniques used to recover more oil from a reservoir after the oil flow driven by natural pressure has diminished or stopped.

Equilibrium moisture. The moisture content of a coal sample that has been saturated with water and then equilibrated in an atmosphere of 100% relative humidity. The value is presumed to represent the moisture that could be held in the coal if all of the pores were filled or held in the coal in its natural state in the seam.

Equilibrium reaction. A chemical reaction that is reversible and that ultimately comes to a state in which there is no net change in the amounts of reactants and products.

Epigenetic. Minerals in coals that precipitated from water, growing in cracks or cleats.

Exothermic. Any process, but usually a chemical one, which liberates heat as it occurs.

Fire boss. A mine employee charged with testing areas of an underground coal mine for the accumulation of flammable gases before the other miners are allowed into the mine.

Firedamp. A gaseous mixture dominantly comprised of methane that accumulates in underground coal mines. Firedamp is extremely dangerous because it is flammable and some concentrations in air are explosive.

Fixed carbon. The carbonaceous solid portion of a coal sample that remains after the volatile matter has been driven off. It is calculated as part of the proximate analysis by subtracting the sum of moisture, volatile matter, and ash from 100.

Float-sink test. A test used to assess the outcome of coal cleaning in which a pulverized sample of a coal is mixed with a liquid having a specific gravity intermediate between the carbonaceous portion and the minerals. A coal-rich, low-mineral fraction floats and a mineral-rich fraction sinks. The amounts of each fraction and their ash yields provide information to guide coal preparation strategies.

Flue-gas desulfurization. The removal of sulfur oxides from the flue gases of a boiler to prevent their being emitted to the environment.

Fly ash. Ash of particle sizes small enough to be entrained in the gaseous combustion products in a boiler so they are carried up and out of the boiler with the gases, rather than falling to the bottom.

Forms-of-sulfur. The ways in which sulfur can be incorporated in coals, which include being chemically bonded to the macromolecular framework of the coal as organic sulfur, being present in pyrite or related minerals such as marcasite, and being present in various sulfate minerals such as jarosite. The determination of the proportion of sulfur present in each of these forms constitutes a forms-of-sulfur analysis.

Fouling. The accumulation of ash as agglomerated or sintered deposits on the heat-exchange surfaces of coal-fired boilers.

Fracking. An informal term for hydraulic fracturing, which involves breaking apart gas- or oil-bearing shales using water at very high pressures, usually in horizontal wells through the shale.

Friability. A property of solid materials such as coals that expresses their tendency to resist abrasion.

Froth flotation. A process for fine-coal cleaning, in which air is blown through a slurry of fine coal particles in water. Carbon-rich particles float and can be removed from the top, while mineral-rich particles sink to the bottom of the vessel.

Fuel cell. An electrochemical device in which the energy of a chemical reaction is converted into electricity and to which the reactants are fed continuously so that it never goes "dead," unlike a battery.

Fuel NO_x. A mixture of nitrogen oxides produced by the reaction of nitrogen atoms chemically bonded into the molecular framework of the coal react with oxygen.

Fuel ratio. The ratio of fixed carbon to volatile matter in a coal; an indicator of its combustion behavior.

Functional group. An atom or, more commonly, a group of atoms that are part of a larger molecule, have a characteristic structure, and have a characteristic chemical behavior.

Furnace. The region of a power plant boiler in which most of the combustion of the coal occurs.

Gasifier. A type of reactor in which coals are converted into gaseous products via reaction with steam and oxygen or steam and air.

Geochemical phase. The phase of coalification in which kerogens are converted to coals by the effects of natural geothermal temperatures and tectonic pressures.

Grade. Classification of coals on the basis of their mineral matter content or ash yield.

Grindability. A property of coals (or other solids) determined in a laboratory test but which indicates the ease or difficulty of grinding a coal in industrial equipment.

Gymnosperm. The class of seed plants that includes the conifers.

Hazardous air pollutants. Commonly known as HAPs; a listing of substances considered to be human health hazards in the environment, including a number of trace elements that occur in coals.

Heat rate. The amount of heat from coal required to generate 1 kilowatt-hour of electricity.

Heat-recovery coke oven. A device for making metallurgical coke inside which the volatile components of the coal are burned to provide the heat necessary for the coking process. Also known as a *non-recovery coke oven*, because no byproducts are recovered.

Higher heating value. The calorific value calculated on the basis of the products of combustion being carbon dioxide and liquid water; the value includes the heat of condensation of steam to liquid water.

Hot metal. The product of a blast furnace consisting of molten iron containing small amounts of dissolved carbon and other elements. If allowed to solidify instead of being processed immediately into steel, also known as *pig iron*.

Humic acid. Large, dark-colored macromolecular solids of ill-defined structure, soluble in aqueous alkalis and precipitated if the solution is acidified. Formed during diagenesis of organic matter.

Hydrogen donors. Compounds or structural fragments of a coal that readily transfer hydrogen atoms to reactive radicals produced in the thermal decomposition of the coal. The classic example of a hydrogen-donor compound is tetralin.

Hydroliquefaction. A process for the conversion of coals into high yields of liquids by the addition of hydrogen.

Hydrophilic. "Water-loving," an important property of coal and mineral surfaces in fine coal cleaning.

Hydrophobic. "Water-fearing" or "water-avoiding," also used in the context of characterizing coal and mineral surfaces.

Indirect liquefaction. A process that produces liquid fuels from coals by first converting the coal to a mixture of hydrogen and carbon monoxide (synthesis gas) and then, in a separate step, reacting the gas to form liquids.

Indirect solar conversion. A strategy for electricity generation in which the heat in solar radiation is used to produce steam, which then drives a turbine connected to a generator.

Induced seismicity. Earthquakes caused by human activities such as fracking.

Inorganic affinity. Said of any of the inorganic elements in coals that tend to affiliate primarily with the mineral matter.

Inorganic constituents. A general term that includes any of the minerals present in coals as well as those elements that are present as ion-exchangeable ions or are coordinated to atoms in the macromolecular framework.

Integrated gasification combined cycle. A strategy for the generation of electricity in which a coal is gasified to produce a fuel that is burned inside a combustion turbine, after which the hot exhaust gases are used to generate steam for operating a steam turbine. Both turbines are connected to generators, providing two sources of electricity and improving the efficiency of the plant.

Intercalation. A process by which small atoms are chemically inserted between the atomic layers of a solid, such as the insertion of lithium atoms or ions between the layers of carbon atoms in graphite in lithium-ion batteries.

Ion exchange. A chemical process in which ions of one element present in an aqueous solution are interchanged with ions of another element held on a porous solid substrate, such as a low-rank coal.

Kerogen. Brown or black high-molecular-weight solids insoluble in alkalis, non-oxidizing acids, and organic solvents. They represent the end point of diagenesis.

Kinetic isotope effect. In a chemical reaction, the lighter (lower atomic weight) of two isotopes will participate in the reaction at a slightly faster rate than the heavier isotope.

Lepidodendron. Extinct, tree-like plants from the Carboniferous, roughly resembling gigantic versions of modern club mosses.

Liquefaction. In general, the conversion of a solid, such as coal, into liquids in one or more chemical steps (i.e., not simple melting).

Lithotype. Distinct bands appearing in coal seams, differing in appearance, properties, and maceral content.

Longwall mining. A method of underground coal mining in which a machine cuts or slices coal from the face of a seam; for efficient operation of the equipment, the seam face needs to be hundreds of meters long, hence the name "longwall."

Lower heating value. The calorific value calculated on the basis of the products of combustion being carbon dioxide and steam, not including the heat of condensation of steam.

Maceral. Coalified remnants of the original plant matter, distinguishable by optical microscopy and analogous to the grains of minerals that comprise a rock.

Mesocarbon microbeads. Solid, roughly spherical particles of very high carbon content, usually less than 1 millimeter diameter, formed by careful heat treatment of mesophase.

Mesophase. A form of matter that can be thought to be liquid-like, in the sense of being able to flow and assume the shape of its container, but also to be solid-like, in having some intermediate- or long-range ordering of the molecules.

Metallurgical coke. A coke suitable for use in a blast furnace.

Methanation. The chemical process by which methane is produced from the reaction of hydrogen with carbon monoxide. The purpose is often to produce a substitute natural gas.

Methylene insertion. A chemical step crucial for the growth of chains of carbon atoms during the Fischer-Tropsch reaction. In this step a methylene carbon atom, $-CH_2-$, inserts into the bond between a methyl carbon, $-CH_3$, and the catalyst surface to form a —CH_2CH_3 structure. Further methylene insertions can lead to chains of five or more carbon atoms, which would be liquids at ordinary conditions.

Mineral. A naturally occurring inorganic substance characterized by a specific chemical composition, crystal form, and internal structure.

Miner's asthma. A chronic shortness of breath resulting from long-term inhalation of coal or rock dust.

Model compound. Any organic compound having a composition and molecular structure similar to a portion of the structural framework of coals and therefore that can be used to study how that portion of the coal structure might react with a selected reagent. In principle, the model compound should be much easier to study in the laboratory than a solid coal.

Moisture. With specific reference to the proximate analysis of coals, moisture is determined by the weight loss observed by heating the coal to temperatures slightly above 100°C (the boiling point of water).

Molecular sieve. A porous solid, such as an activated carbon, in which the pore diameters are of about the same size as some molecules, thereby separating substances on the basis of their molecular sizes or shapes.

Native. Used to indicate a coal that is still in its natural state and that has not yet undergone thermal or chemical treatments.

Natural greenhouse effect. The warming of Earth caused by the carbon dioxide and water vapor that occur naturally in the atmosphere.

Negative emissions technology. Any of several processes designed to remove more carbon dioxide from the atmosphere in a given time period than is emitted by human activities.

Oil agglomeration. A process of fine coal cleaning in which coal particles are mixed with an oil-based product such as diesel fuel. Coal particles are wetted by the oil and clump together. Mineral particles are not wetted and can be removed as a separate stream.

α-Olefins. A family of industrially useful organic chemicals in which there is a double bond between the first and second carbon atoms of the molecule, such as pentene, CH_2=$CHCH_2CH_2CH_3$,

Once-through catalyst. In direct coal liquefaction or related processes, a catalyst that is introduced to the reactor with the raw materials and separated from the products but then discarded rather than being recycled.

Organic affinity. Said of those inorganic elements in coals that appear to be affiliated mainly with the carbonaceous portion of a coal rather than the mineral matter.

Organic sulfur. The form of sulfur in coals incorporated in the molecular framework of the coal by carbon–sulfur covalent bonds.

Outburst. The sudden breakage of a coal seam caused by the high accumulated pressure of methane in the coal.

Overburden. In surface mining, the amount of sediments lying on top of the coal seam of interest. The overburden must be removed to access the coal.

Oxyfuel combustion. A process in which coal is burned in an atmosphere containing higher concentrations of oxygen than found in ordinary air. Doing so gives combustion products with higher concentrations of carbon dioxide and lower concentrations of nitrogen, making it somewhat easier to capture the CO_2.

Oxygenate. Broadly, organic compounds containing one of more atoms of oxygen, such as alcohols or aldehydes, and which are useful as fuel additives or in the production of chemical products.

Oxygen-blown. Said of coal gasifiers in which oxygen is fed to the unit as pure O_2 rather than as a component of air.

Paper coal. An uncommon form of coal having a laminated structure looking somewhat like sheets or leaves of paper.

Parasitic loss. Energy produced and used inside a power plant or other facility to operate equipment and which therefore is not available to be sold to consumers.

Particulate matter. Fine particles of solids carried out of a combustor or gasifier and consisting of fly ash along with particles of partially reacted coal and soot.

Peat. Type III kerogen, consisting of newly accumulated plant material and partially decomposed organic matter.

Permissible explosives. Various explosives that can be used in underground coal mining and that are characterized by only a short spurt of flame. A tiny, short-lived flame is much less likely to ignite combustible or explosive gases in the mine.

Petrographic analysis. Examination of coals by optical microscopy to identify the macerals in the sample and quantify their relative proportions.

Photovoltaics. Conversion of the energy in light—usually sunlight—into electricity.

Pig iron. The solidified product of smelting iron ore in a blast furnace, primarily iron with small amounts of carbon, silicon, and a few other elements.

Pitch mining. A method of underground coal mining applied to very steep seams in which the miners normally work upward into the seam.

Plastic range. For coals that pass through a fluid stage during heating, the temperature interval between the softening temperature and the resolidification temperature.

Pneumoconiosis. Generally, lung disease caused by chronic inhalation of irritants.

Porosity. The fraction of the total volume of a solid that is occupied by pores or void spaces.

Post-combustion strategies. Methods for reducing emissions from coal combustion by capturing potential pollutants from the flue gases after the coal has burned.

Pre-combustion strategies. Methods for preventing the formation of potential pollutants in coal combustion by reducing their amount in the fuel before it is burned (e.g., by coal cleaning).

Proximate analysis. A set of techniques for characterizing and classifying coals by treating all of the components as if there were only four: moisture, volatile matter, and ash, which are measured directly, and fixed carbon, which is calculated by subtracting the sum of the other three from 100.

Pyritic sulfur. One of the ways in which sulfur is incorporated into coals; in this case as the mineral pyrite or the related marcasite.

Pyrolysis. Generally, any reaction for the breaking of chemical bonds by heat.

Rank. Classification systems for coals that express the extent to which a particular coal has progressed away from the original plant matter and toward the ultimate end point, graphite.

Reactive dissolution. Any process in which a solvent appears to be dissolving components of coals but most likely is reacting with the coals to create molecular fragments that then dissolve in the solvent.

Reflectance. A property of any solid that indicates the fraction of light reflected from the surface; in coal science, this is of particular interest for characterizing vitrinite macerals. The reflectance relates to the aromaticity and size of aromatic ring systems, so is an excellent way of indicating rank.

Renewables. A family of energy sources that are, in principle, inexhaustible and include solar, wind, tidal, geothermal, hydroelectricity, and biomass.

Resolidification temperature. In the Gieseler fluidity determination, the temperature at which the fluidity of the sample has dropped to 20 dial divisions per minute.

Retrogressive reactions. In coal conversion processes, reactions in which radicals formed from the coal react with each other to form new, strong carbon–carbon bonds instead of the desired reaction with hydrogen to form carbon–hydrogen bonds.

Reversible system. Any system in which the conditions, such as temperature and pressure, internal to the system differ from the external values by only an infinitesimal amount.

Roof bolting. A safety measure in underground coal mining in which thin rock strata in the mine roof are bolted together to make a thicker, stronger beam that is less likely to fall.

Room and pillar system. A method of underground coal mining in which a portion of the coal is removed, creating rectangular void spaces in the seam, but leaving the rest of the coal in place to help support the roof and overlying rocks.

Run-of-mine. Used to describe coal samples exactly as they came from the mine, with no special treatment such as drying or coal cleaning.

Salt coals. Coals containing high amounts of sodium chloride (common salt) and which usually are undesirable for utilization because of potential equipment corrosion problems.

Scrubber. The principal device used to capture sulfur oxides in combustion flue gases, operating by using a water slurry of lime or limestone to react with the SO_x.

Selective catalytic reduction. A technique for reducing the concentration of nitrogen oxides in flue gases by reaction with nitrogen-containing compounds such as ammonia or urea over a catalyst. The nitrogen compounds are converted to harmless nitrogen gas.

Semicoke. A highly carbonaceous solid formed as the initial product of resolidification of the fluid produced from coking coals; further heating in the coke oven converts the semicoke to metallurgical coke.

Shaft mine. An underground coal mine in which miners and their equipment are lowered vertically into the mine and coal is removed, roughly resembling an elevator shaft in a building.

Shale revolution. The dramatic changes in the energy economy by the enormous amounts of oil and gas made available from shales by fracking.

Silicosis. A lung disease caused by long-term exposure to and inhalation of dust from silicate and aluminosilicate rocks.

Slacking. The breaking of pieces of coals, usually low-rank coals, as they lose moisture.

Slagging. The formation of molten ash on the internal surfaces of coal-fired boilers, usually very undesirable.

Slagging gasifiers. Various kinds of coal gasifiers that operate at temperatures high enough to cause to ash to melt; it is then removed from the gasifier as a liquid.

Slope mine. An underground coal mine in which the entry slopes downward, as distinct from a shaft mine in which the entry is vertical.

smelting. Treating the ore of a metal with chemical reduction process to obtain the free metal.

Softening temperature. In the Gieseler fluidity test for characterizing caking coals, the temperature at which the fluidity first reaches the value of 20 dial divisions per minute. In the ash fusibility test, the temperature at which the test specimen has fused to a lump in which the height is equal to the width of the base.

Solvent extraction. Any process in which coals are treated with various solvents to dissolve or extract as much of the carbonaceous material as possible or to selectively isolate certain components of the sample, such as waxes or resins.

Solvent swelling. A phenomenon exhibited by coals, and characteristic of crosslinked polymers, in which exposure to a solvent causes the solid to swell to some extent, but not dissolve.

Spontaneous combustion. The ignition of coal in a stockpile or bunker without apparent human intervention, caused by such phenomena as slow oxidation or release of heat of wetting.

Squeeze. A problem in underground coal mining in which the weight of roof rock crushes the coal in the seam, possibly collapsing the roof onto the floor, causing injuries or fatalities.

Stereoisomer. Compounds having the same chemical composition but differing in the three-dimensional arrangement of their constituent atoms in space.

Stinkdamp. Hydrogen sulfide accumulated in the atmosphere of an underground coal mine.

Stripping ratio. In a surface coal mine, the ratio of the thickness of the overburden to the thickness of the coal seam.

Sulfatic sulfur. Sulfur incorporated in coals as various sulfate minerals, such as iron or calcium sulfates. A high concentration of iron sulfates suggests that the coal sample may have been oxidized extensively.

Supercritical fluid. A fluid in which distinct gas and liquid phases do not exist.

Surface area. Normally taken to be the total area of all surfaces that might be accessible to gases or liquids in a coal sample, including the external surface and the surfaces of all the pore walls.

Syngenetic. Minerals that formed in coals at the same time as the coal itself was forming.

Temperature of maximum fluidity. During the heating of a coking coal, this is the temperature at which the fluidity of the plastic mass reaches a maximum value, or the viscosity of the mass is at a minimum. Different coals have different temperatures at which the fluidity is a maximum, and different fluidities at the maximum temperature.

Thermal NO_x. Nitrogen oxides that formed from a reaction of nitrogen in the air with oxygen at the high temperatures of a combustion system.

Thermal pollution. Introduction of undesirably large amounts of heat into the environment by, for example, returning heated water from a condenser directly back into the local water source.

Throw-away catalysts. Inexpensive substances, often a waste from some other process, that can be used one time in a coal conversion reaction and then discarded. Examples include mill scale and red mud.

Timbering. The use of thick, heavy pieces of wood in an underground coal mine to support the sides and roof of the workings.

True density. The density of coal as measured in helium, thought to be "true" because it is presumed that helium is capable of penetrating all pores in the sample, thereby indicating the density only of the carbonaceous material itself.

Type. Coal classification based on the kinds and amounts of the maceral components of the sample, relying on petrographic analysis.

Ultimate analysis. Direct determination of the amounts of carbon, hydrogen, nitrogen, and sulfur in a coal sample, and calculation of the amount of oxygen by subtracting the sum of the other four elements from 100.

Unconventional. Said of oil or gas deposits that are held in rocks of low porosity and low permeability or are of high density and viscosity. In either case they are not able to be recovered by the traditional methods of the industry.

Vascular plant. Plants characterized by the presence of tubes, channels, or similar structures for transport of water and nutrients through the plant.

Volatile matter. The weight loss experienced by a coal sample under the specific conditions of the volatile matter test, which involve heating at 950°C for 7 minutes.

Volatiles. A general term for the gases and vapors driven off a coal sample by heating, regardless of specific temperature, heating rate, or time.

Washability. A term used to indicate how amenable a particular coal would be to removal of mineral matter by the conventional methods of coal cleaning.

Water gas. A useful, inexpensive fuel gas produced by reacting coals with steam.

Water–gas shift. The reaction of carbon monoxide with steam to produce carbon dioxide and hydrogen, or vice versa. The water–gas shift is an equilibrium, so can be run in either direction to produce any desired gas composition with these four components.

Water-tube boiler. The principal device for producing high-temperature, high-pressure steam in a power plant, in which water or steam is enclosed in pipes or tubes heated from the outside by the combustion reactions.

Whitedamp. An accumulation of carbon monoxide in an underground coal mine. This gas is a lethal poison.

Wind turbine. A device that uses the kinetic energy of wind to operate a generator and produce electricity.

Working fluid. Any gas or liquid used to operate a turbine, usually for the purpose of turning a generator to produce electricity.

Yellow boy. A thick, gelatinous precipitate formed in acid mine water, characterized by its yellow-orange color.

Bibliographic Essay

For readers who are interested in probing much more deeply into the topics mentioned in this book, in my professional work I usually reach for one or more of these four books as a starting point: Norbert Berkowitz, *An Introduction to Coal Technology*. Now 40 years on, some topics are out of date, but nonetheless it remains the best short introduction to the field. *Chemistry of Coal Utilization*, edited by Homer Lowry, first appeared as two volumes in 1945, with a supplemental volume in 1963. A second supplemental volume, edited by Martin Elliott, was issued in 1981. Collectively known simply as "Lowry," these books sought to collect most of what was known about coal science and utilization at the time. It is unfortunate that no one has stepped up to prepare a third supplemental volume to capture the work done since the 1980s by the large worldwide cohort of coal scientists in the past era of generous support for coal research and process development, scientists who are mostly already into retirement. In the field of coal science, *the* book is *Coal: Typology—Physics—Chemistry—Constitution*, by D. W. Van Krevelen. A third edition appeared in 1993, greatly expanded relative to its predecessors. The most recent of the "big four" is *Clean Coal Engineering Technology*, by Bruce Miller. It is well written and meticulously detailed. The focus is on combustion processes, but since they account for more than 90% of coal consumption around the world, Miller's book is an excellent source of information. These books all presume that the reader has a background equivalent to university-level studies in the physical sciences or engineering.

For coal classification following the procedures established by the American Society for Testing and Materials the source is the *Annual Book of ASTM Standards*, particularly Volume 05.05, *Gaseous Fuels; Coal and Coke*. A great many books on coal, including this one, of course, excerpt some of the ASTM standards, but it never hurts to go straight to the fount to be sure of the details. New editions are issued annually, but the individual standards are changed very slowly—rightfully so—so that earlier editions remain quite useful.

The Industrial Revolution began in Britain thanks to an abundance of coal, iron ore, and limestone. Lewis Dartnell, in his book *Origins*, discusses—among other topics—why and how coal formed in Britain. Charles Lyell's *Travels in North America*, recently reissued in a two-volume paperback set, provides an interesting view of what was known, or speculated, about coal formation nearly two centuries ago. In addition, readers who are interested in history will find much detail about life in the United States and Canada in the early nineteenth century. A much more recent book, Leonard Wilson's *Lyell in America*, also provides information from letters written by Mary Lyell. The development of coal science and technology in the nineteenth century is also interestingly discussed in *Scientists and Swindlers*, by Paul Lucier. The book discusses the development of scientific consulting in coal and petroleum, as well as the early attempts at converting coals into gaseous and liquid fuels. Readers with some background in chemistry can learn more about coalification in my earlier book, *Chemistry of Fossil Fuels and Biofuels*. An excellent source is Peter Given's "An Essay on the Organic Geochemistry of Coal," which appears in Volume 3 of *Coal Science*, edited by Martin Gorbaty, John Larsen, and Irving Wender.

I am not aware of a biography of Marie Stopes, who deserves to be much better known both for her achievements as a coal scientist and as a crusader for women's rights and family planning. She figures prominently in Michael Boulter's *Bloomsbury Scientists*, which discusses the activities and accomplishments of a group of artists and scientists living in Bloomsbury in the 1910s and 1920s. Readers interested in getting more background on the instrumental techniques applied to the study of coal structure will find that all modern textbooks of organic chemistry cover at least NMR and IR (usually not diffraction). Virtually all have the same title: *Organic Chemistry*. I find the one by John McMurry to be fairly helpful. For a general introduction to the field of organic chemistry, texts written even several decades ago can be good sources, often at reasonable prices on the used-book market. The book by Robert Morrison and Robert Boyd is quite good. After all these years, benzene is still aromatic and methane is still a gas. The specific applications of the instrumental methods to coal are discussed by van Krevelen. The professional literature on coal petrography begins with *Stach's Textbook of Coal Petrology*, with contributions from many leading petrographers. *Stach's Textbook* last appeared in 1982. A revised and updated book (1998), *Organic Petrology*, again brought together a team of experts led by G. H. Taylor.

The two volumes of *Annals of Coal Mining and the Coal Trade* (Robert Galloway) present an extensive record of coal mining in Britain, as it was practiced in the nineteenth century. It provides an appalling catalog of explosions, fires, outbursts, child labor, and other aspects of the "bad old days" of underground mining. George Orwell's *The Road to Wigan Pier* recounts a visit to a mine in Wigan in the mid-1930s, discussing the working conditions in the mine as well as the social and living conditions of the miners. An episode mostly neglected in the immense literature of World War II is chronicled in *Called Up, Sent Down*, by Tom Hickman. In the desperate days of the war, Britain actually sent military draftees to work in coal mines. Hickman details what it was like to work in the mines of that era. The evolutionary development of engines for pumping water, culminating in James Watt's true steam engine, is discussed in my earlier book, *Energy and Society*, which also discusses some of the environmental impacts of mining. Additional discussion of mining is in another of my previous books, *Coal: The Energy Source of the Past and Future*.

Mining is surely the only facet of coal science and technology featured in fiction. Novels featuring the heroic endeavors of organic geochemists or the epic struggle to save the world from boiler fouling seem to be scarce. Of the many novels with a mining theme, the two that I have particularly found memorable are Emile Zola's *Germinal* and Richard Llewelyn's *How Green Was My Valley*. The latter may have resonated because there was a great deal of Welsh heritage in the anthracite-mining region of Pennsylvania where I grew up.

Unfortunately, one of the features of the anthracite region that has become notorious is the Centralia mine fire. Among the growing number publications on this sad situation is the excellent *The Day the Earth Caved In*, by Joan Quigley. It is both interesting and informative.

Probably the central book in the field of coal preparation is in fact *Coal Preparation*, edited by Joseph Leonard. Good additional information is available in the books by Berkowitz and Lowry mentioned in the first paragraph. A comprehensive review of coal preparation, with a primary focus on South African practice, has been published under the aegis of the South African Coal Processing Society and is titled *Coal Preparation in South Africa*.

My *Energy and Society* has an extensive introductory-level discussion of the evolution of boilers and issues relating to coal use in electricity generation. Bruce Miller's *Clean Coal Engineering Technology* book covers many aspects of production of electricity from coal, including emissions controls and research and development activities for generating electricity with near-zero emissions. For decades the Babcock and Wilcox Company has published *Steam: Its Generation and Use*. It is an encyclopedic treatise incorporating a great amount of detail on coal combustion, steam generation, and issues relating to emission control. The book is now in its 42nd edition, which must be some sort of record for books relating to coal. The environmental issues regarding use of coals in power plants are also treated in *Energy and Society* and in *Clean Coal Engineering Technology* at introductory and professional levels, respectively.

Donald Cardwell's *The Norton History of Technology* provides interesting historical background on Abraham Darby and the blast furnace. Many books on the history of the eighteenth and nineteenth centuries discuss the impact of the Industrial Revolution, in which coke and blast furnaces had major roles. Among such books are *The Triumph of the Middle Classes*, by Charles Morazé, and, more recently, Gavin Weightman's *The Industrial Revolutionaries*. Environmental issues associated with coke plants are discussed in the first chapter of *When Smoke Ran Like Water*, whose author, Devra Davis, grew up in Donora, Pennsylvania, a coke and steel town once notorious for its pollution. An excellent history of coke-making is R. A. Mott's essay, "The Triumphs of Coke," published as part of the 50-year jubilee of the Coke Oven Managers' Association in 1965. Fortunately, this otherwise obscure publication can be read online at https://www.yumpu.com/cn/document/read/30544421/the-triumphs-of-coke-by-ra-mott-dsc-fric-flnslf-to-.

Berkowitz and van Krevelen both have much useful information on the chemistry and technology of coke making. My earlier *Chemistry of Fossil Fuels and Biofuels* also discusses this topic. A comprehensive and detailed monograph on production of metallurgical coke is *Coke: Quality and Production*, by Roger Loison, Pierre Foch, and André Boyer.

The bumpy start of what is now the Dakota Gasification plant is discussed in my *Coal: The Energy Source of the Past and the Future*, along with some history of early developments in making gas from coal. Paul Lucier's *Scientists and Swindlers* also talks about coal gas in the nineteenth century. At a more technical level, *Chemistry of Fossil Fuels and Biofuels*, plus Lowry and Berkowitz have chapters devoted to gasification. The "go-to" book, *Gasification*, by Christopher Higman and Maaretn van der Burgt, presents an excellent, thorough discussion of gasification, including thermodynamics, kinetics, gasification processes, and their applications.

Much of the chapter on synthesis gas derives, with modifications, from two of my earlier books *Chemistry of Hydrocarbon Fuels* and *Chemistry of Fossil Fuels and Biofuels*. Higman and van der Burgt's *Gasification* provides a very useful discussion of applications of synthesis gas in the production of fuels and chemicals. An extensive discussion on the promise and potential of methanol is the book *Beyond Oil and Gas: The Methanol Economy*, by the Nobel Laureate George Olah and his colleagues Alain Goeppert and Surya Prakash.

Interesting background on the work of Friedrich Bergius and Carl Bosch is available in *The Crime and Punishment of I. G. Farben*, by Joseph Borkin, and in *Buna Rubber*, by Frank Howard. Papers by Anthony Stranges provide excellent histories of the German efforts to develop coal liquefaction. They include "Friedrich Bergius and the Rise of the German Synthetic Fuel Industry" (*Isis*, vol. 75, pp. 642–667, 1984) and "The Conversion of Coal to Petroleum: Its German Roots" (*Fuel Processing Technology*, vol. 16, pp. 205–225, 1987). Numerous history books discuss the importance of the coal-to-liquids effort in Germany during World War II and the impact of the American precision bombing campaign against these plants. The book *Fire and Fury*, by Randall Hansen, is a recent and particularly useful addition to such literature.

Scientific background on liquefaction is available in Berkowitz, Lowry, Van Krevelen, and *Chemistry of Fossil Fuels and Biofuels*. James Speight's *The Chemistry and Technology of Coal* provides information on a large variety of liquefaction processes. The US Department of Energy Report, "Summary Report of the DOE Direct Liquefaction Process Development Campaign of the Late Twentieth Century" (report number DOE/PC 93054-94) by Frank Burke and his colleagues provides an excellent summary of work done, particularly at Wilsonville, in the last big effort on liquefaction. The National Academies report *Liquid Transportation Fuels from Coal and Biomass*, edited by Michael Ramage and David Tilman, is the most up-to-date review of this topic, at least from the perspective of the US energy economy. It covers technological, economic, and environmental issues.

My book *Energy and Society* discusses issues of reserves, resources, and renewable energy. Meghan O'Sullivan's book *Windfall* treats the dramatic shifts in the energy position of the United States and the changes in global politics arising from the sudden abundance of petroleum and natural gas thanks to the shale revolution. Rachel Maddow also discusses the impact of the oil and gas industry in her book *Blowout*, but portrays oil and gas as a "singularly destructive industry." For his part, Dieter Helm, in *Burn Out*, suggests that the real game-changer will derive from the developing internet-of-things and how it will reduce demand not only for oil and gas but also for renewables. Loren Steffy's biography, *George P. Mitchell*, is the story of the person generally regarded to be the father of fracking—a study in perseverance.

The person who might be regarded as the father of climate change, John Tyndall, is the subject of a recent biography, *The Ascent of John Tyndall*, by Roland Jackson. Tyndall made many contributions to science in the Victorian Age. Perhaps someday Eunice Foote will receive her due in print.

One effect that is undeniably attributable to climate change and global warming is that it has released a tsunami of ink in the form of books, magazine articles, and reports. The general background, as a consequence of the energy balance equation, is in *Energy and Society*. A remarkable human perspective of the impact of climate change is the edited collection *The Earth Is Faster Now: Indigenous Observations of Arctic Environmental Change*, by Igor Krupnik and Dyanna Jolly. As observed by the indigenous peoples in the arctic regions, marked changes were already evident 15 years ago.

Climate change has also triggered a tsunami of electrons on numerous websites. The Global Carbon Capture and Storage Institute maintains a comprehensive database of all large-scale carbon capture and storage (CCS) projects around the world, www.globalccsinstitute.com. A more extensive database, which also includes smaller-scale and pilot CCS projects, is maintained by Howard Herzog at MIT, https://sequestration.mit.edu/tools/projects/index. html. A series of web pages of the Global Monitoring Laboratory, US National Oceanic and Atmospheric Administration (NOAA), provides an excellent introductory-level discussion of carbon isotopes, their measurements in the atmosphere, and how the information shows a significant contribution from fossil fuel combustion to atmospheric CO_2 (https://www.esrl.noaa. gov/gmd/ccgg/isotopes/). The latest information on atmospheric concentrations of CO_2 from the Mauna Loa observatory can be found at https://gml.noaa.gov/ccgg/trends/.

Simon Garfield writes of William Henry Perkin and his career in coal-tar chemistry in *Mauve*. A shorter biographical sketch is in Sharon McGrayne's *Prometheans in the Lab*. Perkin also has a role in *The Industrial Revolutionaries*, by Gavin Weightman. William Brock, in *The Norton History of Chemistry*, has interesting historical background on coal tar chemicals. Readers who are also interested in military history will find two of John Mosier's books, *The Myth of the Great War* and *Verdun*, quite good reading but also with bits and pieces of information on chemicals from coal tar.

Good information on coal-tar chemicals is in the first two volumes of Lowry. Beginning in the early 1990s and carrying on for about 15 years, Chunshan Song and I published a series of papers in the professional literature explaining what we perceived to be the advantages for considering coals as raw materials for chemicals (though not from coal tar) and carbon materials. These efforts culminated in a major report issued in 2005 by the International Energy Agency (IEA) Clean Coal Centre, "Premium Carbon Products and Organic Chemicals from Coal" (report number CCC/98), in which we were joined by our colleague John Andrésen. This report should be available through the IEA. Of course the quantitative information on production, market share, and values is long-since obsolete, but the qualitative information is quite sound. I sometimes suspect that what we accomplished was to verify the diplomat George Kennan's adage that "truth prematurely uttered is scarcely better than error." The field of green chemistry continues to progress steadily, with more and more books and monographs. A good introduction to the field is Mike Lancaster's *Green Chemistry: An Introductory Text*.

A useful short introduction to the rare earths is *The Rare Earth Elements*, by J. H. L. Voncken. A greater level of detail, including history, applications, resources, and extraction, is given in the monograph *Extractive Metallurgy of Rare Earths*, written by Nagaiyar Krishnamurthy and Chiranjib Kumar Gupta.

Two companion volumes present collected chapters prepared by authors working in various fields of carbon technology or science, which, taken together, provide a very thorough grounding in carbon materials. They are *Introduction to Carbon Technologies*, edited by Harry Marsh, Ed Heintz, and Francisco Rodríguez-Reinoso; and *Sciences of Carbon Materials*, by Marsh and Rodríguez-Reinoso. Published at the end of the previous century, they are somewhat dated in places, but still very useful. The IEA report on premium carbon products and chemicals discuss graphites, activated carbons, and other carbon products. A more recent book that provides a good survey is Deborah Chung's *Carbon Materials*.

The need for a trained scientific and engineering workforce and the infrastructure to support them is brought out in Vannevar Bush's *Science: The Endless Frontier*. First written in 1945, parts are remarkably relevant today. More recent is the very good *The Scientific Life*, by Steven Shapin. The five books in the *Incerto* series by Nassim Nicholas Taleb are all solidly good, thought-provoking reading. Of the series, I consider the top two to be *The Black Swan* and *Fooled by Randomness*. In thinking about the future, it's good to occasionally check on https://www.world ometers.info/world-population/.

Finally, it's worth meeting little archy and his free-verse poems about mehitabel. Several collections by Don Marquis are occasionally reissued, or might be available from libraries. There *is* a dance or two in the old dame yet.

Index

cerium, 262
lanthanum, 261, 262
neodymium, 262, 287
promethium, 261
scandium, 261, 262
yttrium, 261
reactive dissolution, process, 43
reclamation, mined land, 81
Rectisol process, 168–69, 170–71
reflectance, 50
refractive index, 50n.8
regulations, coal, 4
renewable energy, 245–46, 297–98
    competition, 209
    economy, 210
    renewables, 218, 242, 301
    solar park with wind turbines, 225*f*
    worldwide forms, 227
    See also solar energy; wind-generated
        electricity
Reppe chemistry, 257
reservoir rock, characteristics of, 210–11
resinites, 49, 309
resolidification temperature (T$_r$), 145–46
retreat mining, 68–69
retrogressive reactions, 195
Revelle, Roger, 230–31
Rice University, 283
Richter scale, earthquakes, 217n.6
Rogers, Henry Darwin, 31–32
roof bolting, 71
room and pillar system, 67–68
Russia
    synfuels, 157
    Ukraine and, 299

safety lamps, mining, 73–74
salt coals, 59–60
Sanada, Yuzo, 204
sapropelic coals, 35, 35n.8
Sasol-Lurgi gasifiers, 164, 181–82, 192–93
Sasol Synfuels plant, 176, 181, 186, 188, 255,
    294
    Sasol-2, 181
    Secunda, 206–7
scandium, 261, 262
sclerotinite, 309
Scrivens, Walter, 285–86
scrubbers, 124–27
    capturing SO$_x$, 124–27
    dry, 126

flue-gas desulfurization (FGD), 124
    non-regenerative, 124–25, 126–27
    regenerative, 125
secondary air, 106
second-generation processes, liquefaction,
    201, 202
selective catalytic reduction (SCR), 130
semicoke, 148
semifusinite, 309
Sentinel Butte Formation, 263
shaft mine, underground mine, 67
shale gas plants, 226
shale oil, 211, 211n.2
    fields as "Cowboyistan", 212–14, 213n.4
shale revolution, 209–10, 227–28, 249
    energy economy, 215
    methane, 215–16
Shenhua Group Corporation, direct
    liquefaction plant, 204–5
silicon carbide, 276
silicosis, 77–78
single-wall nanotubes (SWNTs), 284–85
    structure of, 285*f*
size separation, distribution of coal, 89
slacking, 95
slagging
    gasifiers, 162
    molten ash in boiler, 112
slope mine, underground mine, 67
Smalley, Richard, 283
smelting
    coke, 143
    iron ore, 141, 153–54
Smith, Robert Angus, 122
smokestacks, 3, 3*f*
SNG. *See* substitute natural gas (SNG)
soda ash, 240
sodium carbonate, 240
sodium chloride, 59–60
sodium content in coals, boiler-tube fouling,
    112–13
softening temperature, 145–46
    ash fusibility test, 64–65, 65*f*
solar cells, 222–23
solar energy, 222–26
    cost of electricity, 224
    indirect solar conversion, 222
    photovoltaics, 222–24
    pumped storage, 226
    solar park with wind turbines, 225*f*
    solar thermal power stations, 222